FUNDAMENTALS OF STRUCTURAL ANALYSIS

SI UNITS

By

Col. (Retd.) M.K. PANT

Formerly HOD (Civil Engg. Deptt.),
Registrar and Dean Academics WCTM,
Farrukh Nagar, Gurgaon
HOD (Civil Engg. Deptt.) MRIU, Faridabad, WIT, Sohna and
Savera Group of Institutions, Gurgaon

S.K. KATARIA & SONS®

Publisher of Engineering Books

4885/109, Prakash Mahal, Dr. Subhash Bhargav Lane,
Opposite Delhi Medical Association, Daryaganj, New Delhi–110002
Phone: +91-11-43551243, Mobile: +91-9871775858, 7678425209
e-mail: katariabook@gmail.com, katariabooks@yahoo.com
Website: www.skkatariaandsons.com

(e-books, subscriptions and bulk buying also available on our website)

Published by:

S.K. Kataria & Sons®
4885/109, Prakash Mahal, Dr. Subhash Bhargav Lane,
Opposite Delhi Medical Association, Daryaganj, New Delhi–110002
Phone: +91-11-43551243, Mobile: +91-9871775858, 7678425209
e-mail: katariabook@gmail.com, katariabooks@yahoo.com

First Edition: January, 2012; Second Edition: August, 2014; Reprint: 2026

Laser Type Setting:
PLUS COMPUTERS, Shahdara, Delhi

Printed at:
New A.S. Printing Press, Delhi

Preface to the Second Edition

It is indeed a privilege to present this second edition of this popular and highly acclaimed guide book which has been thoroughly revised and enlarged. A good number of exampels and new chapters have also been added keeping in view the structural analysis syllabus of various indian universities.

I express my sincere thanks to the teachers for their valuable suggestions as well as the students for their excellent response to this unique core book on basics of civil engineering.

At the end, author greatfully acknowledges the all out help extended by the publisher M/s S.K. Kataria & Sons in text processing, composing and manuscript and to bringout this second edition in a short period and in decent getup.

It is hoped that this book should become more popular among the students community and by the new additions it contained, also would be of immense use to the teachers fraternity also.

Col. (Retd.) M.K. Pant

Preface to the First Edition

Structural analysis not only forms the integral part of Civil Engineering but is the backbone of Structural Engineering which involves analysis and design of structures. In broad sense Structural Analysis is the instrument with which the Structural Engineer predicts the behaviour of major engineering projects under influence of external loads at the planning and design stage much before they are built.

Even though Architects conceptualize space planning, it is the structural engineer who delivers safe structure to people to reside inside without fear of their collapse.

While writing this book, I have constantly kept in mind the examination requirements of civil engineering students who faced lot of difficulties for preparing themselves adequately for "Theory of structures", hence converged the text more so in the form of a guide book rather than an exhaustive text book on the subject.

To make the book really useful at all levels, it has been written in an easy style as well as with a simple and systematic mathematical approach. To attain this, attempts has been made to avoid complicated procedures of solution, even if it made the solution a bit longer. I owe my personal gratitude to distinguished and prominent authors of numerous standard books on the subject that I consulted and referred to their texts, my colleagues whose opinions I sought and above all the need of my students in various engineering colleges I taught.

The author's thanks are due to his wife Sujata Pant for providing an ideal atmosphere at home without whose cooperation and encouragement this book would have never been materialised. Although every care has been taken to check mistakes and misprints, yet it is difficult to claim perfection. Suggestions for improvements shall be thankfully acknowledged.

Col. (Retd.) M.K. Pant

Dedicated to My Beloved
Mother, Mrs. Reba Pant
Who Inspired and Encouraged Me to Accomplish
Impossible Missions in
"Life"

Contents

PART I

PART II

PART I

CHAPTER

1 Introduction

1.1. WHY DOES STUDY OF STRUCTURAL ANALYSIS IS IMPORTANT FOR A CIVIL ENGINEER?

Structural analysis may be defined as the prediction of performance of a given structure under stipulated loads and/or other external effects such as support movements and temperature changes.

Stipulated loads comprise of Axial forces, Shear-forces and Bending moments, Deflections and Support reactions.

Thus the analysis of structure involves determination of these quantities for a structure under given loading conditions. These unknown quantities may be obtained by formulating a suitable number of independent equations which are obtained from three following types of relations or combinations which govern the behaviour of the structure:

(*i*) Conditions for static equilibrium of the internal and external forces acting on the structures.

(*ii*) Stress-strain relation of the material involved.

(*iii*) Geometrical relations between strains and displacement of points on the structure. Analysis of structure is based to a large extent upon the principles of statics. The subject forces that are commonly met with are:

- (*a*) Weight
- (*b*) Tension
- (*c*) Normal reaction due to contact
- (*d*) Friction
- (*e*) Forces due to fluid pressure, wind pressure and pressure from grannular material.

The analysis and design process is both creative and technical and requires a fundamental knowledge of material properties and the laws of mechanics which govern material response. Once a preliminary design of a structure is proposed the structure must then be analysed to ensure that it has its required strength and rigidity.

The loadings are determined from codes and local specifications and the forces in the members and their displacements are found using the theory of structural analysis.

❑❑❑

CHAPTER

2 Statically Determinate and Indeterminate Structures

Before starting the force analysis of a structure, it is necessary to establish the determinancy and stability of the structure. In the real sense an exact analysis of a structure can never be carried out, since estimates always have to be made of the loadings and the strength of the materials composing the structure.

The equilibrium equations provide both the necessary and sufficient conditions for equilibrium. When all the forces in a structure can be determined strictly from these equations, the structure is referred to as "Statically determinate". Structures having more unknown forces than available equilibrium equations are called "Statically indeterminate." We have basic three equations of statics in equilibrium *viz.*

$\Sigma V = 0$, Summation of all Vertical forces acting at a joint = 0

$\Sigma H = 0$, Summation of all Horizontal forces acting at a joint = 0

$\Sigma M = 0$, Summation of all Moments forces acting at a joint = 0

When the number of unknown forces (stress components) or reactions remain within the three conditions of static in equilibrium the system is said to be statically determinate. However, when the number of unknown forces (stress components) or relations exceed the number of conditions of equilibrium, the system said to be statically indeterminate or redundant. In such cases, the equation of static equilibrium must be supplemented by the equation of compatibility of deformation. The compatibility conditions relate the displacement/slope of the structure so that its various parts fit together.

When the internal forces cannot be evaluated by the conditions of static equilibrium alone, the structure is called **"Internally indeterminate"** whereas the structure may be connected to foundation in such a way that all the reactions cannot be evaluated from the equations of static equilibrium alone, and such structures are called **"Externally Indeterminate."** The later structures are also called **"Hyper Static Structures"**.

Types of Indeterminancies in a Structural System

Two types of indeterminancies, the static and the kinematic may be used to describe a structural system. If a structure is statically indeterminate, the additional equation needed to solve for the unknown forces are obtained by relating the applied loads and reactions to the displacements or slopes at different points on the structure. These equation which are referred as compatibility equations, must be equal to the Degree of Indeterminancy of the structure.

To determine the static Indeterminancy or static determinancy of a structure, following procedure may be followed.

1. Draw the Free Body Diagram: Whenever the equations of equilibrium are applied, it is first necessary to draw a free body diagram of the structure of its members. The selected member is isolated from its supports and surroundings and its outlined shape drawn. All the external and internal forces must be shown that act on the member. Before drawing the Free Body Diagram, it is essential to understand (*a*) Type of support (Connection), (*b*) Reactions at support and, (*c*) Number of unknowns as explained diagrammatically below: (Fig. 2.1)

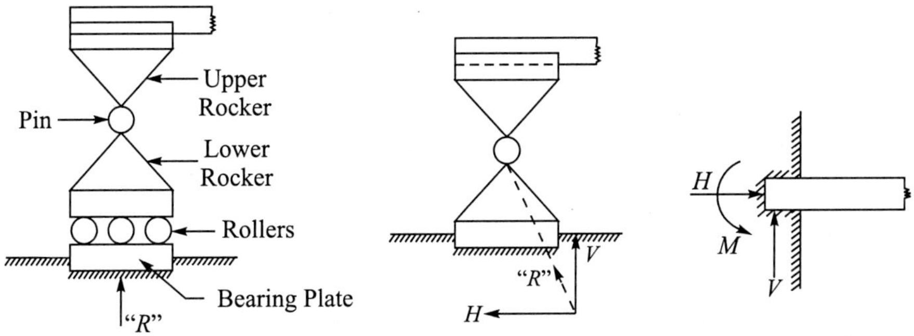

(a) Roller Support (b) Hinged-Support (c) Fixed Support

Fig. 2.1

Table 2.1

Configuration of Reactions at Supports :

Type of Support	Symbolic-Representation	Forces Acting
1. Roller		R "One Reaction"
2. Hinge/Pinned		V, H "Two Reactions"
3. Fixed		M, V, H "Three Reactions"

2. Count the number of support reactions, "*r*".
3. Count the Total No. of internal forces "*i*"– that can be transmitted by internal hings and internal rollers of the structure. "2" in case of each internal hinge and "1" incase of each roller.
4. Determine the total number of unknowns *viz.* $(r + i)$.
5. Count the number of rigid members or portions "*n*" contained in the structure. Since each portion should be in equilibrium, and there are three equations of static equilibrium, the number of equilibrium equations available for the entire structure is "$3n$".
6. Determine if the structure is statically determinate of indeterminate by comparing $(r + i)$ to the total number of equilibrium equations.

   If $(r + i) = 3n$, The structure is statically determinant.

   If $(r + i) > 3n$, The structure is statically indeterminate.

   The procedure will get more clear on solving the numerical problems dealt ahead.

2.1. MAIN DIFFERENCE BETWEEN STATICALLY DETERMINATE AND INDETERMINATE STRUCTURES

Table 2.2

Statically Determinate Structures	Statically Indeterminate Structures
1. Conditions of equilibrium are sufficient to fully analyse the structure.	1. Condtions of equilibrium are insufficient to analyses the structure fully.
2. The B.M. at a section or the force in any member is independent of the material of the components of the structure.	2. The B.M. at a section or the force in any member depends upon the material of the components of the structure.
3. The B.M. at a section or the force in any member is independent of the x-see areas of the components.	3. The B.M. at a section or the force in any member is fully dependent of the x-see areas of the components.
4. No stresses are caused due to temperature changes.	4. Stresses are caused due to temperature changes.
5. No stresses are caused due to lack of fit.	5. Stresses are caused due to lack of fit.
Examples of Determinate Structures:	**Examples of Indeterminate Structures:**
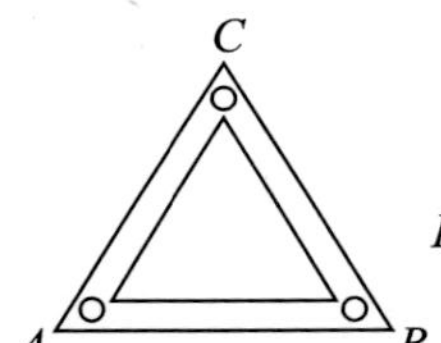 "Plane-Truss" 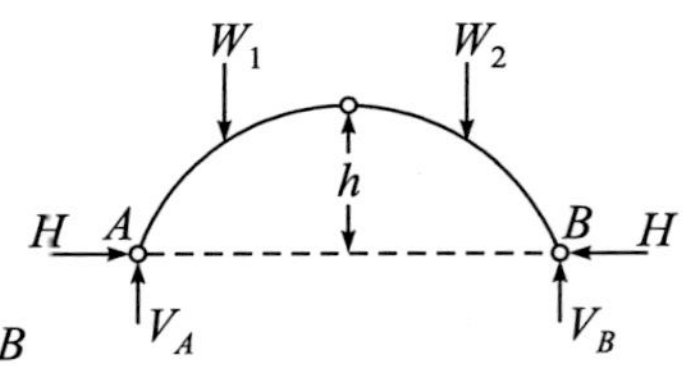 "Three Hinged Arch" 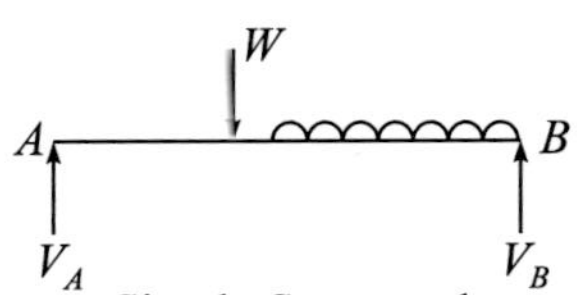 Simply Supported Beam	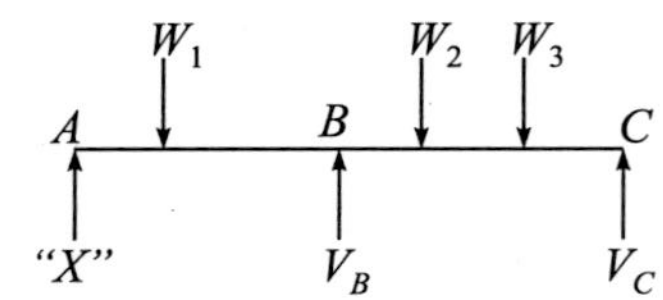 "Contineous Beam" 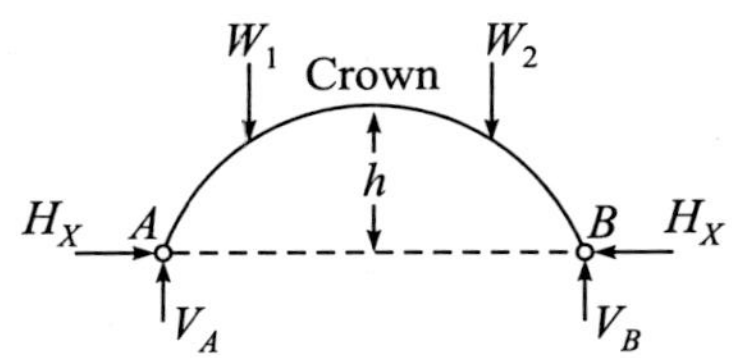 "Two Hinged Arch" 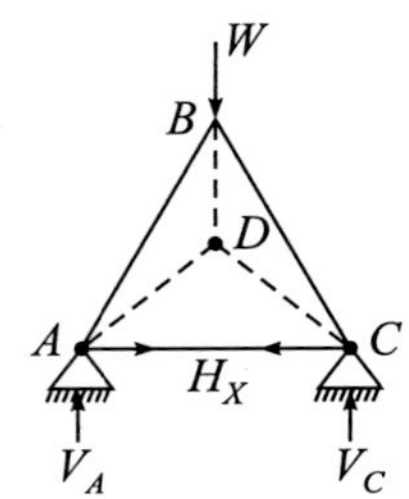 "Space-frame"

The "Kinematic-Indeterminancy" of a structure refers to the number of independent components of joint displacement with respect to specified set of axes, also called the "Degree of Freedom" (DOF). For example, a joint in a space frame will have six-independent components of displacements (3 Translation and 3 Rotation), and will thus have 6 DOF.

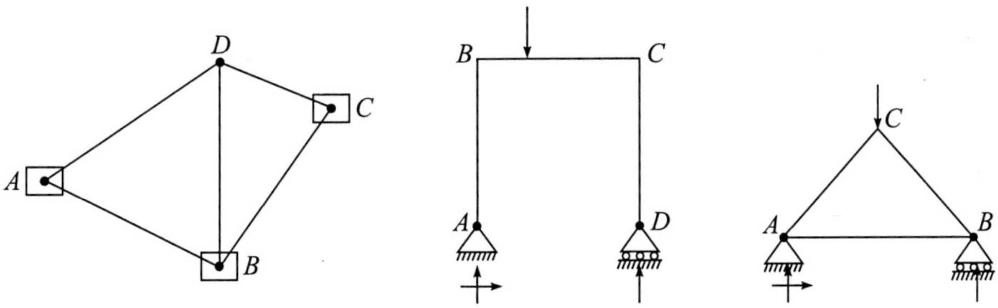

"Space-frame" "Plane-frame" "Plane-truss"

Fig. 2.2

In an another example a plane frame will have three component of displacements at joint (two translations and one rotation) and thus will have 3 DOF.

A plane truss joint, however will have two degree of freedom, both translations.

The Degree of Freedom (DOFs) in a structure are also called "Kinematic-Indeterminancy".

2.2. KINEMATIC-INDETERMINANCY

Thus, in a structure, if "j" represents numbers of joints, "N" represents joint displacement or No. of supports and "C" the number of constraints displacements, then the "Kinematic-Indeterminancy" (Ik)

$$\text{"}I_k\text{"} = Nj - C$$

It should be noted that for Planer-structures:

(*a*) In case of Fixed Supports, "C" = 3

(*b*) In case of Hinged/Pinned Support, "C" = 2

(*c*) In case of Roller Support, "C" = 1

Also if axial deformations are neglected, the displacement components in Y-Direction will be zero and same will be in X-Direction and thus gets reduced to one independent displacement only.

We will now consider some examples of beams, pin-connected structures and framed structures to show how to classify determinancy of a structure.

Example 1: *From the beams shown below, state whether they are statically determinate or indeterminate.*

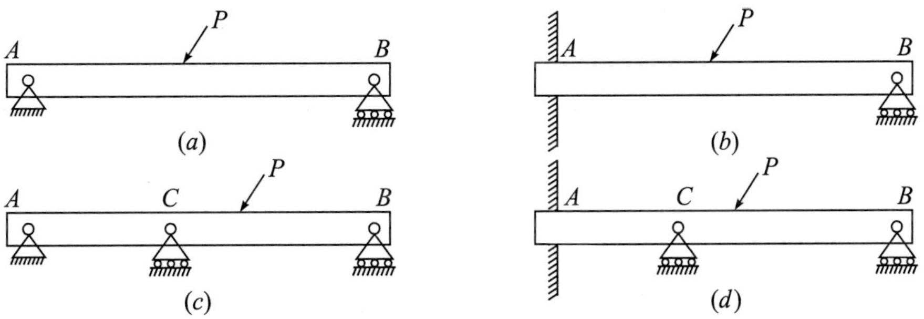

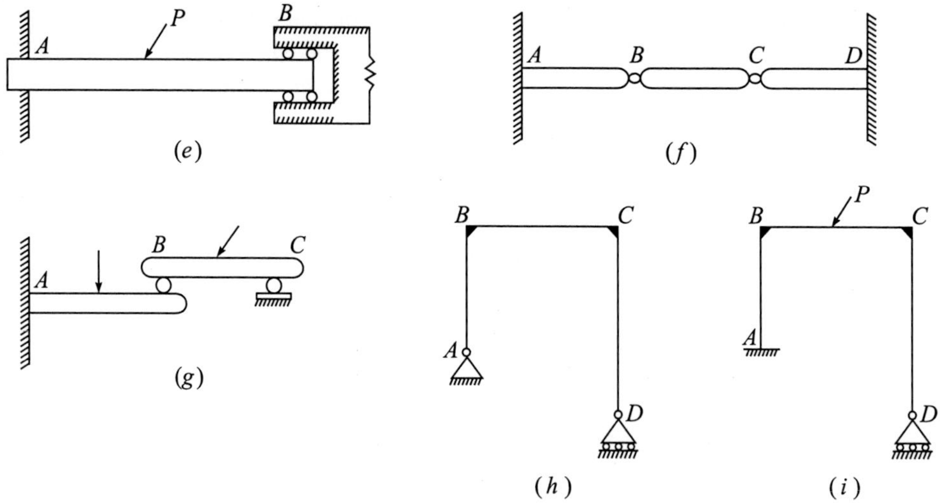

Fig. 2.3

Solution: (*i*) First draw the Free Body Diagram for each beam.

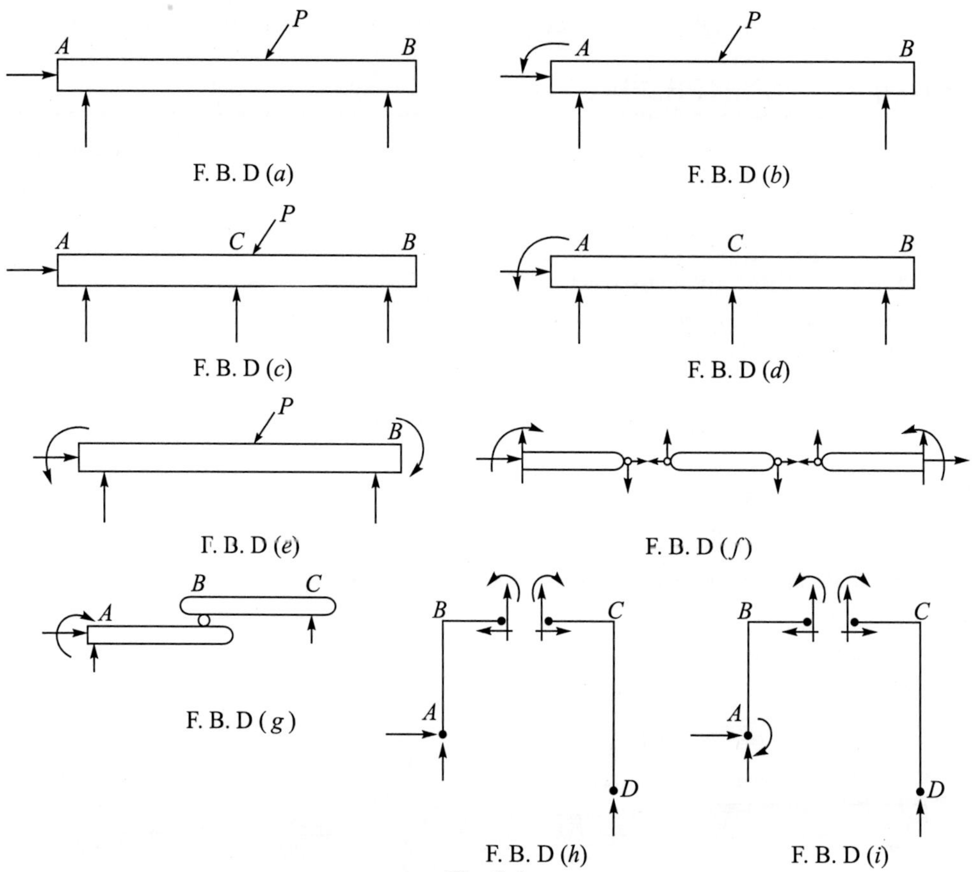

Fig. 2.4

(*ii*) Tabulate the results as under:

Table 2.3

Structure Fig. No.	Number of Reactions "r"	Number of Internal Forces "i"	Total (r + i)	Number of Parts "n"	Total number of Equations of Equilibrium Available (3 × n)	Classification and Degree of Freedom (r + i) = 3n
(*a*)	3	–	3	1	3 × 1 = 3	3 = 3 (Determinate Structure)
(*b*)	4	–	4	1	3 × 1 = 3	4 = 3 (Indeterminate Structure) DOF = 01
(*c*)	4	–	4	1	3 × 1 = 3	4 = 3 (Indeterminate Structure) DOF = 01
(*d*)	5	–	5	1	3 × 1 = 3	(Indeterminate Structure) DOF = 02
(*e*)	5	–	5	1	3 × 1 = 3	(Indeterminate Structure) DOF = 02
(*f*)	6	4 (At *B* and *C*)	10	3	3 × 3 = 9	(Indeterminate Structure)
(*g*)	4	1 (At *B*)	5	2	3 × 2 = 6	(Determinate Structure)
(*h*)	3	3	6	2	3 × 2 = 6	(Determinate Structure)
(*i*)	5	3	8	2	3 × 2 = 6	(Indeterminate Structure) DOF = 02

Example of Kinematic Indeterminancy:

Determine the Kinematic indeterminancy of the structures shown below:

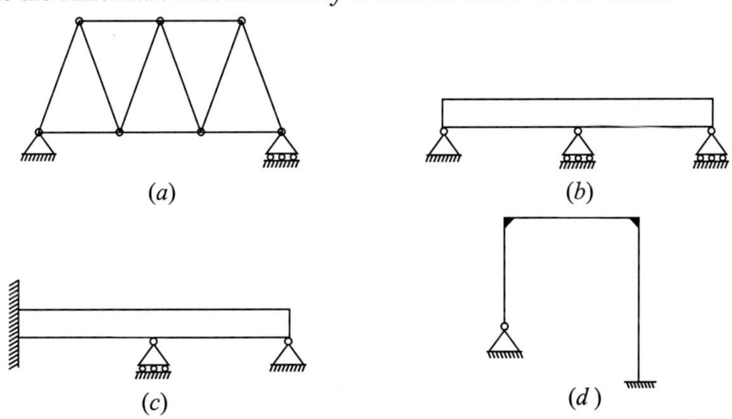

Fig. 2.5

Solution: The results of the Kinematic-Indeterminancy are tabulated as under:

Table 2.4

Structure Figure	Number of Restraints "C"	Number of Joints "j"	Total D.O.F. (N × j) N = No. of Supports	Kinematic Indeterminancy Ik = Nj – C
(*a*)	2 + 1 = 3	7	2 × 7 = 14	$I_k = 14 - 3 = 11$
(*b*)	2 + 1+ 1 = 4	3	3 × 3 = 9	$I_k = 9 - 4 = 5$
(*c*)	3 + 1 + 2 = 6	3	3 × 3 = 9	$I_k = 9 - 6 = 3$
(*d*)	2 + 3 = 5	4	3 × 4 = 12	$I_k = 12 - 5 = 7$

SUMMARY

- When all the forces in a structure can be determined from the conditions of static equilibrium, such structures are called determinate structures.
- When the internal forces cannot be evaluated by the conditions of static equilibrium, the structure is called indeterminate structure.
- Two types of indeterminancies, the static and kinematic may be used to describe a structural system.
- The kinematic-indeterminancy of a structure refers to the number of independent components of joint displacement with respect to specified set of axes, called the Degree of Freedom (DOF).
- To determine the static Determinancy Indeterminancy of a structure, a set procedure has to be followed:
  (*a*) Draw the Free Body Diagram.
  (*b*) Count the number of support reactions "*r*".
  (*c*) Count the total no. of internal forces "*i*".
  (*d*) Determine the total no. of unknowns viz. $(r + i)$
  (*e*) Take the total no. of equilibrium equations as $3n$
  If $(r + i) = 3n$. The structure is Determinate.
  If $(r + i) > 3n$. The structure is Indeterminate.

OBJECTIVE TYPE QUESTIONS

1. Before conducting force analysis of a structure, it is necessary to establish:
(*a*) Determinancy of structure
(*b*) Indeterminancy of structure
(*c*) Determinancy and stability of structure

2. The equations of statics in equilibrium are :
(*a*) $\Sigma V = 0$, $\Sigma H = 0$, $\Sigma M = 0$ (*b*) ΣV and $\Sigma H = 0$ (*c*) None above

3. The reactions at Hingle support are:
(*a*) 1 (*b*) 2 (*c*) 3

4. The reactions at Fixed support are:
(*a*) 1 (*b*) 3 (*c*) 2

5. A plane truss joint, will have
   (*a*) Two degree of freedom (DOF)
   (*b*) Three degree of freedom (DOF)
   (*c*) One degree of freedom (DOF)

Answers: 1. (*c*), 2. (*a*), 3. (*b*), 4. (*b*), 5. (*a*).

EXERCISE FOR PRACTICE

1. Classify if the structure is determinate or indeterminate

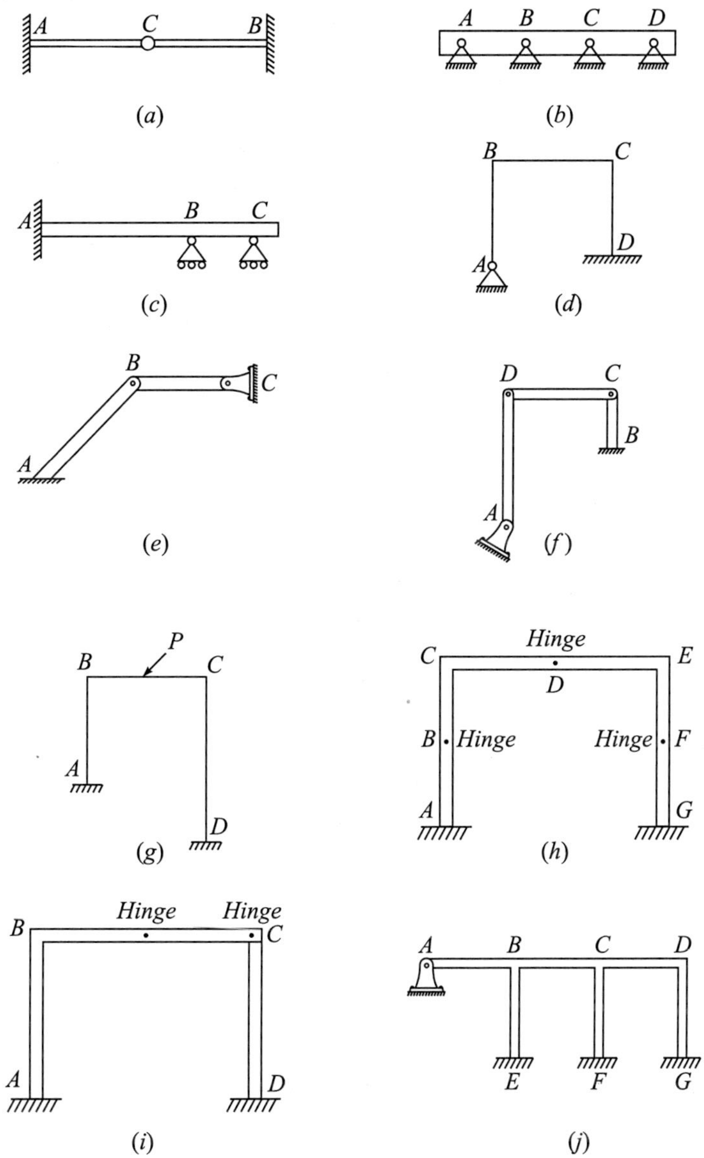

Fig. 2.6.

CHAPTER

3 Stress and Strains

3.1. STRESS

Whenever a body is subjected to an external force, it tends to undergo deformation (*i.e.*, change in shape or dimension). Due to cohesion between the molecules of the body, the body resists deformation. This internal resistance is equal and opposite to the applied force. This resisting force per unit area is known as stress or intensity of stress. Hence, the internal resistance per unit area of cross-section offered by a body against the deformations called '**Stress**'.

Mathematically, stress may be defined as the force per unit area, *i.e.*,

Stress, $\sigma = \dfrac{P}{A}$,

where P = Load or force acting on the body.

A = Cross-sectional area of the body.

In S.I. units, stress is expressed as N/m^2 or N/mm^2. N/m^2 is also called Pascal (Pa).

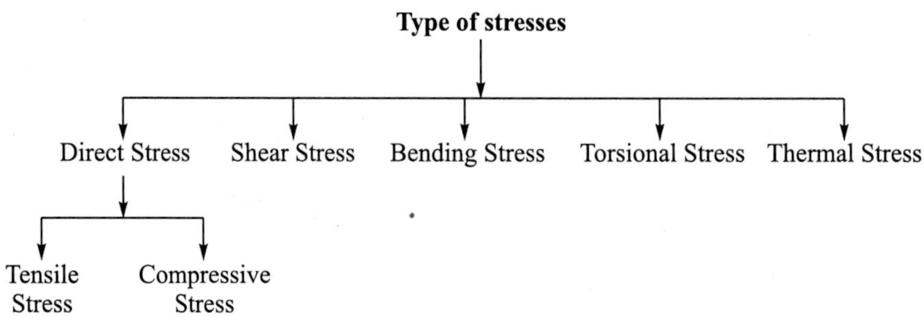

3.1.1 Direct Stress

Direct stress is that which acts perpendicular to the cross-section of the member. It is also known as Normal Stress. The direct or normal stress is either tensile or compressive in nature according to the nature of loading.

(*i*) **Tensile Stress:** When two equal and opposite pulls are applied to a member, the stress induced in a member is known as 'Tensile Stress'. The member subjected to tensile stress is known as "Tie".

Let an axial pull "P" is applied on a member of cross-sectional area "A", then Tensile Stress

$$\text{“}\sigma_t\text{”} = \frac{P}{A}$$

(*ii*) **Compressive Stress:** When two equal and opposite pushes are applied to a member, the stress induced in a member is known as 'Compressive Stress'. The member subjected to compressive stress is known as 'Strut'.

Let an axial push '*P*' is applied on a member of X-sec area "*A*", then the Compressive Stress

$$\text{“}\sigma_c\text{”} = \frac{P}{A}$$

3.1.2 Shear Stress

When two equal and opposite forces are acting tangentially to the cross-sectional, the stress induced in a member is known as Shear-Stress. Due to these two equal and opposite forces, the member tends to shear off across the section. The figure below shows the failure of a rivet due to shear. Beam webs are subjected to shear stress.

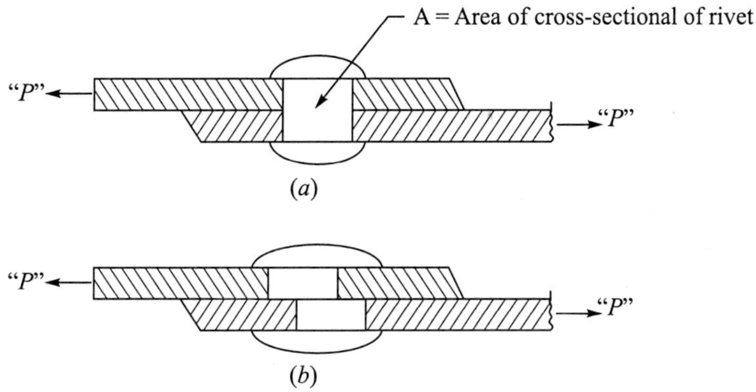

Fig. 3.1

Consider a block *ABCD* with face *CD* fixed to the surface as shown in the Fig. above. When a force '*P*' is applied tangentially, the block takes the shape *A, B, C, D.*

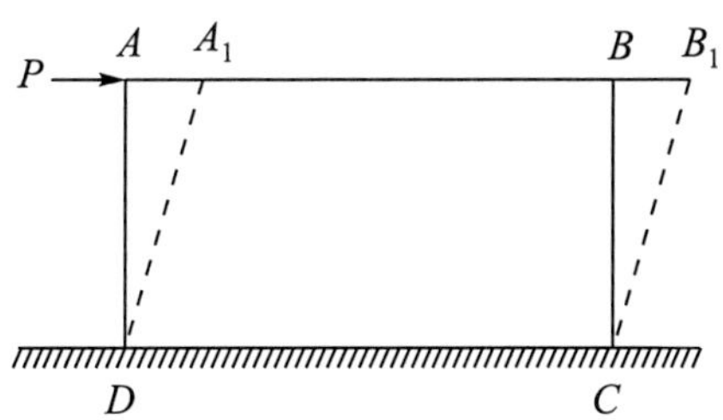

Fig. 3.2

If *A* = Shear area (*CD* × width of block), then Shear Stress

$$\text{“}\tau\text{”} = \frac{\text{Shear Force}}{\text{Shear Area}} = \frac{P}{A}$$

3.1.3 Bending Stress

When a beam is loaded with some external loads, the B.Ms and S.Fs are setup at all the sections of the beam. The B.Ms at a section tends to bend or deflect the beam. The material of the beam will offer resistance or stresses to bending. The stresses produced to resist the B.Ms are known as Bending stress.

3.1.4 Torsional Stress

The Shear Stress set up in the shaft when equal opposite torques (Twisting moments '*T*') are applied to the ends of a shaft about its axis, is called Torsional Stress.

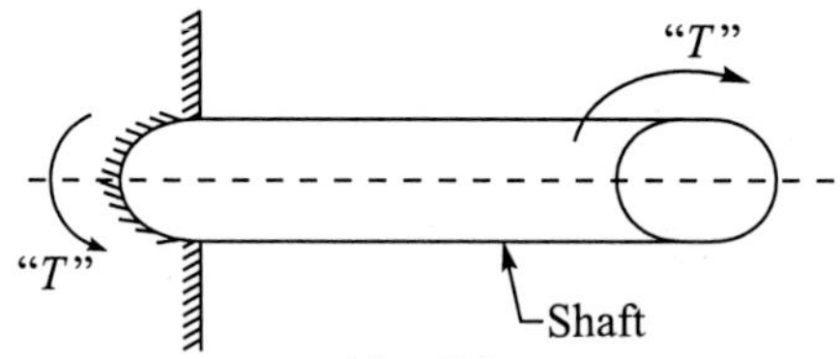

Fig. 3.3

3.1.5 Thermal Stress

When the temperature of a material is changed, its dimension also undergoes change. If this change in dimension is prevented, then a stress is set up in the material, which is called a thermal stress.

3.1.6 Strain

It is the measure of the deformation caused by the loaded body. The ratio of change in dimension of the body to the original dimension is known as Strain.

$$\therefore \qquad \text{Strain "}e\text{"} = \frac{\text{Change in dimension}}{\text{Original dimension}}$$

Remember, strain is a ratio, as such it has no units.

"Type of Stresses"

Tenoile-Strain

$$\text{"}e_t\text{"} = \frac{\delta l}{l}$$

(δl — Increase in length; l — Original length)

Compressive Strain

$$\text{"}e_c\text{"} = \frac{\delta l}{l}$$

(δl — Decrease in length; l — Original length)

Shear Strain

It is the measure of the angle through which a body is distorted under the action of S.F.

$$\text{"}e_s\text{"} = \tan\phi = \phi \quad (\phi \text{ being very small})$$

$$= BB_1/BC$$

Volumetric Strain

$$\text{"}e_v\text{"} = \frac{\text{Change in volume}}{\text{Original volume}} = \frac{\delta v}{v}$$

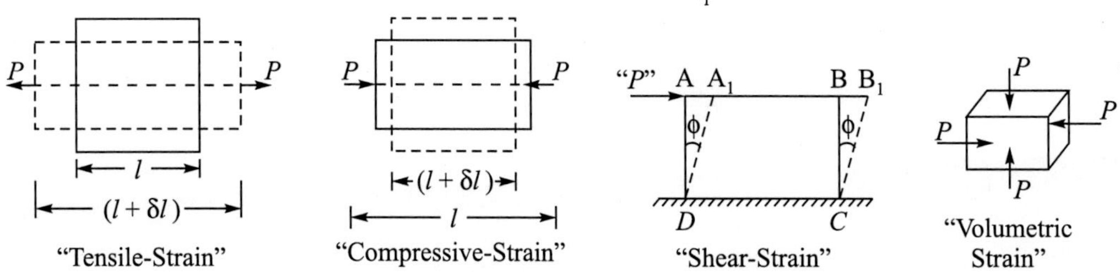

Fig. 3.4.

3.1.7 Hooke's Law

Robert Hooke in 1676, discovered experimentally, a basic law in elasticity. It states that– When a material is loaded within elastic limit, the stress is proportional to strain produced by the stress.

i.e., Stress $\propto$ Strain

or $$\frac{\text{Stress}}{\text{Strain}} = \text{a constant}$$

This constant is known as co-efficient of elasticity or modulus of elasticity.

The value of modulus of elasticity depends upon:

(*i*) The nature of the material of the body.

(*ii*) The manner in which the body is deformed.

There are three moduli of elasticity corresponding to the three types of strains:

(*a*) Young's Modulus of Elasticity (E)

(*b*) Modulus of Rigidity (C)

(*c*) Bulk Modulus of Elasticity (K)

These three modulii of elasticity are also known as '**Elastic Constants**'. These have different values for different materials.

3.1.8 Young's Modulus of Elasticity

The ratio between tensile stress and tensile strain or between Compressive Stress and Compressive Strain is called Young's modulus of elasticity. This is denoted by "*E*".

$$\therefore \quad \text{"}E\text{"} = \frac{\text{Tensile stress}}{\text{Tensile strain}} \text{ or } \frac{\text{Compressive stress}}{\text{Compressive stress}}$$

$$= \frac{\sigma_t}{e_t} \text{ or } \frac{\sigma_c}{e_c}$$

The units of Young's modulus of elasticity are N/m^2 or N/mm^2.

3.1.9 Modulus of Rigidity or Shear Modulus

The ratio of Shear Stress to the Shear Strain is known as Shear modulus of elasticity or modulus of rigidity. This is denoted by "*C*" or "*N*" or "*G*".

$$C = \frac{\text{Shear stress}}{\text{Shear strain}} = \frac{\tau}{e_s}$$

The units of modulus of rigidity are N/m^2 or N/mm^2.

3.1.10 Bulk Modulus of Elasticity

The ratio of normal stress to the volumetric strain is called Bulk or Volume modulus of elasticity. It is denoted by "*K*".

$$K = \frac{\text{Normal stress}}{\text{Volumetric strain}} = \frac{\sigma}{e_v}$$

The units of Bulk modulus of elasticity are N/m^2 or N/mm^2.

3.1.11 Poisson's Ratio

The ratio of lateral strain to the longitudinal strain is called Poisson's Ratio. It is denoted by "m " or $\frac{1}{m}$.

$$\therefore \qquad \mu \text{ or } \frac{1}{m} = \frac{\text{Lateral strain}}{\text{Longitudinal strain}}$$

Poisson's ratio is the ratio of the two strains, hence it has **'no units'**. The value of "*m*" for most of the materials lies between 3 and 4. Hence the values of "*m*" varies from 0.25 to 0.33.

Stress-Strain Curve for a Mild Steel:

The salient features of this important curve are:

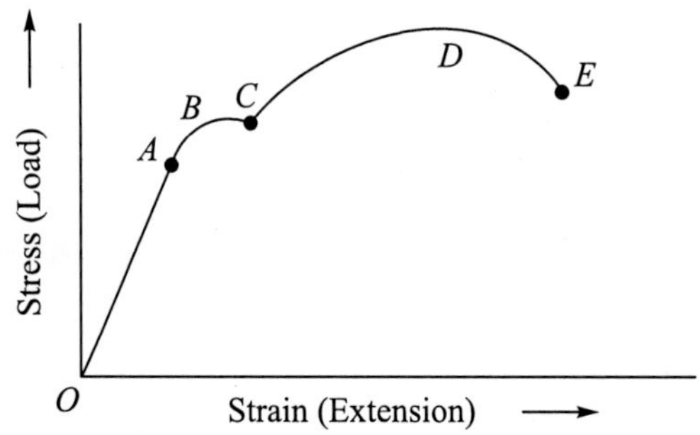

Fig. 3.5

1. Proportional Limit: Point '*A*' is the limit of proportionality. From the origin *O* to a point *A*, stress-strain diagram is a straight line *i.e.*, Stress is proportional to strain. Hooke's law holds good upto this point. Beyond this point, the stress is no longer proportional to the strain. Thus proportional limit is the max. stress at which stress remains directly proportional to strain.

2. Elastic Limit: Point *B* is the elastic limit stage. Between *A* and *B* although the strain increases slightly more than the stress, yet the material is elastic, *i.e.*, on removal of the load, the material will regain its original shape and size. If the material is stressed beyond the point '*B*', the plastic deformation will take place. The value of stress corresponding to point '*B*', upto which material is completely elastic in behaviour is called Elastic Limit.

3. Yield Point: "*C*" is the yield point between *B* and *C*, the strain increases more quickly than the stress. At the yield point, the dial which gives the reading of the load suddenly becomes stationary for few seconds. In tensile test the extensometer is removed at or before the yield point and further extension is measured by scale. The steel shows a well defined yield point.

4. Maximum Stress Point: Beyond the point "*C*", the load again starts increasing but the elongation now increases at much faster rate than the load. As the test is continued, a point of max. Stress is reached at *D*. The stress at this point is called Ultimate stress. At the point '*D*', the bar begins to form a neck or waist.

5. Breaking or Fracture Point: The reduction in area at the neck leads to a drop in the load. After point '*D*', extension remains continuous even with lesser loads and ultimately fracture occurs at point '*E*'. The stress corresponding to point '*E*' is called Breaking stress.

3.2. PRINCIPAL STRESSES AND STRAINS

3.2.1 Principal Planes

It has been observed that at any point in a strained material, there are three planes, mutually perpendicular to each other, which carry direct stresses only and no shear stresses. A little consideration will show that out of these three direct stresses, one will be max, the other minimum and the third an intermediate between the two. These particular planes, which have no shear stress, are known as "Principal Planes".

3.2.2 Principal Stresses

The magnitude of Direct stress, across a Principal plane is known as Principal Stress.

Under the concept of Principal Stresses and Principal Strains – the three planes along *X-X*, *Y-Y* and *Z-Z* axis (mutually perpendicular to each other) carry direct stresses only and no shear stress.

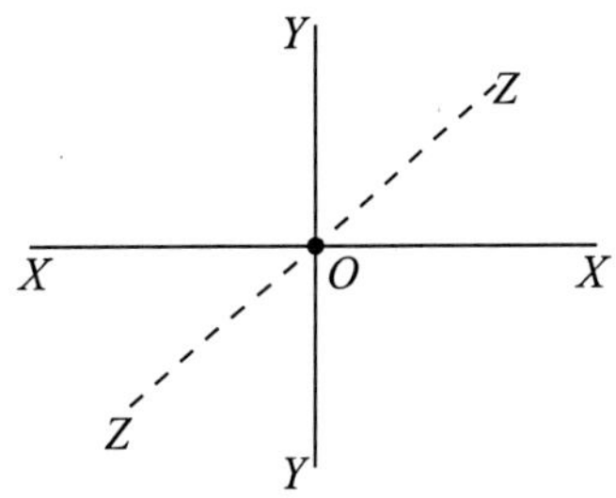

Fig. 3.6

Methods for the Stresses on an oblique section of a body: Following two methods are adopted for the determination of stresses on an oblique section of a strained body.

1. Analytical method
2. Graphical Method

1. Analytical Method: Following two cases will be discussed under this method.

(*a*) A body subjected to a direct stress in One Plane.

(*b*) A body subjected to direct stresses in two mutually perpendicular directions.

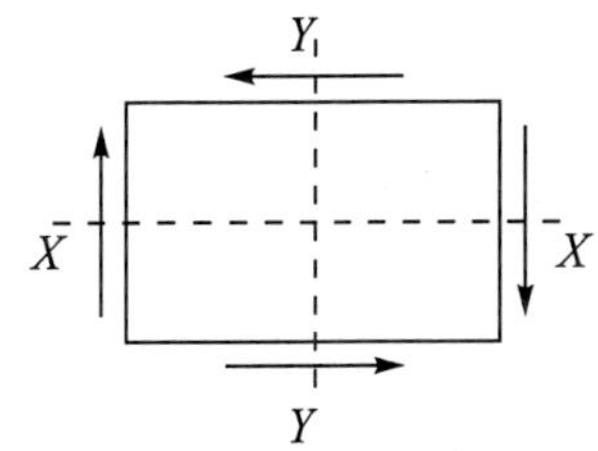

Fig. 3.7

3.2.3 Sign Conventions for Analytical Method

1. All the tensile stresses and strains are taken as positive, where as all the compressive stresses and strains are taken as negative.
2. The shear stress which tends to rotate the element in the clockwise direction is taken as positive, whereas that which tends to rotate in an anticlockwise direction as negative.

In the above Fig. 3.7, the shear stress on the vertical faces (*X–X* axis) is taken as positive, whereas the shear stress on the horizontal faces (*Y–Y* axis) is taken as negative.

3.2.4 Stresses on an Oblique Section of a Body Subjected to a Direct Stress in One Plane

Consider a rectangular body of uniform cross-sectional area and unit thickness subjected to a direct tensile stress along *X–X* as shown in the Fig. 3.8. Let us now consider on oblique section *AB* inclined with the *X–X* axis.

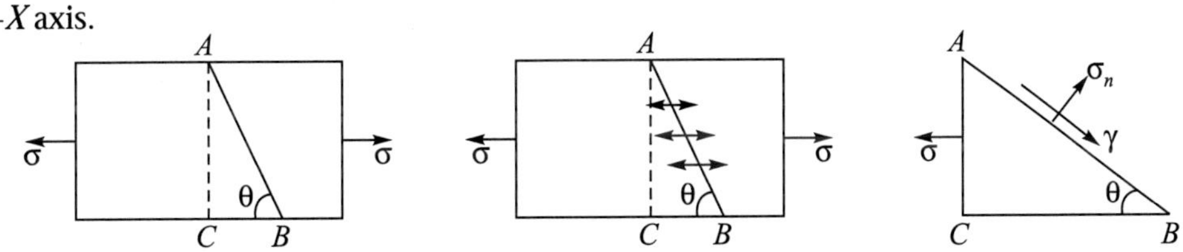

Fig. 3.8

Let σ = Tensile stress across the face *AC* and

θ = Angle, which the oblique section *AB* makes with *BC, i.e.*, with the *X – X* axis in the clockwise direction.

Consider the equilibrium of an element or wedge *ABC* whose free body diagram is shown above. We know that the horizontal force acting on the face *AC*,

$$P = \sigma \cdot AC \ (\leftarrow)$$

Resolving the force $\perp$ to the section *AB*

$$\mathbf{P_n = P \sin\theta = \sigma \cdot AC \sin\theta} \quad \text{...(3.1)}$$

Next, resolving the force tangential to the section *AB*,

$$\mathbf{P_t = P \cos\theta = \sigma \cdot AC \cos\theta} \quad \text{...(3.2)}$$

We know that the normal stress across the section *AB*,

$$\sigma_n = \frac{P_n}{AB} = \frac{\sigma \cdot AC \sin\theta}{AB} = \frac{\sigma \cdot AC \sin\theta}{\dfrac{AC}{\sin\theta}} = \sigma \sin^2\theta$$

$$\text{“}\boldsymbol{\sigma_n}\text{”} = \frac{\sigma}{2}(1 - \cos 2\theta) = \frac{\sigma}{2} - \frac{\sigma}{2}\cos 2\theta \quad \text{...(3.3)}$$

and shear stress across these sections *AB*,

$$\tau = \frac{P_t}{AB} = \sigma \frac{AC\cos\theta}{AB} = \frac{\sigma \cdot AC \cos\theta}{\dfrac{AC}{\sin\theta}} = \sigma \sin\theta \cdot \cos\theta$$

$$\text{“}\tau\text{”} = \frac{\sigma}{2}\sin 2\theta \quad \text{...(3.4)}$$

It will be interesting to know from above equations, that the Normal stress across the section *AB will be* max, when $\sin^2\theta$, = 1, or $\sin\theta = 1$ or $\theta = 90°$.

Similarly the shear stress across the section *AB* will be max when $\sin 2\theta = 1$, or $2\theta = 90°$ or 270° or in other words, the sharer stress will be max on the planes included at 45° and 135° with the line of action of the tensile stress.

$\therefore$ Max. shear stress when $\theta = 45°$

i.e.,
$$\tau_{max} = \frac{\sigma}{2}\sin 90° = \frac{\sigma}{2} \times 1 = \frac{\sigma}{2}$$

and max. shear stress, when $\theta = 135°$

$$\tau_{max} = -\frac{\sigma}{2}\sin 270° = -\frac{\sigma}{2}(-1) = \frac{\sigma}{2}.$$

It is thus obvious that the magnitude of max. shear stress is half of the tensile stress. Now the resultant stress may be found from the relation:

$$\sigma_R = \sqrt{\sigma_n^2 + \tau^2}$$

3.2.5. Stresses on an Oblique Section of a Body Subjected to Direct Stresses in Two Mutually Perpendicular Directions

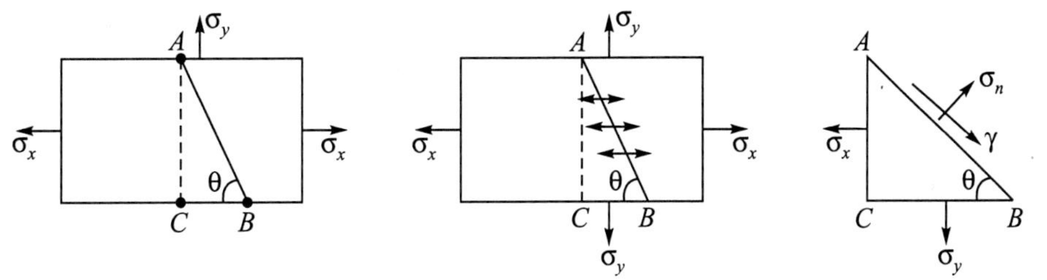

Fig. 3.9

Consider a rectangular body of uniform cross-sectional area and unit thickness subjected to direct tensile stresses in two mutually ⊥ directions along $X–X$ and $Y–Y$ axes as shown in above Fig. 3.9. Next consider an oblique section AB inclined with $X–X$ axis.

Let σ_x = Tensile stress along $X–X$ axis.

σ_y = Tensile stress along $Y–Y$ axis.

θ = Angle which the oblique see AB makes with $X–X$ axis in the clockwise direction.

First of all consider the equilibrium of the wedge ABC. We know that horizontal force acting on the face AC.

$$P_x = \sigma_x \cdot AC \ (\leftarrow)$$

and vertical force acting on the face BC.

$$P_y = \sigma_y \cdot BC \ (\downarrow)$$

Resolving the forces ⊥ to the section AB,

$$\mathbf{P_n} = P_x \sin\theta + P_y \cos\theta = \boldsymbol{\sigma_x} \cdot \mathbf{AC \sin\theta} + \boldsymbol{\sigma_y} \cdot \mathbf{BC \cos\theta} \qquad ...(3.5)$$

Next resolving the forces tangential to section AB,

$$\mathbf{P_t} = P_x \cos\theta - P_y \sin\theta = \boldsymbol{\sigma_x} \cdot \mathbf{AC \cos\theta} - \boldsymbol{\sigma_y} \cdot \mathbf{BC \sin\theta} \qquad ...(3.6)$$

We know that normal stress across the sections AB,

$$\sigma_n = \frac{P_n}{AB} = \frac{\sigma_x \cdot AC \sin\theta + \sigma_y \, BC \cos\theta}{AB}$$

$$= \sigma_x \cdot \frac{AC \sin\theta}{AB} + \frac{\sigma_y \cdot BC \cos\theta}{AB}$$

$$= \sigma_x \cdot \frac{AC \sin\theta}{\frac{AC}{\sin\theta}} + \frac{\sigma_y \cdot BC \cos\theta}{\frac{BC}{\cos\theta}}$$

$$= \sigma_x \sin^2\theta + \sigma_y \cdot \cos^2\theta = \frac{\sigma_x}{2}(1 - \cos 2\theta) + \frac{\sigma_y}{2}(1 + \cos 2\theta)$$

$$= \frac{\sigma_x}{2} - \frac{\sigma_x}{2}\cos 2\theta + \frac{\sigma_y}{2} + \frac{\sigma_y}{2}\cos 2\theta$$

$$\text{“}\boldsymbol{\sigma_n}\text{”} = \frac{\boldsymbol{\sigma_x + \sigma_y}}{\mathbf{2}} - \frac{\boldsymbol{\sigma_x - \sigma_y}}{\mathbf{2}} \cos 2\theta \qquad ...(3.7)$$

and shear stress across the section *AB*,

$$\tau = \frac{P_t}{AB} = \frac{\sigma_x \cdot AC\cos\theta - \sigma_y \cdot BC\sin\theta}{AB}$$

$$= \frac{\sigma_x \cdot AC\cos\theta}{AB} - \frac{\sigma_y \cdot BC\sin\theta}{AB}$$

$$= \frac{\sigma_x \cdot AC\cos\theta}{\frac{AC}{\sin\theta}} - \frac{\sigma_y \cdot BC\sin\theta}{\frac{BC}{\cos\theta}}$$

$$= \sigma_x \cdot \sin\theta \cdot \cos\theta - \sigma_y \sin\theta \cdot \cos\theta$$

$$= (\sigma_x - \sigma_y)\sin\theta \cdot \cos\theta$$

$$\text{“}\tau\text{”} = \frac{\sigma_x - \sigma_y}{2}\sin 2\theta \qquad \text{...(3.8)}$$

It will be interesting to know from the eqs. (3.4), (3.3), the shear stress across the see *AB* will be max., when sin 2θ = 1, or 2θ = 90°, or θ = 45°.

$$\therefore \qquad \tau_{max} = \frac{\sigma_x - \sigma_y}{2}$$

and the resultant stress

$$\sigma_R = \sqrt{\sigma_n^2 + \tau^2}$$

3.2.6 Graphical Method for Stress on an Oblique Section of a Body (Mohr's Circle of Stresses)

Sign Conventions for Graphical Method: 1. The angle is taken with reference to the *X-X* axis. All the angle traced in the anticlockwise direction to the *X-X* axis are taken as negative, whereas those in the clockwise direction as positive. The value of angle θ, untill and unless mentioned is taken as positive and drawn clockwise.

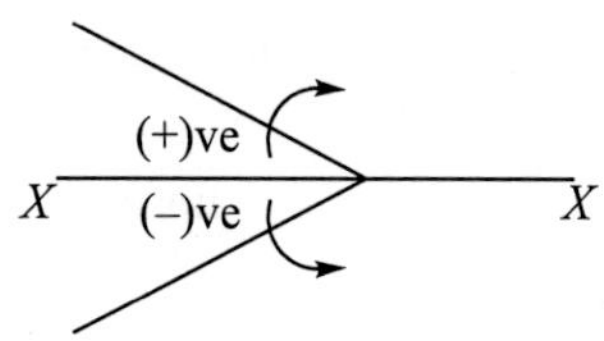

Fig. 3.10

2. The measurements above *X-X* axis and to the right of *Y-Y* axis are taken as positive, whereas those below *X-X* axis and to the left of *Y-Y* axis as negative.

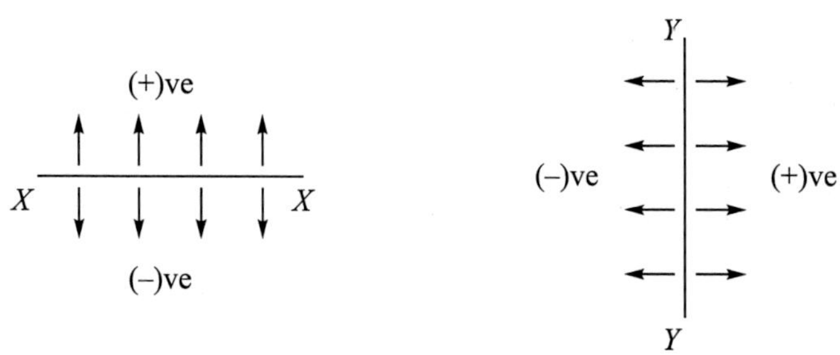

Fig. 3.11

3.3. MOHR'S CIRCLE FOR STRESSES ON AN OBLIQUE SECTION OF A BODY SUBJECTED TO A DIRECT STRESS IN ONE PLANE

Consider a rectangular body of uniform cross-sectional area and unit thickness subjected to a direct tensile stress along *X-X* axis as shown in Fig. 3.12. Now let us consider an oblique section *AB* inclined with *X-X* axis, on which we are required to find out the stresses as shown in Fig. 3.12.

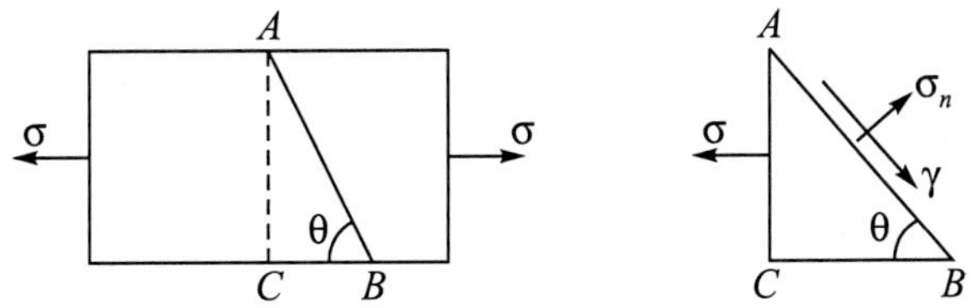

Fig. 3.12

Let σ = Tensile stress, in *X-X* direction.

θ = Angle which the oblique-section *AB* makes with the *X-X* axis in clockwise direction.

First of all, consider the equilibrium of the wedge *ABC*. Now draw the Mohr's circle of stresses as discussed below (refer Fig. 3.13):

(*a*) Select some suitable point "*O*" and through it draw a horizontal line *XOX*.

(*b*) Cut off *OJ* equal to the tensile stress (σ) to some suitable scale and towards right (because σ is tensile). Bisect *OJ* at "*C*".

Now the point "*O*" represents the stress system on plane *BC* and the point '*J*' represents the stress system on plane *AC*.

(*c*) Now with "*C*" as centre and radius equal to *CO* or *CJ* draw a circle. It is known as Mohr's circle for stresses.

(*d*) Now through *C* draw a line *CP* making an angle 2θ with *CO* in the clockwise direction meeting the circle at '*P*'. The point '*P*' represents the section *AB*.

(*e*) Through '*P*' draw $PQ \perp OX$. Join *OP*.

(*f*) Now *OQ*, *QP* and *OP* will give the normal stress, shear stress and resultant stress respectively to the scale. The angle *POJ* is called the angle of obliquity (θ).

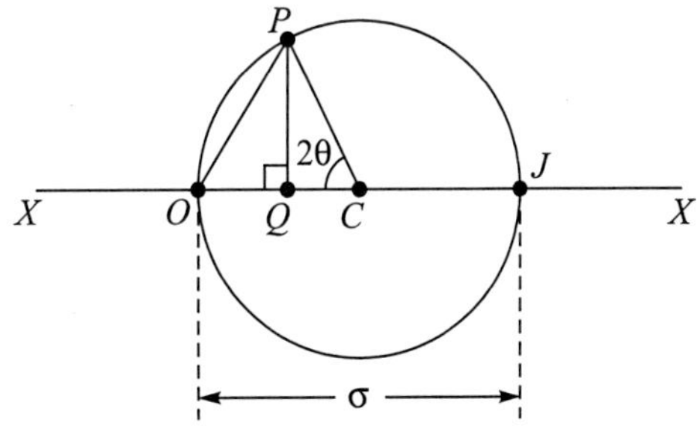

Fig. 3.13

3.4. MOHR'S CIRCLE FOR STRESSES ON AN OBLIQUE SECTIONS OF A BODY SUBJECTED TO DIRECT STRESSES IN TWO MUTUALLY PERPENDICULAR DIRECTIONS

Consider a rectangular body of uniform cross-sectional area and unit thickness subjected to direct tensile stress in two mutually perpendicular directions along *X-X* and *Y-Y* axis as shown in Fig. 3.14. Now let us consider an oblique section *AB* incline with *X-X* axis on which we are required to find out the stresses.

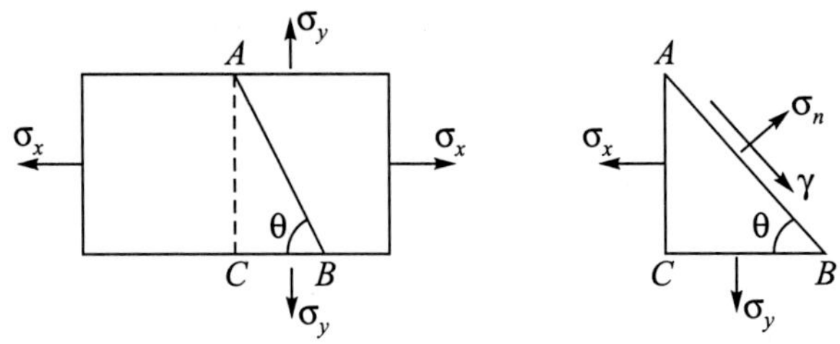

Fig. 3.14

Let σ_x = Tensile stress in *X-X* direction

σ_y = Tensile stress in *Y-Y* direction

θ = Angle which the oblique section *AB* makes with *X-X* axis in clockwise direction.

First of all consider the equilibrium of the wedge *ABC*. Now draw the Mohr's circle of stress as discussed below (Refer (Fig. 3.15):

(*a*) Take some suitable point '*O*' and draw a horizontal line *OX*.

(*b*) Cut of *OJ* and OK equal to the tensile stresses σ_x and σ_y to some suitable scale towards right (because both the stresses are tensile). The point "*J*" represents the stress system on plane *AC* and the point '*K*' represents the stress system on plane *BC*. Bisect *JK* at "*C*".

(*c*) With '*C*' as centre and radius equal to *CJ* draw a circle. It is known as Mohr's circle of stresses.

(*d*) Now through *C*, draw a line *CP* making an angle of 2θ with *CK* in clockwise direction meeting the circle at '*P*'. The point '*P*' represents the stress system on the section *AB*.

(*e*) Through *P*, draw $PQ \perp$ to the line *OX*. Join *OP*.

(*f*) Now *OQ*, *QP* and *OP* will give the normal stress, shear stress and resultant stress respectively to the scale. Similarly *CM* or *CN* will give the max. shear stress to the scale. The angle *POC* is called the angle of obliquity.

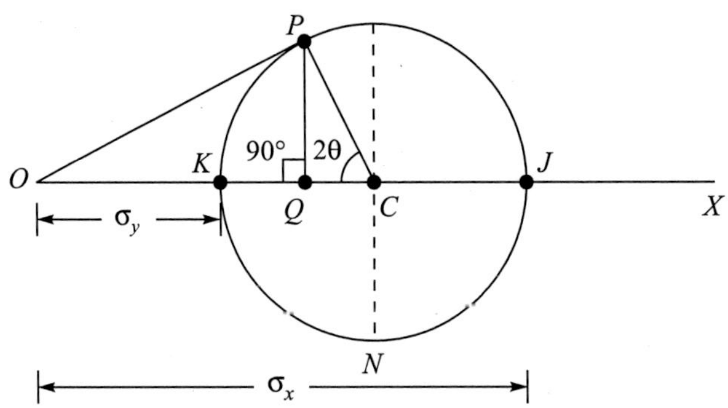

Fig. 3.15

PROBLEM ON PRINCIPAL STRESSES

Example 3.1: *A wooden bar is subjected to a tensile stress of 5 MPa. What will be the values of normal and shear stresses across a section which makes and angle of 25° with the direction of the tensile stress.*

Solution: Given: Tensile stress (σ) = 5 MPa, and angle made by section with the direction of tensile stress (θ) = 25°

(*a*) Normal stress across the section,

$$\sigma_n = \frac{\sigma}{2} - \frac{\sigma}{2}\cos 2\theta = \frac{5}{2} - \frac{5}{2}\cos(2 \times 25°)$$

$$= 2.5 - 2.5 \cos 50° = 2.5 - 1.607 = 0.89 \text{ MPa}$$ **Ans.**

(*b*) Shear stress across the section,

$$\text{“}\tau\text{”} = \frac{\sigma}{2}\sin 2\theta = \frac{5}{2}\sin(2 \times 25°)$$

$$= 2.5 \sin 50° = 2.5 \times 0.766$$

$$= 1.915 \text{ MPa}$$ **Ans.**

Example 3.2: *Two wooden pieces 100 mm × 100 mm in Cross-Sectional are joined together along a line AB, as shown in Fig. 3.16. Find the Max. force (P), which can be applied if the shear stress along the joint AB is 1.3 MPa.*

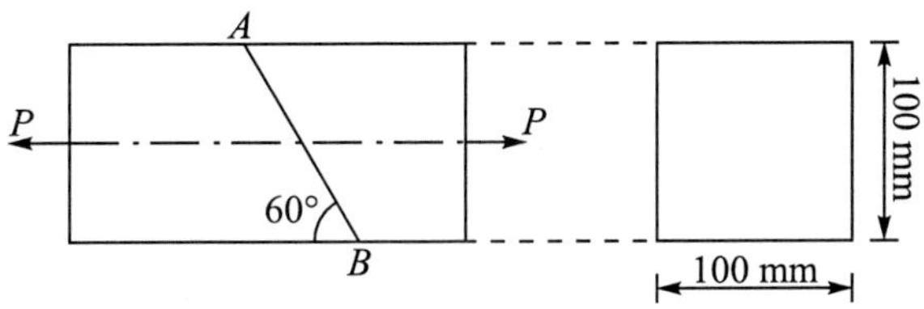

Fig. 3.16

Solution: Given:

Cross-Sectional area = 100 mm × 100 mm

θ = 60°

Permissible shear stress “τ” = 1.3 MPa = 1.3 N/mm^2

Let σ = Safe tensile stress in the member.

Area of Cross-sectional of wooden remember,

$$\text{“}A\text{”} = 100 \times 100 = 10000 \text{ mm}^2$$

and shear stress “τ”,

$$1.3 = \frac{\sigma}{2}\sin(2 \times 60°) = \frac{\sigma}{2}\sin 120° = \frac{\sigma}{2} \times 0.866$$

or $$1.3 = 0.433\,\sigma$$

∴ $$\sigma = \frac{1.3}{0.433} = 3.0 \text{ N/mm}^2$$

∴ The Max. Axial force “P” = σ × A = 3 × 10000 = 30 kN **Ans.**

Example 3.3: *A tension member is formed by connecting two wooden members 200 mm × 100 mm as shown in Fig. 3.17. Determine the safe value of force (P), if* **(MDU–2008)**

$$\sigma_n = 0.5 \text{ MPa and } \gamma = 1.25 \text{ MPa}$$

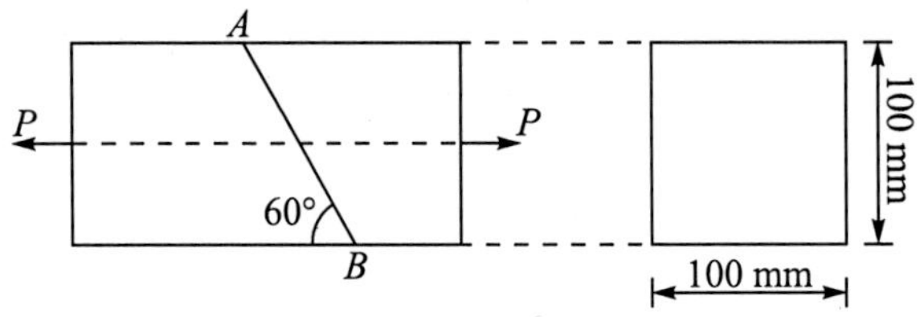

Fig. 3.17

Solution: Given:

Cross-Sectional area = 200 mm × 100 mm

$\theta = 60°$

$\sigma_n = 0.5$ MPa and $\tau = 1.25$ MPa

Let σ = Safe stress in the joint in N/mm^2

We have, Cross-sectional area of member

"A" = 200 × 100 = 20,000 mm^2

We also know that normal stress (σ_n),

$$0.5 = \frac{\sigma}{2} - \frac{\sigma}{2}\cos 2\theta = \frac{\sigma}{2} - \frac{\sigma}{2}\cos(2 \times 60°)$$

$$= \frac{\sigma}{2} - \frac{\sigma}{2}\cos 120° = \frac{\sigma}{2} - \frac{\sigma}{2}(0.5) = 0.75\,\sigma$$

$$\therefore \quad \sigma = \frac{0.5}{0.75} = 0.67 \text{ N/mm}^2 \quad \ldots(i)$$

and shear stress "τ",

$$1.25 = \frac{\sigma}{2}\sin 2\theta = \frac{\sigma}{2}\sin(2 \times 60°) = \frac{\sigma}{2}\sin 120°$$

$$= \frac{\sigma}{2} \times 0.866 = 0.433\,\sigma$$

$$\therefore \quad \sigma = \frac{1.25}{0.433} = 2.89 \text{ N/mm}^2 \quad \ldots(ii)$$

From the above two values, we find that the safe stress is least of the two values *i.e.*, 0.67 N/mm^2.

∴ The safe value of the force

$P = \sigma \times A = 0.67 \times 20000 = 13400$ N = 13.4 kN **Ans.**

Example 3.4: *A point in a strained material is subjected to two mutually perpendicular tensile stresses of 200 MPa and 100 MPa. Determine the intensities of normal, shear and resultant stresses on a plane inclined at 30° with axis of minor tensile stress.*

Solution: Given:

Tensile stress along X-X axis

$\sigma_x = 150$ MPa

Tensile stress along Y-Y axis

$\sigma_y = 100$ MPa

$\theta = 30°$

Normal stress on the inclined plane,

$$\sigma_n = \frac{\sigma_x + \sigma_y}{2} - \frac{\sigma_x - \sigma_y}{2}\cos 2\theta$$

$$= \frac{200+100}{2} - \frac{200-100}{2}\cos(2\times 30^\circ)$$

$$= 150 - (50 \times 0.5) = 125 \text{ MPa}$$ **Ans.**

Shear stress on the inclined plan

$$'\tau' = \frac{\sigma_x - \sigma_y}{2}\sin 2\theta = \frac{200-100}{2}\sin 60^\circ$$

$$= 50 \sin 60^\circ = 50 \times 0.866 = 43.3 \text{ MPa}$$ **Ans.**

Resulted stress on the inclined plane

$$= \sigma_R = \sqrt{\sigma_n^2 + \tau^2} = \sqrt{(125)^2 + (43.3)^2}$$

$$= 132.3 \text{ MPa.}$$ **Ans.**

Example 3.5: *The stresses at a point of a machine component are 150 MPa and 50 MPa both tensile. Find the intensities of normal, shear and resultant stress on a plane inclined at an angle of 55° with the axis of major tensile stress. Also find the magnitude of Max. shear stress in the component.*

(KU–2008)

Solution: Given:

Tensile stress along *X-X* Axis (σ_x) = 150 MPa

Tensile stress along *Y-Y Axis* (σ_y) = 50 MPa

θ = 55°

Normal stress on the inclined plane,

$$\sigma_n = \frac{\sigma_x + \sigma_y}{2} - \frac{\sigma_x - \sigma_y}{2}\cos 2\theta$$

$$= \frac{150+50}{2} - \frac{150-50}{2}\cos(2\times 55^\circ)$$

$$= 100 - 50\cos 110^\circ = 100 - 50(-0.342)$$

$$= 117.1 \text{ MPa}$$ **Ans.**

Shear stress on the inclined plane,

$$\text{“}\tau\text{”} = \frac{\sigma_x - \sigma_y}{2}\sin 2\theta = \frac{150-50}{2}\times \sin(2\times 50^\circ)$$

$$= 50 \sin 110^\circ = 50 \times 0.9397 = 47 \text{ MPa}$$ **Ans.**

Resultant stress on inclined plane

$$\sigma_R = \sqrt{\sigma_n^2 + \tau^2} = \sqrt{(117.1)^2 + (47.0)^2} = 126.2 \text{ MPa}$$ **Ans.**

Magnitude of the max. shear stress in the component

$$\tau_{max} = \pm\frac{\sigma_x - \sigma_y}{2} = \pm\frac{150-50}{2} = \pm 50 \text{ MPa}$$ **Ans.**

Example 3.6: *The stresses at a point in a component are 100 MPa (tensile) and 50 MPa (Compressive). Determine the magnitude of the normal and shear stresses on a plane inclined at an angle of 25° with tensile stress. Also determine the direction of the resultant stress and the magnitude of the Max. intensity of shear stress.*

Solution: Given:

Tensile stress along *X-X* axis (σ_x) = 100 MPa.

Compressive stress along *Y-Y* axis (σ_y) = (–) 50 MPa (–ive sign due to compression)

$\theta = 25°$

Normal stress on the inclined plane,

$$\sigma_n = \frac{\sigma_x + \sigma_y}{2} - \frac{\sigma_x - \sigma_y}{2} \cos 2\theta$$

$$= \frac{100 + (-50)}{2} - \frac{100 - (-50)}{2} \cos (2 \times 25°)$$

$$= 25 - 75 \cos 50° = 25 - (75 \times 0.6428)$$

$$= -23{,}21 \text{ MPa}$$ **Ans.**

Shear stress on the inclined plane

$$\text{“}\tau\text{”} = \frac{\sigma_x - \sigma_y}{2} \sin 2\theta = \frac{100 - (-50)}{2} \sin (2 \times 25°)$$

$$= 75 \sin 50 = 75 \times 0.766 = 57.45 \text{ MPa}$$ **Ans.**

Direction of resultant stress,

Let “θ” be the angle which the resultant stress makes with *X-X* axis.

$$\tan \theta = \frac{\tau}{\sigma_n} = \frac{57.45}{-23.21} = (-)\ 2.4752 \quad \text{or} \quad \text{“}\theta\text{”} = (-)\ 68°$$ **Ans.**

Magnitude of max. shear stress,

$$\tau_{max.} = \pm \frac{\sigma_x - \sigma_y}{2} = \pm \frac{100 - (50)}{2} = \pm 75 \text{ MPa}$$ **Ans.**

Example 3.7: *An element in a strained body is subjected to a tensile stress of 150 MPa and a shear stress of 50 MPa tending to rotate the element in a anticlockwise direction. Find:*

(i) The magnitude of the normal and shear stress on a section inclined at 40° with the tensile stress and

(ii) the magnitude and directions of max. shear stress that can exist on the element.

Solution: Given:

Tensile stress along horizontal *X-X* axis

$\sigma_x = 150$ MPa.

Shear stress $\gamma_{xy} = (-)\ 50$ MPa (– sign due to anticlockwise)

$\theta = 40°$

(*i*) Magnitude of the normal and shear stress on the see:

$$\sigma_n = \frac{\sigma_x}{2} - \frac{\sigma_x}{2} \cos 2\theta - \gamma_{xy} \sin 2\theta$$

$$= \frac{150}{2} - \frac{150}{2}\cos(2\times 40^{o}) - (-50)\sin(2\times 40^{o})$$

$$= 75 - (75\times 0.1736) + (50\times 0.9848)$$

$$= 75 - 13.02 + 49.24 = 111.22 \text{ MPa.}$$ **Ans.**

and shear stress on the section

$$\text{“}\tau\text{”} = \frac{\sigma_x}{2}\sin 2\theta - \gamma_{xy}\cos 2\theta$$

$$= \frac{150}{2}\sin(2\times 40°) - (-50)\cos(2\times 40°)$$

$$= (75\times 0.9848) + (50\times 0.1736)$$

$$= 73.86 + 8.68 = 82.54 \text{ MPa.}$$ **Ans.**

(*ii*) Magnitude and direction of the max. shear stress that can exist on the element.

$$\tau_{max} = \pm\sqrt{\left(\frac{\sigma_x}{2}\right)^2 + \tau_{xy}^{\ 2}} = \pm\sqrt{\left(\frac{150}{2}\right)^2 + (-50)^2}$$

$$= \pm 90.14 \text{ MPa}$$ **Ans.**

Let θ_x = Angle which plane of max. shear stress makes with *X-X* axis.

$$\tan 2\theta_s = \frac{\sigma_x}{2\tau_{xy}} = \frac{150}{2\times 50} = 1.5$$

or $2\theta_s = 56.3°$

$\therefore$ $\theta_s = 28.15°$ or $118.15°$ **Ans.**

Example 3.8: *A plane element in a boiler is subjected to tensile stresses of 400 MPa on one plane and 150 MPa on the other at right angles to the former. Each of the stresses is accompanied by a shear stress of 100 MPa such that when associated with the minor tensile stress tends to rotate the element in anticlockwise direction. Find:*

(a) Principal stresses and their directions.

(b) Max. shearing stresses and the direction of the plane on which they act.

Solution: Given:

Tensile stress along *X-X* axis (σ_x) = 400 MPa

Tensile stress along *Y-Y* axis (σ_y) = 150 MPa

and shear stress τ_{xy} = (–) 100 MPa (– sign due to anticlockwise)

(*a*) Principal stresses and their directions

$$\sigma_{max} = \frac{\sigma_x + \sigma_y}{2} + \sqrt{\left(\frac{\sigma_x - \sigma_y}{2}\right)^2 + \tau_{xy}^2}$$

$$= \frac{400 + 150}{2} + \sqrt{\left(\frac{400 - 150}{2}\right)^2 + (-100)^2}$$

$$= 275 + 160.1 = 435.1 \text{ MPa}$$ **Ans.**

and minimum principal stress,

$$\sigma_{min} = \frac{\sigma_x + \sigma_y}{2} - \sqrt{\left(\frac{\sigma_x - \sigma_y}{2}\right)^2 + \tau_{xy}^2}$$

$$= \frac{400+150}{2} - \sqrt{\left(\frac{400-150}{2}\right)^2 + (-100)^2}$$

$$= 275 - 160.1 = 114.9 \text{ MPa}$$ **Ans.**

Let θ_p = Angle which plane of principal stress makes with *X-X* axis,

$$\tan 2\theta_p = \frac{2\tau_{xy}}{\sigma_x - \sigma_y} = \frac{2\times 100}{400-150} = 0.8$$

or $2\theta_p = 38.66°$

or $\theta_p = 19.33°$ or $109.33°$ **Ans.**

(*b*) Max. shearing stress and their direction.

$$\tau_{max} = \sqrt{\left(\frac{\sigma_x - \sigma_y}{2}\right)^2 + \tau_{xy}^2} = \sqrt{\left(\frac{400-150}{2}\right)^2 + (-100)^2}$$

$$= 160.1 \text{ MPa}$$ **Ans.**

Let θ_s = Angle which plane of max. shearing stress makes with *X-X* axis.

$$\tan 2\,\theta_s = \frac{\sigma_x - \sigma_y}{2\tau_{xy}} = \frac{400-150}{2\times 100} = 1.25 \quad \text{or} \quad 2\,\theta_s = 51.34°$$

or $\theta_s = 25.76°$ or $115.67°$ **Ans.**

Example 3.9: *A point in the strained material is subjected to the stresses as shown in Fig. 3.18.*

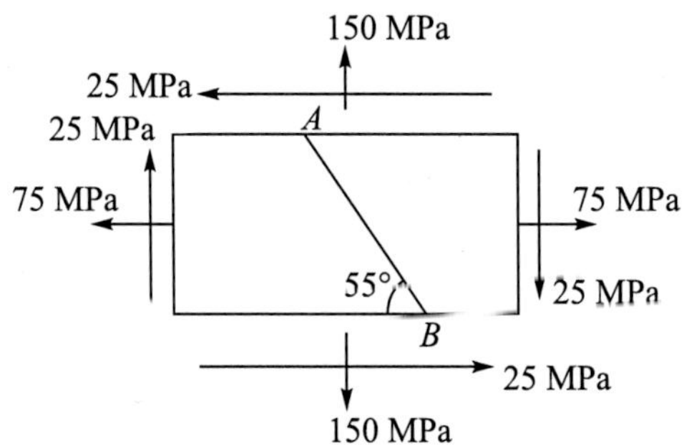

Fig. 3.18

Find normal and shear stress on the See. AB.

Solution: Given:

Tensile stress along horizontal *X-X* Axis,

$$\sigma_x = 75 \text{ MPa}$$

Tensile stress along vertical *Y-Y* Axis,

$$\sigma_y = 150 \text{ MPa}$$

Shear stress $\tau_{xy} = 25$ MPa

$\theta = 55°$

Normal stress on section *AB*,

$$\sigma_n = \frac{\sigma_x + \sigma_y}{2} - \frac{\sigma_x + \sigma_y}{2}\cos 2\theta - \tau_{xy}\sin 2\theta$$

$$= \frac{75+150}{2} - \frac{75-150}{2}\cos(2\times 55°) - 25\sin(2\times 55°)$$

$$= 112.5 + 37.5\cos 110° - 25\sin 110°$$

$$= 112.5 + 37.5 \times (-0.342) - 25\,(0.9397)$$

$$= 112.5 - 12.83 - 23.49 = 76.18 \text{ MPa}$$ **Ans.**

Shear stress on section *AB*,

$$\tau = \frac{\sigma_x - \sigma_y}{2}\sin 2\theta - \tau_{xy}\cos 2\theta$$

$$= \frac{75-150}{2}\sin(2\times 55°) - 25\cos(2\times 55°)$$

$$= -37.5\sin 110° - 25\cos 110°$$

$$= -37.5 \times 0.9397 - 25 \times (-0.342)$$

$$= (-)\ 26.69 \text{ MPa}$$ **Ans.**

Example 3.10: *A machine component is subjected to the stresses as shown in Fig. 3.19.*

(MDU–2009)

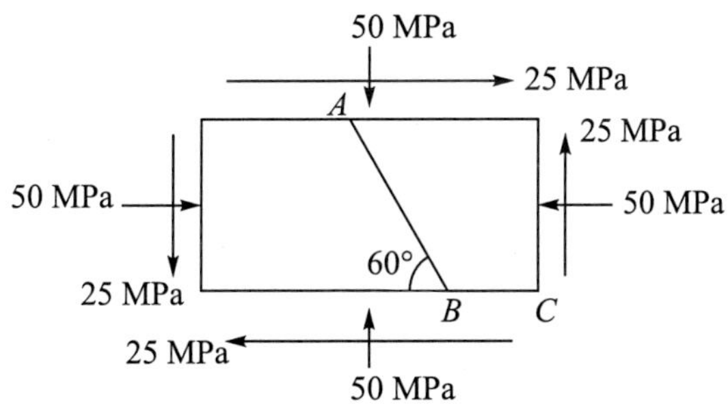

Fig. 3.19

Find the normal and shearing stresses on the section AB inclined at an angle of 60° with X-X axis. Also find the resultant stress on the section.

Solution: Given:

Compressive stress along horizontal *X-X* axis (σ_x)

$= (-)\ 100$ MPa (– sign due to compressive stress)

Compressive stress along vertical *Y-Y* axis (σ_y)

$= (-)\ 50$ MPa (– sign due to compressive stress)

Shear stress

$\tau_{xy} = (-)\ 25$ MPa (– sign the to anticlockwise)

$\theta = 60°$

Normal stress on the section *AB*.

$$\sigma_n = \frac{\sigma_x + \sigma_y}{2} - \frac{\sigma_x - \sigma_y}{2} \cos 2\theta - \tau_{xy} \sin 2\theta$$

$$= \frac{-100 + (-50)}{2} - \frac{100 - (-50)}{2} \cos (2 \times 60°) - [-25 \sin (2 \times 60°)$$

$$= -75 + 25 \cos 120° + 25 \sin 120°$$

$$= -75 + [25 \times (-0.5)] + (2.5 \times 0.866)$$

$$= (-)\ 65.85 \text{ MPa}$$

Ans.

Shear stress on see *AB*,

$$\tau = \frac{\sigma_x - \sigma_y}{2} \sin 2\theta - \tau_{xy} \cos 2\theta$$

$$= \frac{-100 - (-50)}{2} \sin (2 \times 60°) - [-25 \cos (2 \times 60°)$$

$$= -25 \sin 120° + 25 \cos 120°$$

$$= -25 \times 0.866 + [25 \times (-0.5)]$$

$$= (-)\ 34.15 \text{ MPa}$$

Ans.

Resultant stress on section *AB*,

$$\sigma_R = \sqrt{\sigma_n^2 + \tau_\gamma^2} \sqrt{(-65.85)^2 + (-34.15)^2}$$

$$= 74.2 \text{ MPa}$$

Ans.

Example 3.11: *A circular bar is subjected to an axial pull of 100 kN. If the max. intensity of shear stress on any oblique plane is not to exceed 60 MN/m², determine the diameter of the bar.*

Solution: Axial pull, $P = 100$ KN

Max. intensity of shear stress,

$$\tau_{max} = 60 \text{ MN/m}^2$$

Diameter of the bar, "*d*",

Cross-sectional area of the bar

$$A = \frac{\pi}{4} \times d^2$$

But the max. shear stress produced,

$$\tau_{max} = \frac{P}{2A}$$

$$60 \times 10^6 = \frac{100 \times 1000}{2 \times \left(\frac{\pi}{4}\right) \times d^2}$$

$$\text{"}d\text{"} = \left[\frac{100 \times 1000 \times 4}{2 \times \pi \times 60 \times 10^6}\right]^{\frac{1}{2}} = 0.0326 \text{ m}$$

or $$\text{"}d\text{"} = 32.6 \text{ mm}$$

Ans.

PROBLEMS ON HOOKE'S LAW AND ELASTIC CONSTANTS

Example 3.1: *A steel rod 1 m long and 20 mm × 20 mm in Cross-sectional area is subjected to a tensile force of 40 KN. Determine the elongation of the rod, if modulus of elasticity for the rod is 200 GPa.*

Solution: We know that,

Stress, $\sigma = \dfrac{P}{A}$

and Strain 'Σ' $= \dfrac{\sigma}{E} = \dfrac{P}{AE}$ $\quad (\because \ \sigma = \dfrac{P}{A})$

and deformation "δl" $= \Sigma \times l = \dfrac{P}{AE} \times l$

Given, $l = 1\text{m} = 1 \times 10^3$ mm

$$A = 20 \times 20 = 400 \text{ mm}^2$$

$$P = 40 \text{ KN} = 40 \times 10^3 \text{ N}$$

$$E = 200 \text{ GPa} = 200 \times 10^3 \text{ N/mm}^2$$

$$\therefore \quad \delta l = \frac{P}{AE} \times l = \frac{(40 \times 10^3) \times (1 \times 10^3)}{400 \times (20 \times 10^3)} = 0.5 \text{ mm} \qquad \textbf{Ans.}$$

Example 3.2: *In an experiment, a steel specimen of 13 mm φ was found to elongate 0.2 mm in a 200 mm gauge length when it was subjected to a tensile force of 26.8 KN. If the specimen was tested within the elastic range, what is the value of Young's modulus for the steel specimen?*

Solution: Given:

"d" = 13 mm, $\delta l = 0.2$ mm

"l" = 200 mm, $P = 26.8$ KN

Let 'E' be the Young's modulus for the steel specimen

We know that,

$$A = \frac{\pi \times d^2}{4} = \frac{\pi \times (13)^2}{4} = 132.73 \text{ mm}^2$$

$$\delta l = 0.2 = \frac{P}{AE} \times l = \frac{26.8 \times 20}{132.73E} = \frac{40.38}{E}$$

$$\therefore \quad \text{"}E\text{"} = \frac{40.38}{0.2} = 201.9 \text{ KN/mm}^2 = 201.9 \text{ GPa} \qquad \textbf{Ans.}$$

Example 3.3: *A hollow cylinder 2 m long has an outside diameter of 50 mm and inside diameter of 30 mm. If the cylinder is carrying a load of 25 KN, find the stress in the cylinder. Also find the deformation of the cylinder, if the value of modulus of elasticity for the cylinder material is 100 GPa.*

Solution: Given:

Length (l) = 2 m = 2×10^3 mm

Outside diameter 'D' = 50 mm,

Inside diameter 'd' = 30 mm

Load 'P' = 25 KN = 25×10^3 N

"E" = 100 GPa = 100×10^3 N/mm^2

Stress in the cylinder:

We know that X-see area of the hollow-cylinder

$$\text{“}A\text{”} = \frac{\pi}{4} \times (D^2 - d^2) = \frac{\pi}{4} \times [(50)^2 - (30)^2]$$

$$= 1257 \text{ mm}^2$$

and Stress in cylinder

$$\text{“}\sigma\text{”} = \frac{P}{A} = \frac{25 \times 10^3}{1257} = 19.9 \text{ MPa}$$ **Ans.**

Deformation of the cylinder,

$$\text{‘}\delta l\text{’} = \frac{P \times l}{A.E} = \frac{(25 \times 10^3) \times (2 \times 10^3)}{1257 \times (100 \times 10^3)} = 0.4 \text{ mm}$$ **Ans.**

Example 3.4: *A hollow steel tube 3.5 m long has external dia. of 120 mm. In order to determine the internal diameter, the tube was subjected to a tensile load of 400 KN and extension was measured to be 2 mm. If the modulus of elasticity for the tube material is 200 GPa, determine the internal dia. of the tube.*

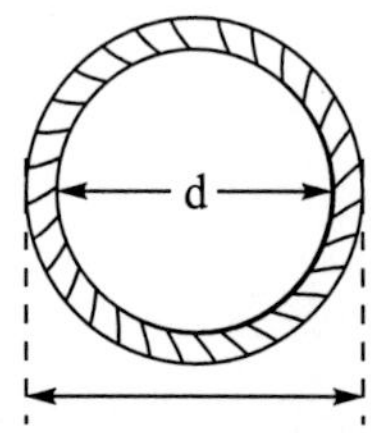

Fig. 3.20

Solution: Given:

$$\text{“}l\text{”} = 3.5 \text{ m} = 3.5 \times 10^3 \text{ mm}$$

$$\text{“}D\text{”} = 120 \text{ mm}, \text{“}P\text{”} = 400 \text{ KN} = 400 \times 10^3 \text{ N}$$

$$\text{“}\delta l\text{”} = 2 \text{ mm}, \text{‘}E\text{’} = 200 \text{ GPa} = 200 \times 10^3 \text{ N/mm}^2$$

Let ‘d’ = Internal dia. of the tube in mm.

We know that,

$$\text{“}A\text{”} = \frac{\pi}{4} \times [(120)^2 - (d)^2] = 0.7854\,[(120)^2 - (d)^2]$$

and extension of the tube (δl),

$$2 = \frac{P \times l}{A \times E} = \frac{(400 \times 10^3) \times (3.5 \times 10^3)}{0.7854\,[(120)^2 - (d)^2] \times 200 \times 10^3}$$

$$= \frac{8913}{14400 - d^2}$$

$\therefore$ $$28800 - 2d^2 = 8913 \quad \text{or} \quad 2d^2 = 28800 - 8913 - 19887$$

$\therefore$ $$d^2 = \frac{19887}{2} = 9943.5$$

or $$\text{“}d\text{”} = 99.71 \text{ mm}$$ **Ans.**

Example 3.5: *A rod 20 mm in diameter and 2 m long is subjected to an axial pull of 35 KN. If the Young's modulus of elasticity of the material of the rod is 210 GN/m², determine:*

(i) the stress

(ii) the strain

(iii) the elongation of the rod.

Solution:

$$"l" = 2 \text{ m}$$
$$'d' = 20 \text{ mm}$$
$$A = \frac{\pi d^2}{4} = \frac{\pi}{4} \times 20^2 = 314.16 \text{ mm}^2$$
$$= 314.16 \times 10^{-6} \text{ m}^2$$

Axial pull $P = 35 \text{ kN} = 35000 \text{ N}$

Young's modulus of elasticity,

$$E = 210 \text{ GN/m}^2 = 210 \times 10^9 \text{ N/m}^2.$$

(*i*) Stress, $\sigma = \frac{P}{A} = \frac{35000}{314.16 \times 10^{-6}} = 111.41 \times 10^6 \text{ N/m}^2$

$= 111.41 \text{ N/mm}^2$ **Ans.**

(*ii*) Strain, $e = \frac{\sigma}{E} = \frac{111.41 \times 10^6}{210 \times 10^9} = 0.00053$ **Ans.**

(*iii*) Elongation, $\delta l = l \times \text{strain}$ $\quad (\because \text{ Strain} = \frac{\delta l}{l})$

$= 2 \times 0.00053 = 0.00106 \text{ m} = 1.06 \text{ mm}$ **Ans.**

Example 3.6: *Compute the stress in steel rod, if it is subjected to an elongation of 0.09% of its length. E for steel = 2×10^5 N/mm^2.*

Solution:

$$E = 2 \times 10^5 \text{ N/mm}^2$$
l = original length of rod.

$\therefore$ Elongation, $\delta l = \frac{0.09}{100} \times l$

or $\frac{\delta l}{l} = \frac{0.09}{100}$

$\therefore$ Strain $= \frac{\delta l}{l} = \frac{0.09}{100} = 0.0009$

Stress $= E \times \text{Strain}$ $\quad \left[\because E = \frac{\text{Stress}}{\text{Strain}}\right]$

$= 2 \times 10^5 \times 0.0009 = 180 \text{ N/mm}^2$ **Ans.**

Example 3.7: *A hollow cast iron cylinder 4 m long, 300 mm outer diameter and thickness of metal 50 mm is subjected to a central load on the top when standing, straight. The stress produced is 75000 KN/m^2. Assume Young's modulus for cast-iron as 1.5×10^8 KN/m^2 and find: (i) magnitude of the load, (ii) longitudinal strain produced, and (iii) total decrease in length.*

Solution: $D = 300 \text{ mm} = 0.3 \text{ m}$

$$t = 50 \text{ mm} = 0.05 \text{ m}$$

$$l = 4 \text{ m}$$
$$\sigma = 75000 \text{ kN/m}^2$$
$$E = 1.5 \times 10^8 \text{ kN/m}^2$$
Inner diameter "d" $= D - 2t = 0.3 - 2 \times 0.05 = 0.2$ m

(*i*) Magnitude of the load "P"

$$\sigma = P/A$$

or $$P = \sigma \times A = 75000 \times \frac{\pi}{4}(D^2 - d^2)$$
$$= 75000 \times \frac{\pi}{4}(0.3^2 - 0.2^2)$$

∴ "P" = 2945.2 kN. **Ans.**

(*ii*) Longitudinal strain produced, e:

$$e = \frac{\text{Stress}}{E} = \frac{75000}{1.5 \times 10^8} = 0.0005$$ **Ans.**

(*iii*) Total decrease in length, δl:

$$\text{Strain} = \frac{\text{Change in length}}{\text{Original length}} = \frac{\delta l}{l}$$
$$0.0005 = \frac{\delta l}{4}$$
$$\delta l = 0.0005 \times 4 = 0.002 \text{ m} = 2 \text{ mm}$$

Hence decrease in length = 2 mm **Ans.**

Example 3.8: *A steel wire 2 m long and 3 mm in diameter is extended by 0.75 mm when a weight w is suspended from the wire. If the same weight is suspended from a brass wire, 2.5 m long and 2 mm in diameter, it is elongated by 4.64 mm. Determine the modulus of elasticity of brass if that of steel be* 2.0×10^5 *N/mm²*.

Solution: Given:

$$E_s = 2.0 \times 10^5 \text{ N/mm}^2$$
$$l_b = 2.5 \text{ m}, d_b = 2 \text{ mm}, \delta l_b = 4.64 \text{ mm}$$

Modulus of elasticity of brass, E_b,

From Hooke's Law, $$\delta l = \frac{Pl}{A \times E},$$

(*a*) For steel wire,

$$\delta l_s = \frac{P \times l_s}{A_s \times E_s}$$

or $$0.75 = \frac{P \times (2 \times 1000)}{\left(\frac{\pi}{4} \times 3^2\right) \times 2.0 \times 10^5}$$

or $$P = 0.75 \times \left(\frac{\pi}{4} \times 3^2\right) \times 0.2 \times 10^5 \times \frac{1}{2000} \quad ...(i)$$

(*b*) For brass wire,

$$\delta l_b = \frac{P \times l_b}{A_b \times E_b}$$

$$4.64 = \frac{P \times (2.5 \times 1000)}{\left(\frac{\pi}{4} \times 2^2\right) \times E_b}$$

or $$P = 4.64 \times \left(\frac{\pi}{4} \times 2^2\right) \times E_b \times \frac{1}{2500} \quad ...(ii)$$

Equating eqs. (*i*) and (*ii*),

$$0.75 \times \left(\frac{\pi}{4} \times 3^2\right) \times 2.0 \times 10^5 \times \frac{1}{2000}$$

$$= 4.64 \times \left(\frac{\pi}{4} \times 2^2\right) \times E_b \times \frac{1}{2500}$$

$\therefore$ $$E_b = 0.909 \times 10^5 \text{ N/mm}^2$$ **Ans.**

Example 3.9: *A steel bar is 900 mm long, its two ends are 40 mm and 30 mm in diameter and length of each rod is 200 mm. The middle portion of the bar is 15 mm in diameter and 500 mm long. If the bar is subjected to an axial tensile load of 15 KN. Find its total extension.*

Take $E = 200$ GN/m^2 $(1 \text{ G} = 10^9)$

Solution:

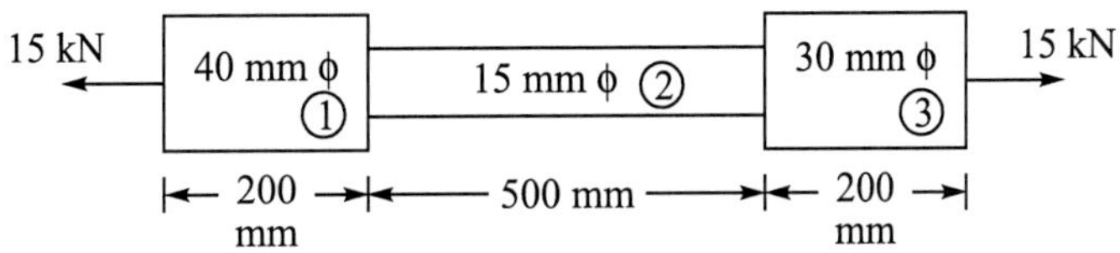

Fig. 3.21

Load $P = 15$ kN.

Area, $A_1 = \frac{\pi}{4} \times 40^2 = 1256.6 \text{ mm}^2 = 0.001256 \text{ m}^2$

Area, $A_2 = \frac{\pi}{4} \times 15^2 = 176.7 \text{ mm}^2 = 0.0001767 \text{ m}^2$

Area, $A_3 = \frac{\pi}{4} \times 30^2 = 706.8 \text{ mm}^2 = 0.0007068 \text{ m}^2$

Length $l_1 = 200$ mm $= 0.2$ m,

$l_2 = 500$ mm $= 0.5$ m

and $l_3 = 200$ mm $= 0.2$ m

Total extension of the bar:

Let δl_1, δl_2 and δl_3 be the extensions in the parts 1, 2 and 3 of he steel bar respectively,

Then, $$\delta l_1 = \frac{Pl_1}{A_1 E}, \delta l_2 = \frac{Pl_2}{A_2 E}, \delta l_3 = \frac{Pl_3}{A_3 E}$$

$$\left[\because \; E = \frac{\sigma}{e} = \frac{P/A}{\frac{\delta l}{l}} = \frac{P \cdot l}{A \cdot \delta l} \quad or \quad \delta l = \frac{Pl}{AE}\right]$$

Total extension of the bar,

$$\delta l = \delta l_1 + \delta l_2 + \delta l_3$$

$$= \frac{Pl_1}{A_1 E} + \frac{Pl_2}{A_2 E} + \frac{Pl_3}{A_3 E} = \frac{P}{E}\left[\frac{l_1}{A_1} + \frac{l_2}{A_2} + \frac{l_3}{A_3}\right]$$

$$= \frac{15 \times 10^3}{200 \times 10^9}\left[\frac{0.20}{0.001256} + \frac{0.50}{0.0001767} + \frac{0.26}{0.0007068}\right]$$

$$= 0.0002454 \text{ m} = 0.2454 \text{ mm}$$ **Ans.**

Example 3.10: *A brass bar is subjected to axial forces as shown in Fig. 3.22. Find the total change in the length of the bar.* **(KU–2010)**

Take $E = 1 \times 10^5 \text{ N/mm}^2$.

Solution:

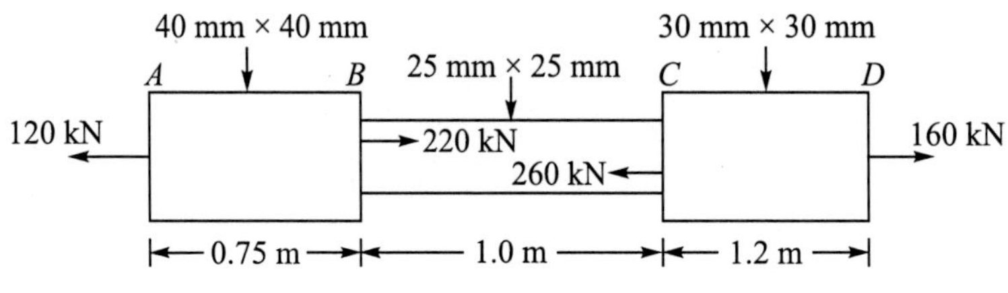

Fig. 3.22

Part AB:

Area, $A_1 = 40 \times 40 = 1600 \text{ mm}^2$

Length, $l_1 = 0.75 \text{ m} = 750 \text{ mm}$

This section of the bar is subjected to a tensile force of 120 kN.

Increase in length of *AB*,

$$\delta l_1 = \frac{\sigma_1 l_1}{E} = \frac{P_1 l_1}{A_1 E} = \frac{120 \times 1000 \times 750}{1600 \times 1 \times 10^5} = 0.56 \text{ mm}$$

Part BC:

Area, $A_2 = 25 \times 25 = 625 \text{ mm}^2$

Length, $l_1 = 1 \text{ m} = 1000 \text{ mm}$

The section of the bar is subjected to a compressive force (220 – 120) = 100 kN.

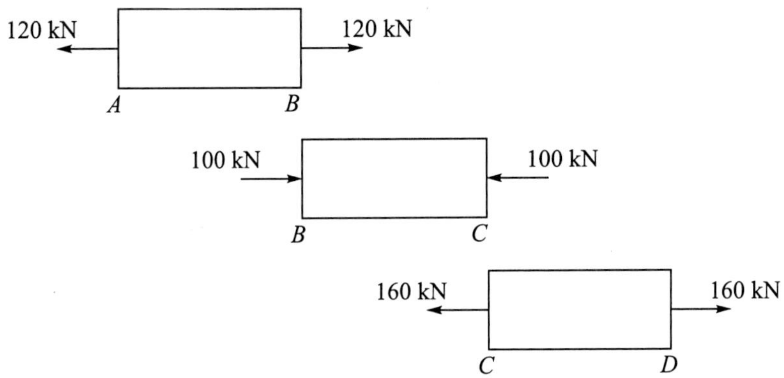

Fig. 3.23

Decrease in length of *BC*,

$$\delta l_2 = \frac{\sigma_2 l_2}{E} = \frac{P_2 l_2}{A_2 E} = \frac{100 \times 1000 \times 1000}{625 \times 1 \times 10^5} = 1.60 \text{ mm}$$

Part CD:

Area, $A_3 = 30 \times 30 = 900 \text{ mm}^2$

Length, $l_3 = 1.2 \text{ m} = 1200 \text{ mm}$

The section of the bar is this part is subjected to a tensile force (260 – 100) = 160 kN.

∴ Increase in length of *CD*,

$$\delta l_3 = \frac{\sigma_3 l_3}{E} = \frac{P_3 l_3}{A_3 E} = \frac{160 \times 1000 \times 1200}{900 \times 1 \times 10^5} = 2.13 \text{ mm}$$

∴ Net change in lenght of the bar

$$\delta l = \delta l_1 - \delta l_2 + \delta l_3$$
$$= 0.56 - 1.60 + 2.13$$
$$= (+)\ 1.09 \text{ mm (elongation)}$$ **Ans.**

Example 3.11: *A brass bar having cross-sectional area of 1000 mm², is subjected to axial forces as shown in Fig. 3.24. Calculate the force 'P' necessary for equilibrium of bar. Determine the total elongation of the bar.*

Take $E = 1.05 \times 10^5 \text{ N/mm}^2$.

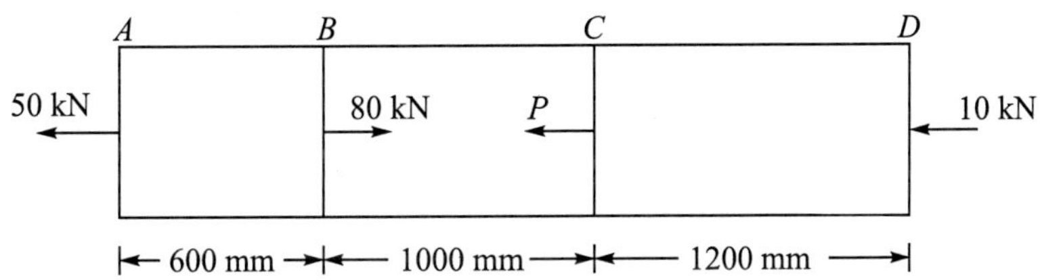

Fig. 3.24

Solution: For equilibrium, the algebraic sum of forces on bar along its axis must be zero.

$$80 = 50 + P + 10$$

∴ "P" = 20 kN.

Part AB:

Area, $A = 100 \text{ mm}^2$

length $l_1 = 600$ mm

The section of the bar in this part is subjected to a tensile force of 50 kN *i.e.*, 50,000 N.

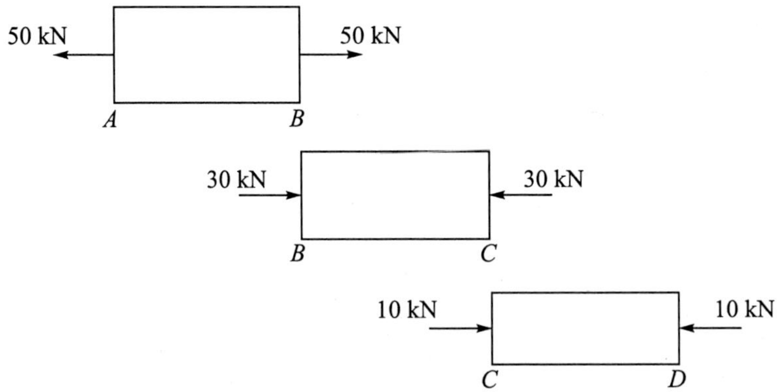

Fig. 3.25

∴ Increase in length of *AB*,

$$\delta l_1 = \frac{P_1 l_1}{AE} = \frac{50,000 \times 600}{1000 \times 1.05 \times 10^5} = 0.286 \text{ mm}$$

Part BC:

Area, $A = 1000 \text{ mm}^2$

Length $l_2 = 1000$ mm

The section of the bar is subjected to a compressive force of (80 – 50) = 30 kN

Decrease in length of *BC*,

$$\delta l_2 = \frac{P_2 \times l_2}{AE} = \frac{30,000 \times 1000}{1000 \times 1.05 \times 10^5} = 0.286 \text{ mm}$$

Part CD:

Area, $A = 1000 \text{ mm}^2$

Length $l_3 = 1200$ mm

The section of the bar is subjected to a compressive force of (30 – 20) = 10 kN.

Decrease in length of *CD*,

$$\delta l_3 = \frac{P_3 l_3}{AE} = \frac{10000 \times 1200}{1000 \times 1.05 \times 10^5} = 0.114 \text{ mm}$$

∴ Total elongation of bar = $\delta l_1 - \delta l_2 - \delta l_3$

= 0.286 – 0.286 – 0.114 = – 0.114 mm **Ans.**

(–) sign shows that there will be decrease in length of bar by 0.114 mm.

Example 3.12: *An R.C.C. column is 400 mm × 400 mm in section. The column is reinforced with 08 steel bars of 20 mm diameter. The column carries a load of 15×10^5 N. Find the stresses in the concrete and steel bars. Take "E" for steel and concrete as 2.1×10^5 N/mm^2 and 1.4×10^4 N/mm^2.*

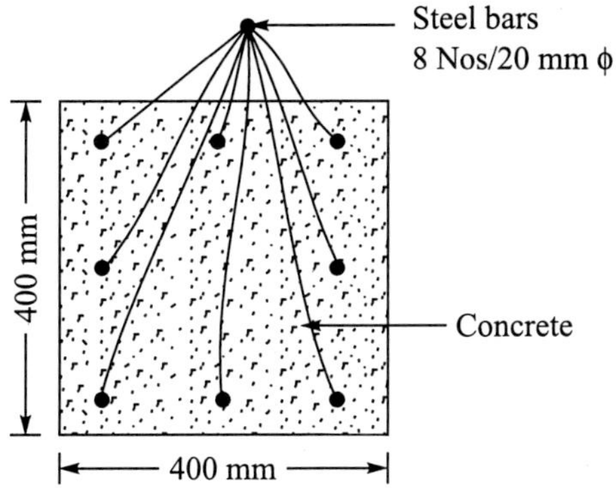

Fig. 3.26

Solution:

Area of steel bars, $A_1 = \frac{\pi}{4} \times (20)^2 \times 8 = 2513.27\ \text{mm}^2$

Area of concrete, $A_2 = A - A_1$

$= 160000 - 2513.27$

$= 157486.73\ \text{mm}^2$

Strain in steel = Strain in concrete

$$\frac{\sigma_1}{E_1} = \frac{\sigma_2}{E_2}$$

or $$\sigma_1 = \frac{\sigma_2 \times E_1}{E_2} = \frac{\sigma_2 \times 2.1 \times 10^5}{1.4 \times 10^4}$$

$\therefore$ $$\sigma_1 = 15\ \sigma_2$$

Total load on column = Load on steel + Load on concrete

$$15 \times 10^5 = \sigma_1 A_1 + \sigma_2 A_2$$
$$= 15\ \sigma_2 \times 2513.27 + \sigma_2 \times 157486.73$$
$$= (37699.05 + 157486.73) \times \sigma_2$$

or $$15 \times 10^5 = 195185.78\ \sigma_2$$

$\therefore$ $$\sigma_2 = \frac{15 \times 10^5}{195185.78} = 7.68\ \text{N/mm}^2$$ **Ans.**

and $$\sigma_1 = 15\ \sigma_2 = 15 \times 7.68 = 115.2\ \text{N/mm}^2$$ **Ans.**

SUMMARY

- Stress $\sigma = \frac{P}{A}$, where P is the applied load and A is the cross sectional area.
- Strain $e = \frac{\delta l}{l}$, where δl is the change in length and l is the original length.

- The stress induced in a body, which is subjected to two equal and opposite pulls, is known as tensile stress.
- The stress induced in a body, which is subjected to two equal and opposite pushes, is known as compressive stress.
- Stress is expressed as N/mm^2 or N/m^2 or GN/m^2 strain is just a number having no units.
- Elasticity is the property of the material by virtue of which it returns back to its original position after the removal of external forces.
- Hooke's law states that within elastic limit, the stress is proportional to strain.
- Factor of safety $= \dfrac{\text{Ultimate stress}}{\text{Working stress}}$
- In case of a composite bar having two or more bars of different lengths, the extension or compression in each bar will be equal and the total load will be equal to the sum of the loads carried by each bar.
- The stresses induced in the body due to change in temperature are known as thermal stresses.

$$\text{Thermal stress} = \alpha\, tE$$
$$\text{Thermal strain} = \alpha\, t$$

OBJECTIVE TYPE QUESTIONS

1. Fill in the blanks:
   (*a*) The internal resistance per unit area is known as ....
   (*b*) The strain has .... units.
   (*c*) The property by virtue of which certain materials return back to their original position after the removal of the external forces is known as ....
   (*d*) The stresses induced in a body due to change in temperature are known as ....
   (*e*) The ratio of lateral strain to linear strain is known as ....
2. Simple stress is often called
   (*a*) Direct stress (*b*) Transverse stress
   (*c*) Total stress
3. When a tensile or compressive force (*P*) acts on a body, the change in its length is given by
   (*a*) $\dfrac{Pl}{AE}$ (*b*) $\dfrac{AE}{Pl}$
   (*c*) $\dfrac{PE}{Al}$ (*d*) $\dfrac{PA}{lE}$
4. Value of Poisson's ratio ranges from
   (*a*) 1 to 2 (*b*) 1 to 5
   (*c*) 0 to 0.5 (*d*) 0.5 to 1.5
5. Limit of proportionality depends upon
   (*a*) type of loading (*b*) type of material
   (*c*) Area of cross-section

Answers: 1. (*a*) stress, (*b*) no, (*c*) elasticity, (*d*) thermal stresses, (*e*) Poisson's ratio
2. (*a*), 3. (*a*), 4. (*c*), 5. (*c*).

EXERCISE FOR PRACTICE

1. Find the axial deformation of the rod shown in Fig. 3.27. **(MDU 2009)**

Dia. of 'A' = 30 mm
Dia. of 'B' = 20 mm
Dia. of 'C' = 10 mm

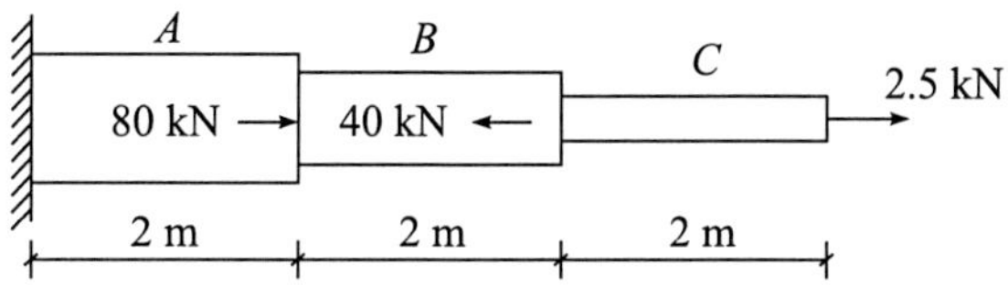

Fig. 3.27

2. A member $ABCD$ is subjected to axial loads P_1, P_2, P_3 and P_4 as shown in Fig. 3.28. Calculate the force P_2 necessary for equilibrium, if P if P_1 = 4500 N, P_3 = 45000 N and P_4 = 13000 N. Determine the total elongation of the member if modulus of elasticity is 2.1×10^6 N/cm². The cross sectional area of AB is 6.25 cm², BC is 25 cm² and CD is 12.5 cm².

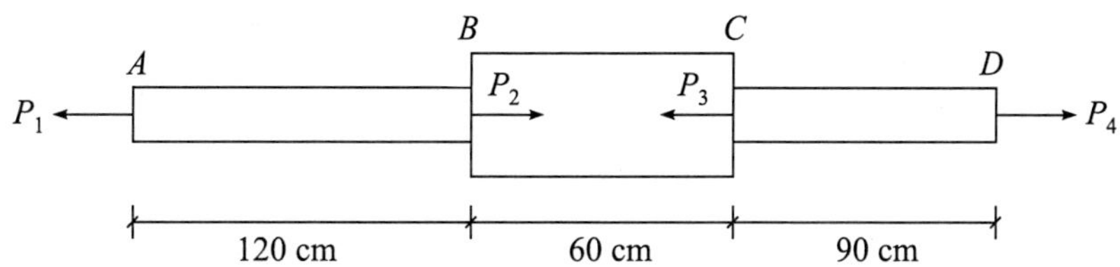

Fig. 3.28

3. A solid steel bar 500 mm long and 70 mm ϕ is placed inside an aluminium tube 75 mm inside dia and 100 mm outside dia. The Aluminium cylinder is 15 mm longer than steel cylinder. An axial load of 600 kN is applied to the bar and cylinder through rigid cover plate on both sides. Find the stresses in steel bar and aluminium tube. Take $E_S = 2.1 \times 10^5$ N/mm² and $E_A = 1.0 \times 10^5$ N/mm². **(MDU 2010)**
4. Draw the Mohr's stress circle for direct stresses of 65 N/mm² (Tensile) and 35 N/mm² (compressive) and estimate the magnitude and direction of the resultant stresses on planes making angles of 20° and 65° with the plane of the first principal stress. Also find the normal and tengential stresses on these planes.
5. A R.C.C. circular column of 400 mm ϕ, has steel bars of 20 mm ϕ embedded in it. Find the max. load which the column can carry, if the stress in steel and concrete are not to exceed 120 MPa and MPa. Take $E_S = 18\ E_C$.

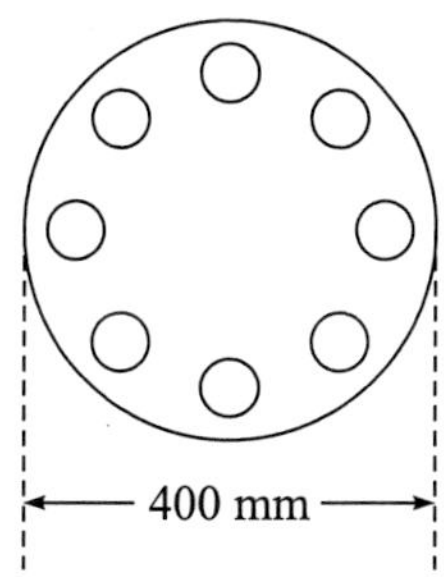

Fig. 3.29

CHAPTER

4 Bending Stresses in Beams

When some external load acts on a Beam, the *S.F.* and *B.M.* are set up at all sections of the beam. Due to the *S.F.* and *B.M.*, the beam undergoes certain deformation. The material of the beam will offer resistance or stresses against these deformations. The stresses produced to resist the *B.M.* are known as bending stresses. The stresses produced to resist the action of shearing force are known as shearing stresses.

4.1. PURE BENDING OR SIMPLE BENDING

The bending of beam not accompanied by any shear force is termed as Pure Bending or Simple Bending.

Figure 4.1 shows the beam *ABCD* with equal overhangs and supported at *B* and *C*. Let a point load '*W*' be applied at each end of the beam. *S.F.* and *B.M.* diagrams are drawn for the given loading. From these diagrams, it is clear that there is no *S.F.* between *B* and *C* but the *B.M.* between *B* and *C* is constant. This means that betwen *B* and *C*, the beam is subjected to a constant *B.M.* only. This condition of the beam between *B* and *C* is known as Pure/Simple bending.

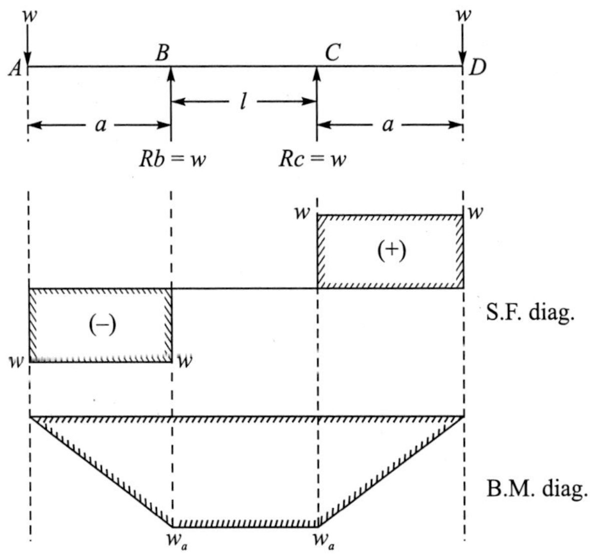

Fig. 4.1

4.1.1. Assumptions in the Theory of Simple Bending

1. The material of the beam is homogeneous and isotropic.
2. A transverse section of a beam, which is plane before bending will remain plans after bending.
3. The value of young's modulus of elasticity is the same in tension and compression.
4. There is no resultant push or pull across the transverse section of the beam.
5. The loads are applied in the plane of bending.
6. The beam material is stressed within its elastic limit and thus obeys Hooke's Law.
7. The radius of curvature of the beam before bending is very large as compared to the transverse dimensions of the beam.
8. Each layer of the beam is free to expand or contract independently of the layer above or below it.

4.2. THEORY OF SIMPLE BENDING (BENDING EQUATION)

Consider a portion (*ABCD*) of the beam which is subjects to *B.M.* "*M*". Let this portion of the beam bends into an arc of a circle with "*O*" as centre under the effect of applied loads a shown in the Fig. 4.2.

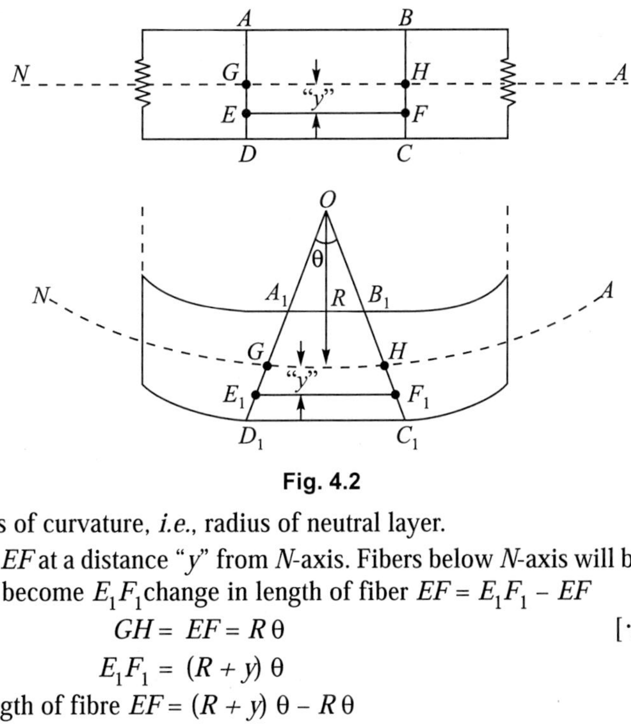

Fig. 4.2

Let "*R*" = Radius of curvature, *i.e.*, radius of neutral layer.

Consider a layer *EF* at a distance "*y*" from *N*-axis. Fibers below *N*-axis will be under tension. After deformation *EF* will become E_1F_1 change in length of fiber $EF = E_1F_1 - EF$

But $\quad GH = EF = R\theta \qquad [\because \text{Arc length} = R \times \theta]$

$\therefore \quad E_1F_1 = (R + y)\,\theta$

$\therefore$ Change in length of fibre $EF = (R + y)\,\theta - R\theta$

$$= R\theta + y\theta - R\theta = y\theta$$

$$\text{Strain in layer } EF,\ e = \frac{\text{Change in length of } EF}{\text{Original length}}$$

$$\text{“}e\text{”} = \frac{E_1F_1 - EF}{EF} = \frac{y\theta}{R\theta} = \frac{y}{R}$$

But strain $= \frac{\text{Stress}}{E} = \frac{\sigma b}{E}$

$\therefore$ $$\frac{\sigma b}{E} = \frac{y}{R} \quad \text{or} \quad \boxed{\frac{\sigma b}{y} = \frac{E}{R}}$$

E and R are constants for a given beam.

Hence stress in any fibre is proportional to the distance of the fibre from the N-axis.

Next consider elementary areas a_1, a_2, a_3 etc., at a distance at a distance of y_1, y_2, y_3 etc., respectively from the N-axis (consider Fig. 4.3).

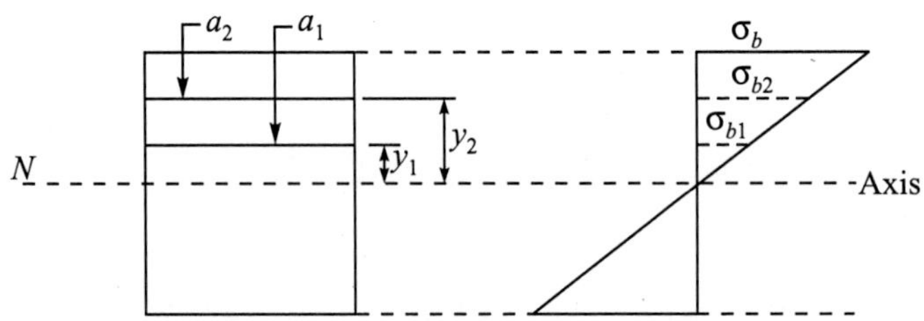

Fig. 4.3

Force on elementary area $a_1 = \sigma b_1 \times a_1$

Force on elementary area $a_2 = \sigma b_2 \times a_2$ and so on.

Moment of all these forces about N-axis,

$$M = \sigma b_1\, a_1\, y_1 + \sigma b_2\, a_2\, y_2 + \sigma b_3\, a_3\, y_3 + \ldots \qquad \ldots(4.1)$$

But $$\frac{\sigma b}{y} = \frac{E}{R} \qquad \text{(Proved Above)}$$

$\therefore$ $$\sigma b = \frac{E}{R} \times y \qquad (E \text{ and } R \text{ are constants})$$

$\therefore$ $$\sigma b_1 = \frac{E}{R} \times y_1$$

$\therefore$ $$\sigma b_2 = \frac{E}{R} \times y_2 \text{ and so on ...}$$

Put values of σb_1, σb_2 etc., in = n. (4.1) above

We get,

$$M = \frac{E}{R}\, a_1 y_1\, y_1 + \frac{E}{R}\, a_2 y_2\, y_2 + \frac{E}{R}\, a_3 y_3\, y_3 + \ldots$$

$$M = \frac{E}{R}\,(a_1\, y_1^2) + \frac{E}{R}\,(a_2\, y_2^2) + \frac{E}{R}\,(a_3\, y_3^2) + \ldots$$

or $$\text{“}M\text{”} = \frac{E}{R}\,[a_1\, y_1^2 + a_2\, y_2^2 + a_3\, y_3^2 + \ldots]$$

$a_1 y_1^2$ is the *M.I.* of a_1 about *N*-axis.

$a_2 y_2^2$ is the *M.I.* of a_2 about *N*-axis. ...... and so on.

$$\therefore \quad M = \frac{E}{R} \times I$$

or

$$\frac{M}{I} = \frac{E}{R}$$

$$\therefore \quad \boxed{\frac{M}{I} = \frac{\sigma b}{y} = \frac{E}{R}}$$, This is called Bending Equation.

where,

M = *B.M.* or *M.R.* of the section in *N*-mm.

I = *M.I.* of the section about *N*-axis in mm^4.

σb = Bending stress at a distance "y" from *N-A*, in *N*/mm^2.

y = Distance of fibre from *N*-axis in mm.

E = Young's modulus of elasticity in *N*/mm^2.

R = Radius of curvature of *N.A.* in mm.

∴ Bending equation is the relationship between the radius of curvature to which beam bends, the *B.M.*, the bending stresses and the *x*-see dimensions of the beam.

Moment of Resistance: When the beam is subjected to the *B.M.*, the section above and below *N.A.* developes internal forces or stresses of opposite nature. These internal stresses or forces have certain moments about *N*-axis. The sum of moments of the internal forces about the *N*-axis is known as the *M.R.*

From bending equation, we have

$$\frac{M}{I} = \frac{\sigma b}{y} \quad \text{or} \quad M = \sigma b \times \frac{I}{y}$$

or

$$M = \sigma b \times z \qquad \left(\text{where } z = \frac{I}{y}\right)$$

where "z" is called the Section modulus.

Sectional Modulus: Section modulus is defined as the ratio of *M.I.* of a section about the *N*-axis to the distance of the outermost layer from the *N*-axis.

It is denoted by, "z"

Mathematically,

$$\text{“}z\text{”} = \frac{I}{y}$$

where

I = *M.I.* about *N*-axis.

y = Distance of the outermost layer from the *N*-axis.

Now, from bending equation,

$$\frac{M}{I} = \frac{\sigma b}{y}$$

$\therefore$ $$M = \sigma b \times \frac{I}{y}$$

or $$\boxed{M = \sigma b \times z}$$

"Shearing-Stresses": In actual practice when a beam is loaded, the *S.F.* at a section always comes into play along with *B.M.* It has been observed that the effect of shearing stress as compared to the bending stress is quite negligible and is of not much importance. But sometimes, the shearing stress at a section assumes much importance in the design criterion.

Shearing stress at a section in a Loaded Beam: Consider a small portion *ABCD* of length "*dx*" of a beam loaded with unifomally distributed load (*udl*) as shown in Fig. 4.4.

We know that when a beam is loaded with *udl*, the *S.F.* and the *B.M.* vary at every point along the length of the beam.

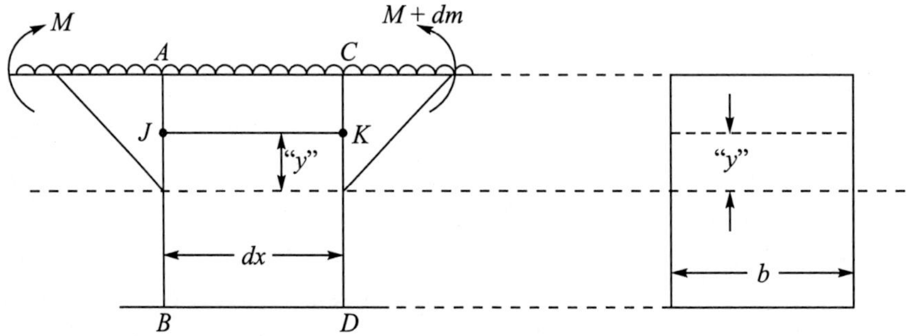

Fig. 4.4

Let $M =$ *B.M.* at *AB*

$M + dm =$ *B.M.* at *CD*

$F =$ *S.F.* at *AB*

$F + dF =$ *S.F.* at *CD* and $I =$ *M.I.* of section about its *N.A.*

Now consider an elementary strip at a distance "*y*" from the *N*-axis as shown in Fig. 4.4.

Let σ = Intensity of bending stress across *AB* at a distance "*y*" from *N*-axis.

and a = *x*-sec. area of the strip.

We have already discussed that,

$$\frac{M}{I} = \frac{\sigma}{y} \quad \text{or} \quad \sigma = \frac{M}{I} \times y$$

Similarly, $$\sigma + d\sigma = \frac{M + dM}{I} \times y$$

where, $\sigma + d\sigma$ = Intensity of Bending stress across *CD*.

We know that the force acting across *AB* = Stress × Area = $\sigma \times a$

$$= \frac{M}{I} \times y \times a \tag{4.2}$$

Similarly, force acting across *CD*

$$= (\sigma + d\sigma) \times a = \frac{M + dM}{I} \times y \times a \tag{4.3}$$

∴ Net unbalance force on the strip,

$$= \frac{M + dM}{I} \times y \times a - \frac{M}{I} \times y \times a$$

$$= \frac{dM}{I} \times y \times a$$

The total unbalance force (F) above the N-axis may be found out by integrating the above equation between 0 and $\frac{d}{2}$.

$$= \int_0^{d/2} \frac{dM}{I} \times a \times y \times dy$$

$$= \frac{dM}{I} \int_0^{d/2} a \cdot y \cdot dy = \frac{dM}{I} A\bar{y} \qquad (4.4)$$

where, A = Area of the beam above N-axis.

and $\bar{y}$ = Distance between the $C.G.$ of the area and the N-axis.

We know that, the intensity of the shear stress

$$\text{“}\tau\text{”} = \frac{\text{Total force}}{\text{Area}}$$

$$= \frac{\frac{dM}{I} \cdot A\bar{y}}{dx \cdot b} \qquad \text{(where “}b\text{” is the width of the beam)}$$

$$= \frac{dM}{dx} \times \frac{A \cdot \bar{y}}{I \times b}$$

$$\text{“}\tau\text{”} = F \times \frac{A\bar{y}}{I \times b} \qquad \left(\text{substitutly } \frac{dM}{dx} = F = S.F.\right)$$

4.3. TORSION OF CIRCULAR SHAFTS

A shafts is said to be in pure torsion, when equal and opposite torques are applied at the two ends of the shaft.

The product of the tangential force acting at the circumference causing twist and the radius of the shaft is called Torque or Turning moment or Twisting moment. It is denoted by "T".

Units of Torque : *N-m* or *N-mm.*

Effect of Torsion on a Shaft: The shaft subjected to a pure torsion has a tendency to shear off at every cross-section, perpendicular to the longitudinal axis of the shaft. Hence the effect of torsion is to produce share stress in the material of the shaft.

Assumptions in the theory of Pure Torsion:

1. The material of the shaft is uniform throughout.
2. The shaft is of uniform circular section throughout.
3. Cross-sections of the shaft, which are plane before twist remain plane after twist.

4. The twist along the length of the shaft is uniform throughout.
5. The distance between any two normal cross-section remain the same after the application of torque.
6. Max. shear stress induced in the shaft due to application of torque does not exceed it elastic limit value.

4.3.1. Formulation of Torsion Equation

Consider a shaft fixed at one end *AA* and free at the end *BB* as shown in the Fig. 4.5.

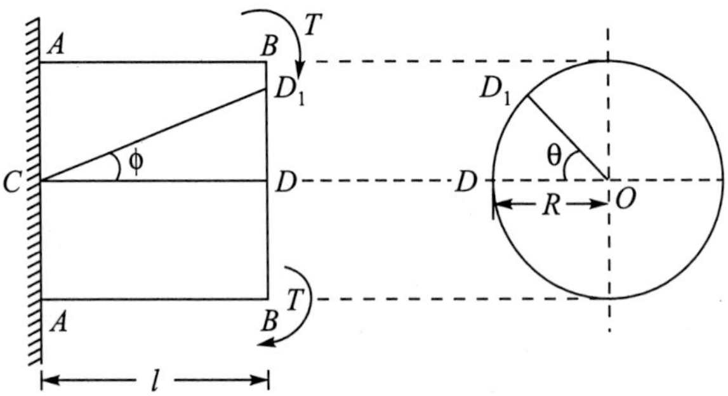

Fig. 4.5

Let T = Torque or Twisting moment.

R = Radius of shaft.

l = Length of shaft.

τ = Shear stress at radius R.

C = Modulus of rigidity of the material of the shaft.

θ = Angle of twist.

I_P = Polar *M.I.* of the shaft.

Let a line CD on the surface of the shaft take a new position CD_1 after the application of torque. The X-sections will be twisted through angle θ and surface by angle ϕ.

$$\text{Shear strain} = \tan\phi = \phi = \frac{DD_1}{CD} \qquad \text{(If } \phi \text{ is very small than } \tan\phi = \phi)$$

$$\text{Arc } DD_1 = OD \times \theta = R\theta \qquad (\because OD = R = \text{Radius of shaft}).$$

$$\therefore \quad \text{Shear strain} = \frac{R\theta}{l} \qquad (4.5)$$

$$\text{Also Shear strain} = \frac{\text{Shear stress}}{\text{Modulus of rigidity}}$$

or
$$\phi = \frac{\tau}{C} \qquad (4.6)$$

From equation (4.1 (*a*)) and (4.2), we get,

$$\frac{\tau}{C} = \frac{R\theta}{l} \quad \text{or} \quad \frac{\tau}{R} = \frac{C\theta}{l} \qquad (4.7)$$

or $$\text{“}\tau\text{”} = \frac{R \times C \times \theta}{l}$$

In a given shaft, under a given torque,

C, θ and l are contents.

$\therefore$ $$\tau \propto R$$

or $$\frac{\tau}{R} = \text{constant} \tag{4.8}$$

From equation (4.8), it is clear that shear stress at any point in the shaft is proportional to the distance of the point from the axis of the shaft. Hence the shear stress is max. at the outer surface and shear stress is zero at the axis of the shaft.

Next consider an elementary circular ring of thickness "*dr*" at a distance "*r*" from the centre as shown is the Fig. 4.6.

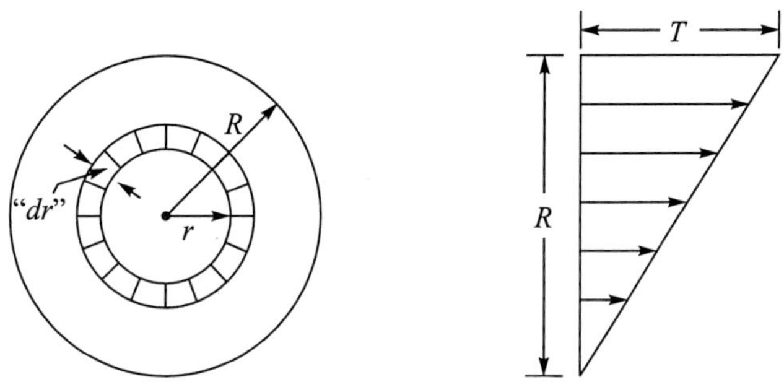

Fig. 4.6

Area of the elementary circular ring = $2\pi r \cdot dr$

Shear stress at a radius "*r*" = $\frac{\tau}{R} \times r$ $\left[\because \frac{\tau}{R} = \text{const.}\right]$

Turning force on the dementry circular ring = Stress × Area

$$= \frac{\tau}{R} \times r \times 2\pi r^2 \cdot dr = \frac{\tau}{R} \times 2\pi r^2 \cdot dr$$

Turning moment due to turning force on the dementry ring.

$$dT = \frac{\tau}{R} \times 2\pi r^2 \cdot dr \cdot r = \frac{\tau}{R} \times 2\pi r^3 \cdot dr$$

$\therefore$ Total turning moment

$$\text{“}T\text{”} = \int dT = \int_0^R \frac{\tau}{R} \times 2\pi r^3 \cdot dr$$

or $$T = \frac{\tau}{R} \times 2\pi \int_0^R r^3 \cdot dr = \frac{\tau}{R} \times 2\,\pi \left|\frac{r^4}{4}\right|_0^R$$

$$= \frac{\tau}{R} \times 2\pi \times \frac{r^4}{4} = \tau \times \frac{\pi}{2} \times R^3$$

or $$T = \tau \times \frac{\pi}{2} \times \left(\frac{D}{2}\right)^3 = \tau \times \frac{\pi}{2} \times \frac{D^3}{8} \qquad \left(\because R = \frac{D}{2}\right)$$

$$= \frac{\pi}{16} \times \tau \times D^3$$

$\therefore$ $$\tau = \frac{16\,T}{\pi\,D^3}$$

Substituting the values of τ in equation (4.7), we get

$$\frac{16\,T}{\pi \times D^3 \times R} = \frac{C\theta}{l}$$

$$\frac{16\,T \times 2}{\pi \times D^3 \times D} = \frac{C\theta}{l}$$

or $$\frac{32\,T}{\pi \times D^4} = \frac{C\theta}{l}$$

But $$\frac{\pi\,D^4}{32} = I_P \text{ (Polar } M.I.\text{)}.$$

$\therefore$ $$\frac{T}{I_P} = \frac{C\theta}{l} \qquad (4.9)$$

From equations (4.8) and (4.9), we get,

$$\boxed{\frac{T}{I_P} = \frac{\tau}{R} = \frac{C\theta}{l}}$$

This equation is known as Torsion equation,

where, T = Twisting moment or torque in N-mm

I_P = Polar *M.I.* of the shaft in mm^4.

τ = Shear stress at radius R in N\mm^2.

R = Radius of shaft in mm.

C = Modulus of rigidity of the material of the shaft in N/mm^2.

θ = Angle of twist.

l = Length of shaft in mm.

Example 4.1: *A simply supported beam of span 5 m has a cross-section 150 mm × 250 mm. If the permissible stress is 10 N/mm², find:*

(*a*) *maximum intensity of uniformly distributed load it can carry.*

(*b*) *maximum concentrated load P applied at 2 m from one end it can carry.*

Solution: Such beam is shown in Fig. 4.7.

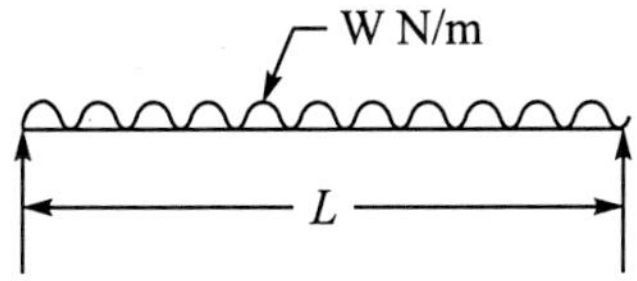

Fig. 4.7.

$$I = \frac{1}{12} bd^3$$

$$y_{max} = \frac{d}{2}$$

$$\therefore \qquad Z = \frac{1}{6} bd^2 = \frac{1}{6} \times 150 \times 250^2 = 1562500 \text{ mm}^3$$

$\therefore$ Moment carrying capacity $M = f_{per}\, Z = 10 \times 1562500$ N-mm

(*a*) If *w* is the intensity of load in N/m units (Fig. 4.12) then maximum moment

$$= \frac{wL^2}{8} = \frac{w \times 5^2}{8} = \frac{w \times 25}{8} \text{ N-mm}$$

$$= \frac{w}{8} \times 25 \times 1000 \text{ N-mm}$$

Equating it to moment carrying capacity, we get maximum intensity of load as

$$\frac{w \times 25}{8} \times 1000 = 10 \times 1562500$$

$$w = 5000 \text{ N/m}$$

$$= 5 \text{ kN/m} \quad \textbf{Ans.}$$

(*b*) If *P* is the concentrated load as shown in Fig. 4.8, then maximum moment occurs under the load and its value

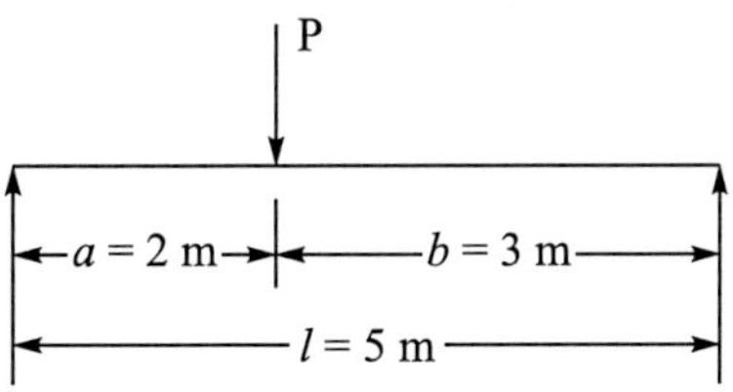

Fig. 4.8.

$$M = \frac{P \times a \times b}{L} = \frac{P \times 2 \times 3}{5}$$

$$= 1.20\, P \text{ kN-m}$$

$$= 1200\, P \text{ N-mm}$$

Equating it to moment carrying capacity we get

$$1200\, P = 10 \times 1562500$$

$$P = 13020.83 \text{ N}$$

$$= 13.02083 \text{ kN} \quad \textbf{Ans.}$$

Example 4.2: *A circular pipe of external diameter 70 mm and thickness 8 mm is used as a simply supported beam over an effective span 2.5 m. Find the maximum concentrated load that can be applied at the centre of the span if permissible stress in tube is 150 N/mm².*

Solution: The circular pipe considered is shown in Fig. 4.9.

$$\text{External dia. } D = 70 \text{ mm}$$

$$\text{Internal dia. } d = 70 - 2 \times 8 = 54 \text{ mm}$$

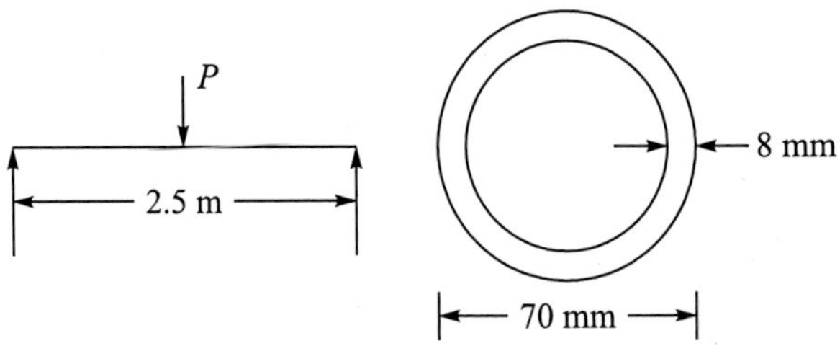

Fig. 4.9.

$$I = \frac{\pi}{64}(70^4 - 54^4)$$

$$= 761195.33 \text{ mm}^4$$

$$y_{max} = \frac{D}{2} = 35 \text{ mm}$$

$$\therefore \quad Z = \frac{I}{y_{max}} = \frac{761195.33}{35}$$

$$= 21748.438 \text{ mm}^3$$

Moment carrying capacity, $M = fZ$

$$= 150 \times 21748.438$$

$$= 3262265.7 \text{ N-mm}$$

Maximum moment in the beam occurs at the mid span and is equal to

$$\frac{PL}{4} = \frac{P \times 2.5}{4} \text{ N} - \text{mm} \qquad \text{where } P \text{ is in the Newton.}$$

$$= \frac{P \times 2.5}{4} \times 1000 \text{ N} - \text{mm}$$

Equating maximum moment to moment carrying capacity we get,

$$3262265.7 = \frac{P \times 2.5 \times 1000}{4}$$

$$P = 5219.63 \text{ N}$$

$$= 5.21963 \text{ kN} \quad \textbf{Ans.}$$

Example 4.3: *A symmetric I-section has flanges of size 180 mm × 10 mm and its overall depth is 500 mm. Thickness of web is 8 mm. It is strengthened with a plate of size 240 mm × 12 mm on compression side. Find the moment of resistance of the section, if permissible stress is 150 N/mm². How much uniformly distributed load it can carry if it is used as a cantilever of span 3 m?*

Solution: The section of beam is as shown in Fig. 4.10. Let y be the distance of centroid from the bottom-most fibre.

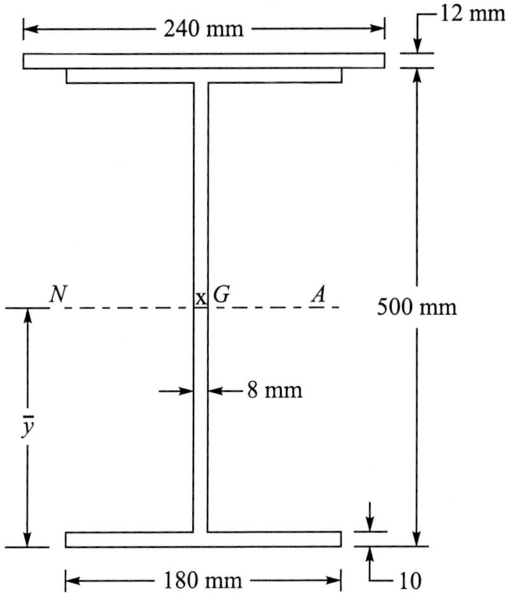

Fig. 4.10.

$$\bar{y} = \frac{\text{Moment of area about bottom fibre}}{\text{Total area}}$$

$$= \frac{240 \times 12 \times 506 + 180 \times 10 \times 495 + 180 \times 10 \times 5 + 480 \times 8 \times 250}{240 \times 12 + 180 \times 10 + 180 \times 10 + 480 \times 8}$$

$$= \frac{3317280}{10320} = 321.442 \text{ mm}$$

$$I = \frac{1}{12} \times 240 \times 12^3 + 240 \times 12(506 - 321.442)^2$$

$$+ \frac{1}{12} \times 180 \times 10^3 + 180 \times 10\,(495 - 321.442)^2$$

$$+ \frac{1}{12} \times 180 \times 10^3 + 180 \times 10\,(5 - 321.442)^2$$

$$+ \frac{1}{12} \times 8 \times 480^3 + 8 \times 480\,(250 - 321.442)^2$$

$$= 4.25952 \times 10^8 \text{ mm}^4$$

$$y_{top} = 512 - 321.442 = 190.558 \text{ mm}$$

$$\therefore \quad y_{max} = \bar{y} = 321.442 \text{ mm}$$

Moment of resistance (Moment carrying capacity)

$$= f_{per} \times Z$$

$$= 150 \times \frac{4.25952 \times 10^8}{321.442} = 1.98769 \times 10^8 \text{ N} - \text{mm}$$

$= 198.769$ kN – mm **Ans.**

Let the load on cantilever be w/m length as shown in Fig. 4.11.

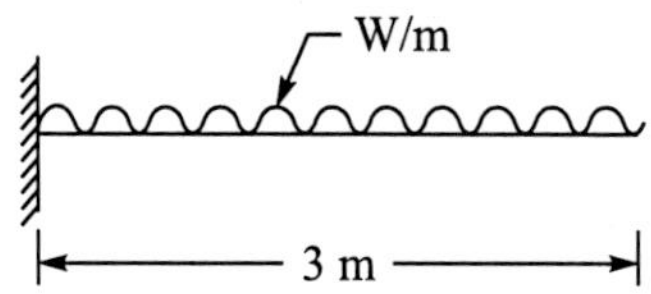

Fig. 4.11.

Then maximum moment produced = $\frac{wL^2}{2}$ kN-m (where w is in kilo Newtons)

$$= w \times \frac{3^2}{2} = 4.5 \text{ w kN-m}$$

Equating moment of resistance to maximum moment, we get maximum load w

$$4.5\,w = 198.769$$

$$w = 44.171 \text{ kN/m} \quad \textbf{Ans.}$$

Example 4.4: *A cast iron beam has an I-section with top flange 80 mm × 40 mm, web 120 mm × 20 mm and bottom flange 160 mm × 40 mm. If tensile stress is not to exceed 30 N/mm² and compressive stress 90 N/mm², what is the maximum uniformly distributed load the beam can carry over a simply supported span of 6 m if the large flange is in tension ?*

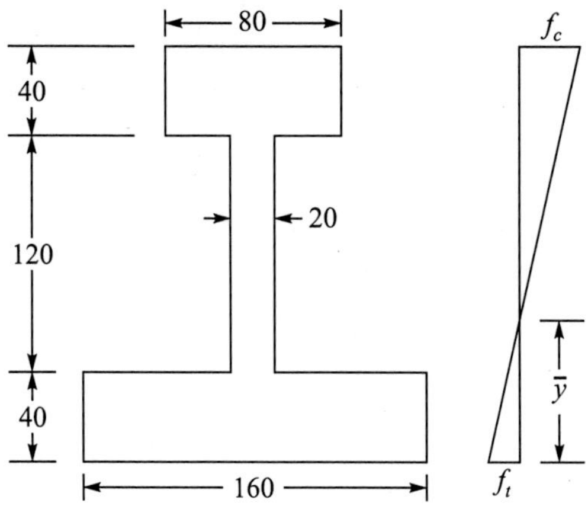

Fig. 4.12.

Solution: The cross section of the beam is as shown in Fig. 4.12. Let $\bar{y}$ be the distance of centroid (hence Neutral axis) from bottom fibre. Then

$$\bar{y} = \frac{\Sigma a_i y_i}{A}$$

$$= \frac{80 \times 40 \times 180 + 120 \times 20 \times 100 + 160 \times 40 \times 20}{80 \times 40 + 120 \times 20 + 160 \times 40}$$

$$= \frac{944000}{12000} = 78.6667 \text{ mm}$$

$$I = \frac{1}{12} \times 80 \times 40^3 + 80 \times 40\ (180 - 78.6667)^2$$

$$+ \frac{1}{12} \times 20 \times 120^3 + 20 \times 120\ (100 - 78.6667)^2$$

$$+ \frac{1}{12} \times 160 \times 40^3 + 160 \times 40(20 - 78.6667)^2$$

$$= 60138670 \text{ mm}^4$$

Tension occurs at bottom and compression at top. Extreme fibre distances of top and bottom fibres are

$$y_t = \bar{y} = 78.6667 \text{ mm}$$

$$y_c = 200 - \bar{y} = 121.3333 \text{ mm}$$

Moment carrying capacity considering tensile strength

$$= f_{\text{per}} \times \frac{1}{y_t}$$

$$= 30 \times \frac{60138670}{78.6667} = 22934224 \text{ N-mm}$$

$$= 22.934229 \text{ kN-m}$$

Moment carrying capacity from compression stress consideration

$$= f_{\text{per}}\ I$$

$$= 90 \times \frac{60138670}{121.333} = 44608366 \text{ A /mm}$$

$$= 40.608366 \text{ kN-m}$$

Actual moment carrying capacity of the section is smaller of the above two *i.e.*, moment carrying capacity of the section = 22.934229 kN-m.

Max. bending moment in a simply supported beam of span 6 m due to udl of wkN/m is

$$= \frac{w \times 6^2}{8} = 4.5\ w$$

Equating it to the moment carrying capacity, we get load carrying capacity w as

$$4.5\ w = 22.934229$$

$$w = 5.0965 \text{ kN/m} \quad \textbf{Ans.}$$

Example 4.5: *An I-section beam has 200 mm wide flanges and an overall depth 500 mm. Each flange is 25 mm thick and the web is 20 mm thick. At a certain section bending moment is M. Find what percentage of M is resisted by flanges and web.*

Solution: The *I*-section considered is shown in Fig. 4.13

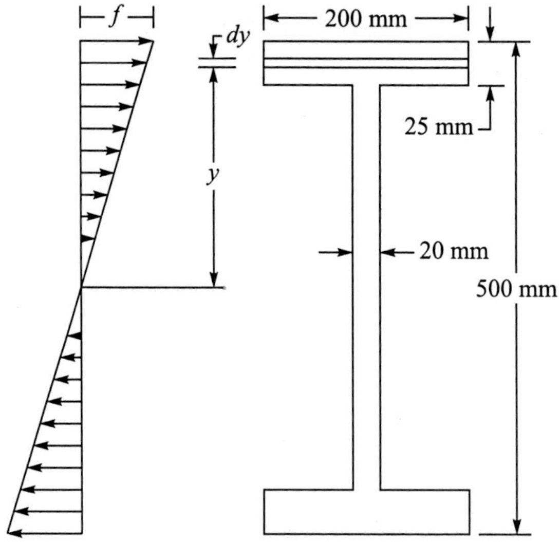

Fig. 4.13.

Due to symmetry centroid and hence neutral axis is at middle of the depth of *I*-section.

Consider an element of thickness *y* at distance *y* from Neutral Axis as shown in Fig. 4.13. Let *f* be maximum bending stress. Since bending stress varies linearly with depth, stress on the element,

$$= \frac{y}{250} f$$

$$\text{Area of the element} = 200\ \text{dy}$$

$$\therefore \quad \text{Force on the element} = \frac{y}{250} f \times 200 dy$$

Its moment of resistance = Moment of resisting force about $N - A$

$$= \frac{y}{250} f \times 200 dy \times y$$

∴ Moment of resistance of top flange

$$= \int_{225}^{250} \frac{200}{250} y^2 f\, dy$$

Due to symmetry moment of resistance of lower flange is also same as that of top flange.

$$\text{Moment of resistance of flanges} = 2 \times \int_{225}^{250} \frac{200}{250} y^2 f\, dy$$

$$= 1.6 \times f \times \left[\frac{y^3}{3}\right]_{225}^{250}$$

$$= 2258333.3\ f$$

Now, moment of inertia of *I*-section is

$$I = \frac{1}{12}\left[200 \times 500^3 - 180 \times 450^3\right]$$

$$= 7.16458 \times 10^8 \text{ mm}^4$$

∴ Moment acting on entire section

$$M = f\frac{I}{y_{max}} \qquad \text{since } f \text{ is the value at } y = 250 \text{ mm}$$

$$= f\frac{7.16458 \times 10^8}{250}$$

$$= 2865833.3\, f$$

∴ Percentage moment resisted by flanges

$$= \frac{2258333.3}{2865833.3} \times 100$$

$$= 78.802 \quad \textbf{Ans.}$$

∴ Percentage moment resisted by web

$$= 100 - 78.802$$

$$= 21.198 \quad \textbf{Ans.}$$

Example 4.6: *Determine the diameter of solid shaft which will transmit 440 kW at 280 rpm. The angle of twist must not exceed one deegree per metre length and the maximum torsional shear stress is to be limited to 40 N/mm². Assume G = 84 kN/mm².*

Solution:

$$P = 440 \text{ kW}$$

$$= 440 \times 10^3 \text{ N-m/sec}$$

$$= 440 \times 10^6 \text{ N-m/sec}$$

$$N = 280 \text{ rpm}$$

$$\theta = 1° = \frac{\pi}{180} \text{ rad}$$

$$L = 1 \text{ m} = 1000 \text{ mm}$$

$$q_s = 40 \text{ N/mm}^2$$

$$G = 84 \times 10^3 \text{ N/mm}^2$$

Now, $$P = \frac{2\pi \times N \times T}{60}$$

∴ $$440 \times 10^6 = \frac{2\pi \times 280T}{60}$$

i.e., $$T = 15006037 \text{ N-mm}$$

and $$J = \frac{\pi d^4}{32}$$

From consideration of shear stress

$$\frac{T}{J} = \frac{q_s}{R}$$

$$\frac{15006037}{\frac{\pi d^4}{32}} = \frac{40}{\frac{d}{2}}$$

$$\therefore \quad \frac{\pi d^3}{16} \times 40 = 15006037$$

$$d = 124.087 \text{ mm}$$

From consideration of angle of twist

$$\frac{T}{J} = \frac{G\theta}{L}$$

$$\frac{15006037}{\frac{\pi d^4}{32}} = \frac{84 \times 10^3 \frac{\pi}{180}}{1000}$$

$$d^4 = 15006037 \times 1000 \times \frac{32}{\pi} \times \frac{1}{84 \times 10^3} \frac{180}{\pi}$$

$$\therefore \quad d = 101.048 \text{ mm}$$

$\therefore$ Minimum diameter of the shaft to be used is 124.087 mm **Ans.**

Example 4.7: *The working condition to be satisfied by a shaft transmitting power are: (i) the shaft must not twist more than 1^0 in a length of 15 times diameter, (ii) the shear stress must not exceed 80 MN/m^2. What is the actual working stress and diameter of the shaft to transmit 736 kW at 200 rpm ? Take shear modulus as 80 GN/m^2.*

Solution:

$$G = 80 \text{ GN/m}^2 = \frac{80 \times 10^9}{10^6} \text{ N/mm}^2$$

$$= 80 \times 10^3 \text{ N/mm}^2$$

$$q_s = 80 \text{ MN/mm}^2 = 80 \text{ N/mm}^2$$

$$P = 736 \text{ kW} = 736 \times 10^3 \text{ Watts}$$

$$= 736 \times 10^6 \text{ N-mm/sec}$$

$$N = 200 \text{ rpm}$$

Now,

$$P = \frac{2\pi NT}{60}$$

$$\therefore \quad 736 \times 10^6 = \frac{2\pi \times 200T}{60}$$

or

$$T = 35.1414 \times 10^6$$

From consideration of angle of twist :

$$\theta = 1° = \frac{\pi}{180} \text{ radians}$$

$$L = 15\,d$$

$$\frac{T}{J} = \frac{G\theta}{L}$$

i.e.,
$$\frac{351414\times10^6}{\frac{\pi d^4}{32}} = 80\times10^3\,\frac{\pi}{180}\times\frac{1}{15d}$$

$$\therefore \quad d^3 = 35.1414\times10^6\times\frac{32}{\pi}\times\frac{1}{80\times10^3}\times\frac{180}{\pi}\times15$$

$$\therefore \quad d = 156.668 \text{ mm}$$

Corresponding stress at surface is

$$\frac{T}{J} = \frac{q_s}{d/2}$$

$$\frac{35.1414\times10^6}{\frac{\pi d^4}{32}(156.668)^4} = \frac{q_s}{\frac{156.668}{2}}$$

$$q_s = 35.1414\times10^6\times\frac{32}{\pi}\times\frac{156.668}{2\times(156.668)^4}$$

$$= 46.542 \text{ N/mm}^2 < 80 \text{ N/mm}^2 \quad \text{OK}$$

∴ Consideration of angle of twist governs the selection of diameter of rod.

Maximum diameter required is 156.661 mm and corresponding shear stress is 46.542 N/mm^2.

Ans.

Example 4.8: *During tests on a sample of steel bar 25 mm in diameter, it is found that the pull of 50 kN produces an extension of 0.095 mm on a length of 200 mm and a torque 200 N-m produces an angular twist of 0.9 degrees on a length of 250 mm. Find the Poisson's Ratio of the steel.*

Solution:

$$d = 25 \text{ mm}$$

$$\text{Pull} = 50 \text{ kN}$$

$$\delta_1 = 0.095 \text{ mm}$$

$$\text{Gauge length} = 200 \text{ m}$$

Using the expression for extension of a bar under direct force, we have

$$0.095 = \frac{50\times10^3\times200}{\frac{\pi}{4}\times25^2\times E}$$

or
$$E = 214440.34 \text{ N/mm}^2$$

Now, from torsion experiment

$$T = 200 \text{ N-m} = 200\times10^3 \text{ N-mm}$$

$$\theta = 0.9 \text{ degrees} = 0.9\times\frac{\pi}{180} \text{ rad}$$

$$L = 250 \text{ mm}$$

and $$J = \frac{\pi}{32}d^4 = \frac{\pi}{32}25^4$$
$$= 38349.52 \text{ mm}^4$$

∴ From the relation $$\frac{T}{J} = \frac{G\theta}{L}$$

we get, $$\frac{200\times10^3}{38349.52} = G\times0.9\times\frac{\pi}{180}\times\frac{1}{250}$$

∴ $$G = 83002.314 \text{ N/mm}^2$$

From the relation between elastic constants,
$$E = 2G(1 + \mu)$$
we get, $$214440.34 = 2 \times 83002.314\ (1 + \mu)$$
∴ $$1 + \mu = 1.29177$$
or $$\mu = 0.29177$$ **Ans.**

Example 4.9: *A hollow circular shaft of 6 m length and inner and outer diameters of 75 mm and 100 mm is subjected to a torque of 10 kN-m. If G = 80 GPa, determine the maximum shear stress produced and the total angle of twist.*

Solution:
$$L = 6 \text{ m} = 6000 \text{ mm}$$
$$d_1 = 100 \text{ mm} \qquad \therefore \quad R = 50$$
$$d_2 = 75 \text{ mm}$$
$$T = 10 \text{ kN-m} = 10 \times 10^6 \text{ N-mm}$$
$$G = 80 \text{ GPa} = 80 \times \frac{10^9}{10^6} \text{ N/mm}^2$$
$$= 80 \times 10^3 \text{ N/mm}^2$$
$$J = \frac{\pi}{32}\left(d_1^4 - d_2^4\right)$$
$$= \frac{\pi}{32}(100^4 - 75^4)$$
$$= 6.711166 \times 10^6 \text{ mm}^4$$

Now, $$\frac{T}{J} = \frac{q_s}{R}$$

i.e., $$\frac{10\times10^6}{6.711166\times10^6} = \frac{q_s}{50}$$

∴ $$q_s = 74.503 \text{ N/mm}^2$$ **Ans.**

and also, $$\frac{T}{J} = \frac{G\theta}{L}$$

∴ $$\theta = \frac{TL}{GJ} = \frac{10\times10^6\times6000}{80\times10^3\times6.711166\times10^6}$$

$$= 0.111754 \text{ rad}$$

$$= 0.111754 \times \frac{180}{\pi} \text{ degrees}$$

$$= 6.403 \text{ degrees} \quad \textbf{Ans.}$$

Example 4.10: *A hollow circular shaft 200 mm external diameter and thickness of metal 25 mm is transmitting power at 200 rpm. The angle of twist over a length of 2 m was found to be 0.5 degrees. Calculate the power transmitted and the maximum shear stress induced in the section. Take modulus of rigidity of material as 84 kN/mm².*

Solution: $d_1 = 200$ mm $\quad \therefore \quad R = 100$ mm

$t = 25$ mm

$\therefore \quad d_2 = 200 - 2 \times 25 \text{ mm} = 150$ mm

$N = 200$ rpm

$$\theta = 0.5° = 0.5 \times \frac{\pi}{180}$$

$L = 2 \text{ m} = 2000$ mm

and $\quad G = 84 \text{ kN/mm}^2 = 84 \times 10^3 \text{ N/mm}^2$

$$J = \frac{\pi}{32}\left(d_1^4 - d_2^4\right) = \frac{\pi}{32}(200^4 - 150^4)$$

$$= 1.07378 \times 10^8 \text{ mm}^4$$

$$\frac{T}{J} = \frac{G\theta}{L}$$

$$\therefore \quad T = GJ\frac{\theta}{L}$$

$$= 84 \times 10^3 \times 1.07378 \times 10^8 \times 0.5 \times \frac{\pi}{180} \times \frac{1}{2000}$$

$$= 39.356 \times 10^6 \text{ N-mm}$$

$$P = \frac{2\pi NT}{60} = \frac{2\pi \times 200 \times 39.356 \times 10^6}{60}$$

$$= 824.27 \times 10^6 \text{ N-mm/sec.}$$

i.e., $\quad P = 824.27$ kW **Ans.**

From the relation $\quad \dfrac{q_s}{R} = \dfrac{G\theta}{L}$

we get $\quad q_s = GR\dfrac{\theta}{L} = 84 \times 10^3 \times 100 \times 0.5 \times \dfrac{\pi}{180} \times \dfrac{1}{2000}$

$$= 36.652 \text{ N/mm}^2 \quad \textbf{Ans.}$$

Example 4.11: *A hollow propeller shaft of a steam ship is to transmit 3750 kW at 240 rpm. If the internal diameter is 0.8 times the external diameter and if the maximum shear stress developed is to be limited to 160 N/mm², determine the size of the shaft.*

Solution:

$$P = 3750 \text{ kW} = 3750 \times 10^6 \text{ N-mm/sec}$$
$$N = 240 \text{ rpm}$$
$$d_2 = 0.8\, d_1$$
$$q_s = 160 \text{ N/mm}^2$$

Now,
$$J = \frac{\pi}{32}\left[d_1^4 - (0.8d_1)^4\right] = \frac{\pi}{32}(1 - 0.8^4)\, d_1^4 = 0.05796 d_1^4$$

Now,
$$P = \frac{2\pi NT}{60}$$

$\therefore$
$$3750 \times 10^6 = \frac{2\pi \times 240 \times T}{60}$$

or
$$T = 1.49207 \times 10^8 \text{ N-mm}$$

From torsion formula
$$\frac{T}{J} = \frac{q_s}{R}$$

we have,
$$\frac{1.49207 \times 10^8}{0.05796 \times d_1^4} = \frac{160}{d_1/2}$$

$$d_1^3 = 8044718.3$$
$$d_1 = 200.37 \text{ mm} \quad \textbf{Ans.}$$
$$d_2 = 0.8d_1 = 160.30 \text{ mm} \quad \textbf{Ans.}$$

SUMMARY

- Pure bending is the bending unaccompanied by any S.F. The stresses produced due to constant bending moment (with zero shear force) are known as bending stresses.
- The bending stress in any layer is directly proportional to the distance of the layer from the *N*-axis. The bending stress on the *N*-axis is zero.
- The moment of resistance offered by the section is known as the strength of the section.
- Total moment of resistance of a composite beam is the sum of the moment of resistance of individual sections.
- The bending equation is given by,

$$\frac{M}{I} = \frac{\sigma_b}{y} = \frac{E}{R}$$

where M = B.M. or M.R.
σ_b = Bending stress
I = M.I. about N.A.
y = Distance of fibre from N.A.
R = Radices of curvature
E = Young's modulus of elasticity

- Section modulus '*Z*' is given by the relation

$$Z = \frac{I}{y}$$

- A shaft is in torsion, when equal and opposite torques are applied at the two ends of a shaft. Strength of the shaft means the max. torque or max. power the shaft can transmit.
- The effect of torsion is to produce shear stress in the material of the shaft, the shear stress is maximum on the surface of the shaft and is zero at the axis of the shaft.
- Torsion equation is given by,

$$\frac{T}{I_p} = \frac{\tau}{R} = \frac{C_\theta}{l}$$

- The torque transmitted by a hollow circular shaft is given by

$$T = \frac{\pi}{16} \times \tau \times \left(\frac{D^4 - d^4}{D}\right)$$

where $D =$ External diameter
$d =$ Internal diameter

- The torque transmitted by a sotid shaft is given by

$$T = \frac{\pi}{16} \times \tau \times D^3$$

where $D =$ Dia. of solid shaft
$\tau =$ Max. shear stress.

- Power transmitted by a shaft

$$P = \frac{2\pi\ NT}{60 \times 1000}\ \text{kW}$$

where $T =$ mean/average torque in Nm
$N =$ rpm.

OBJECTIVE TYPE QUESTIONS

1. The *N*-axis of a section is an axis, at which the bending stress is
(*a*) minimum (*b*) zero (*c*) maximum (*d*) infinity
2. In the theory of simple bending, the bending stress in the beam section varies
(*a*) linearly (*b*) parabolically (*c*) elliptically (*d*) none of above
3. The section modulus of a rectangular section having width (*b*) and depth (*d*) is
(*a*) $\frac{bd}{6}$ (*b*) $\frac{bd^2}{6}$ (*c*) $\frac{bd^3}{6}$ (*d*) $\frac{b^2d}{6}$
4. Strength of a beam is more if its section modulus is
(*a*) Decreased (*b*) Zero (*c*) Increased (*d*) None above
5. The strength of a beam mainly depends on
(*a*) B.M. (*b*) Section modulus
(*c*) Its weight (*d*) C.G. of the section
6. Shear stress at a section in a transversely loaded beam is Max. at
(*a*) Top fibre (*b*) Bottom fibre (*c*) At neutral plane (*d*) None of these.

7. In case of a circular section the section modulus is given as

(*a*) $\frac{\pi d^2}{16}$ (*b*) $\frac{\pi d^3}{16}$ (*c*) $\frac{\pi d^3}{32}$ (*d*) $\frac{\pi d^4}{64}$

8. The shafts are made of

(*a*) Mild steel (*b*) Alloy steel (*c*) Copper alloys (*d*) Any of above.

9. Shafts are designed on the basis of

(*a*) Strength (*b*) Rigidity
(*c*) Either of above (*d*) both (*a*) and (*b*)

10. The angle of twist is .... proportional to the twisting moment.

(*a*) Directly (*b*) Inversely
(*c*) Either (*a*) or (*b*) (*d*) None of above

11. When a solid shaft is subjected to torsion, the shear stress induced in the shaft at its centre is

(*a*) Zero (*b*) Minimum
(*c*) Maximum (*d*) Average

12. For the same material, length and given torque, a hollow shaft wieghs ...... solid shaft

(*a*) Less than (*b*) More than
(*c*) Equal to (*d*) None of above

Answers: 1. (*b*), 2. (*a*), 3. (*b*), 4. (*c*), 5. (*b*), 6. (*c*), 7. (*c*), 8. (*d*), 9. (*d*), 10. (*a*), 11. (*a*), 12. (*a*)

CHAPTER

5 Columns and Struts

Strut: A strut is a slender bar or member in any position other than vertical, subjected to a compressive load and fixed rigidity or hinged or pin connected/jointed at one or both the ends.

Column: A column is a long vertical slender bar or vertical member subjected to an axial compressive load and fixed rigidity at both ends.

Slenderness Ratio: It is the ratio of unsupported length of the column to the minimum radius of Gyration of the Cross-sectional ends of column. It has no units.

5.1. CLASSIFICATION OF COLUMNS

Depending upon the slenderness ratio or length to diameter ratio, columns can be divide into three classes.

5.1.1. Short Columns

Columns which have lengths less than 8 times their respective diameters or slenderness ratio less than 32 are called short columns. When short columns are subjected to compressive loads, their buckling is generally neglected as such the buckling stresses are very small as compared with direct compressive stresses. Therefore, it is assumed that short columns are always subjected to direct compressive stresses only.

5.1.2. Medium Size Columns

The columns which have their lengths varing from 8 times their diameter to 30 times their respective diameters or their slenderness ratio lying between 32 and 120 are called medium size columns or intermediate columns. In these columns, both the buckling as well as direct stresses are of significant values. Therefore, in the design of intermediate columns both these stresses are taken into account.

5.1.3. Long Columns

The columns having their lengths more than 30 times their respective diameters or slenderness ratio more than 120 are called long columns. They are usually subjected to Buckling stress only Direct compressive stress is very small as compared to buckling stress and hence it is neglected.

5.2. OTHER-DEFINITIONS

5.2.1. Buckling Factor

It is the ratio between the equivalent length of the column to the minimum radius of gyration.

5.2.2. Buckling Load

The maximum limiting load at which the column tends to have lateral displacement or tends to buckle is called Buckling or Crippling load. The buckling takes place about the axis having minimum radius of Gyration or least moment of inertia.

5.2.3. Safe Load

It is the load to which a column is actually subjected to and is well below the buckling load. It is obtained by dividing the buckling load by a suitable Factor of Safety (F.O.S)

i.e.,
$$\text{Safe load} = \frac{\text{Buckling/Crippling load}}{\text{F.O.S.}}$$

5.2.4. Factor of Safety (FOS)

The ratio of the critical load to the safe load is known as Factor of Safety.

For Timber, the value of *FOS* is taken as 6.

For Wrought Iron, Mild/Medium steel is taken as 3.

For Cast Iron, Mild/Medium steel is taken as 5.

Failure of a Column: The failure of a column takes place due to any of the following stresses set up in the columns:

(*a*) Direct compressive stresses.

(*b*) Buckling stresses

(*c*) Combination of direct compressive and buckling stresses.

5.2.4.1. Failure of a Short Column

Consider a short column of uniform cross-sectional area *A*, subjects to an axial compressive load *P*, as shown in Fig. 5.1.

Compressive stresses induced $= \frac{P}{A}$. If the compressive load on the short column is gradually increased, a stage will reach when the column will be on the point of failure by crushing. The stress induced in the column corresponding to this load is known as crushing stress (σ_c) and the load is called crushing load P_c.

$$\text{Crushing stress } \text{“}\sigma_c\text{”} = \frac{P_c}{A}$$

Remember all short columns fail due to crushing

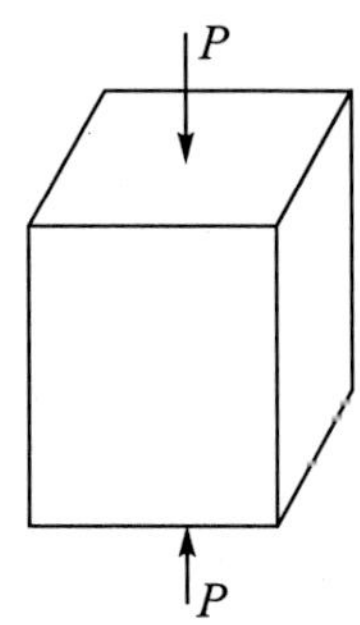

Fig. 5.1

5.2.4.2. Failure of a Long Column

Consider a long column of uniform cross sectional area *A*, subjected to an axial compressive load *P*, as shown in Fig. 5.2. A column is known as long column if the length of the column in comparison to its least lateral dimension, is very large. Such columns do not fail by crushing alone, but also by bending (also known as buckling). The load at which the column just buckles, is known as Buckling load. The buckling load is less than the crushing load for a long column.

Let l = Length of a long column.

P = Buckling load.

A = Cross-sectional area of the column.

e = Max. bending of the column at the centre.

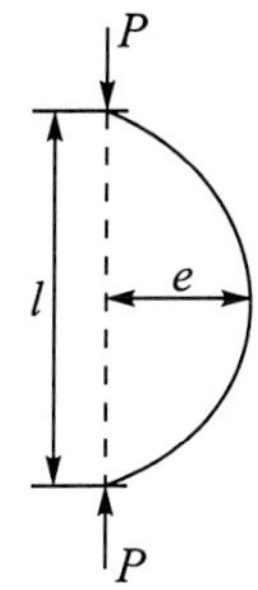

Fig. 5.2

Stress due to direct load $\sigma_d = \frac{P}{A}$

Stress due to bending at the centre of column,

$$\sigma_b = \frac{P.e}{z}$$

where z = Sectional modulus about the axis of bending. Extreme stresses on the mid section are given by:

Max. stress, $\sigma_{max} = \sigma_d + \sigma_b$

Min. stress, $\sigma_{min} = \sigma_d - \sigma_b$

The column will fail when max. stress (σ_{max}) is more than the crushing stress σ_c. But in long columns, the direct compressive stresses are negligible as compared to buckling stresses.

5.2.5. End Conditions

The end conditions of a loaded column can be had in four ways:

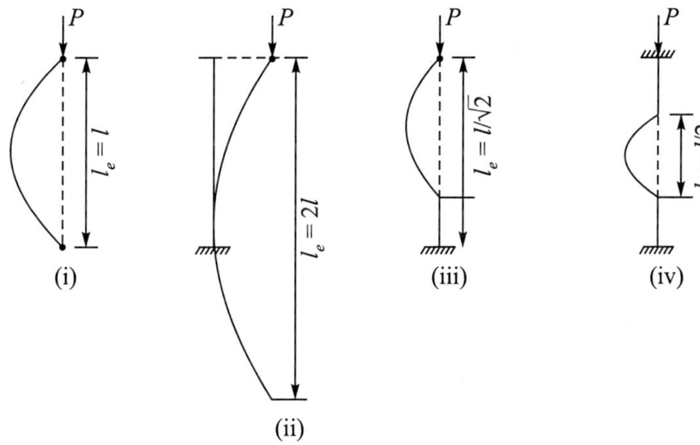

Fig. 5.3

(*i*) Both ends pin jointed or hinged or rounded or free.
(*ii*) One end fixed and other end free.
(*iii*) One end fixed and the other pin jointed.
(*iv*) Both ends fixed.

5.2.6. Equivalent Length (l_e)

Equivalent length of a column is the distance between adjacent points of inflexion. A point of inflexion is found at every column end that is free to rotate and at every point where there is a change of the axis.

Let l = Actual length of column.

and le = Equivalent or Effective length of column.

Hence in case of:

(*i*) **Both Ends Hinged:**

Equivalent length or = Actual length

Effective length $l_e = l$

(*ii*) **One End Fixed and Other End Free:** $l_e = 2l$, the free and sway sideways and the curvature in the length l will be similar to that of the upper half of the simple column.

(*iii*) **One End Fixed and the Other Pin Jointed:** $l_e = \frac{l}{\sqrt{2}}$, between the top of the column and inflexion point.

(*iv*) **Both Ends Fixed:** $l_e = \frac{l}{2}$, the distance between the two inflexion points.

5.3. EULER'S THEORY (FOR LONG COLUMNS)

Assumptions: Following assumptions are made while deriving the Euler's formula:

1. The column is initially straight and of uniform lateral dimension.
2. The compressive load is exactly axial and it passes through the centroid of the col. section.
3. The material of the column is perfectly homogeneous and isotropic.
4. Pin joints are frictionless and fixed ends are perfectly rigid.
5. The weight of the column itself is neglected.
6. The column fails by buckling alone.
7. Limit of proportionality is not exceeded.

5.3.1. Sign Conventions for Bending Moments

A bending moment which bends the columns so as to present convexity towards the initial centre line of the member will be regarded positive Fig. 5.4(*i*).

A bending moment which bends the column as to present concavity towards the initial centre line of the member will be regarded as negative Fig. 5.4(*ii*).

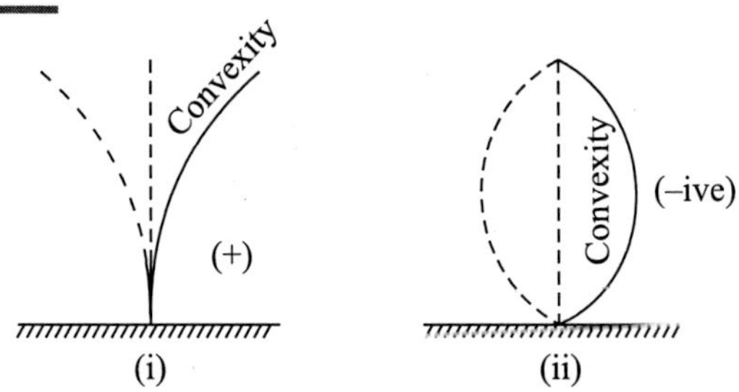

Fig. 5.4

5.3.2. Euler's Formula

Euler's formula is used for calculating the critical load for a column or strut, and is expressed as under:

$$P_e = \frac{\pi^2 EI}{l_e^2}$$

where P_e = Critical load

E = Modulus of elasticity

l_e = Equivalent or Effective length of col. or strut

I = Least *M.I.* of section of the column.

A column of given length, cross section and material will have different values of buckling loads for different end conditions as tabulated below:

Table 5.1

Case	End conditions	Equivalent length (l_e)	Buckling load euler
1.	Both ends hinged/pin jointed or free	"l"	$\frac{\pi^2 EI}{l_e^2} = \frac{\pi^2 EI}{l^2}$
2.	One end fixed, other end free	$2l$	$\frac{\pi^2 EI}{l_e^2} = \frac{\pi^2 EI}{4l^2}$
3.	One end fixed other end pin jointed	$\frac{l}{\sqrt{2}}$	$\frac{\pi^2 EI}{l_e^2} = \frac{2\pi^2 EI}{l^2}$
4.	Both ends fixed or encastered	$\frac{l}{2}$	$\frac{\pi^2 EI}{l_e^2} = \frac{4\pi^2 EI}{l^2}$

5.4. LIMITATIONS FOR THE USE OF EULER'S FORMULA

Following are the limitations due to which Euler's formula is of little practical use:

1. It is applicable to an ideal strut only and in practice there is always crookedness in the column and the load applied may not be exactly co-axial.
2. It takes no account of direct stresses. It means it may give a buckling load for struts, far in excess of load which they can withstand under direct compression.
3. It is only applicable for long columns only.

5.5. DERIVATIONS OF EULER'S FORMULA FOR DIFFERENT END CONDITIONS

Case I: When both the ends are hinged/pinned: Fig. 5.5 shows a column *AB* of length "l", and uniform sectional area *A*, hinged at both the ends *A* and *B*. Let "*P*" be the crippling load at which the column has just buckled. Due to the crippling load, the column will deflect into a curved form *ACB*.

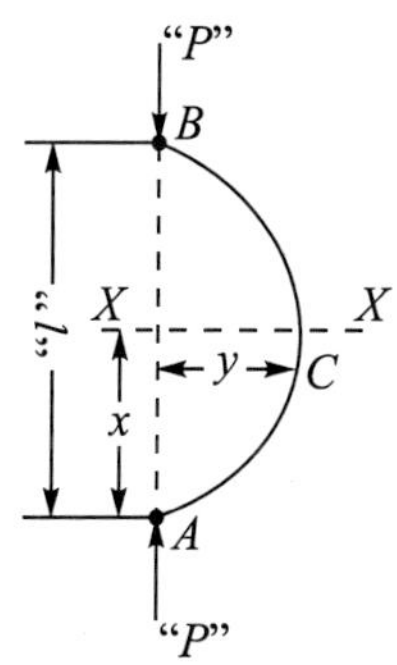

Fig. 5.5

Consider any section *X-X* at a distance x from the end *A*. Let y = Deflection (lateral displacement) at the section.

The moment due to the crippling load at the section = $-P \cdot y$

(–ive sign is taken due to sign conventions)

$$\text{But moment} = EI \frac{d^2 y}{dx^2}$$

Equating the two moments, we get

$$EI \frac{d^2 y}{dx^2} = -P \times y$$

or $$EI \frac{d^2 y}{dx^2} + P \times y = 0$$

or $$\frac{d^2 y}{dx^2} + \frac{P}{EI} \times y = 0 \qquad \text{[Dividing by } EI]$$

The solution of the above differential equation is

$$y = C_1 \cdot \cos\left[x\sqrt{\frac{P}{EI}}\right] + C_2 \cdot \sin\left[x\sqrt{\frac{P}{EI}}\right] \qquad \ldots(5.1)$$

where C_1 and C_2 are the constants of intigration.

The values of C_1 and C_2 are obtained as given below:

(*i*) At $A, x = 0$ and $y = 0$

Substituting these values in eq. (5.1), we get

$$0 = C_1 \cdot \cos 0 + C_2 \cdot \sin 0$$

or $$0 = C_1 \times 1 + 0 \quad \text{or} \quad C_1 = 0$$

(*ii*) At $B, x = l$ and $y = 0$

Substituting these values in eq. (5.1), we get

$$0 = 0 + C_2 \cdot \sin\left[l\sqrt{\frac{P}{EI}}\right]$$

or $$0 = 0 + C_2 \cdot \sin\left[l\sqrt{P/EI}\right] \qquad \ldots(5.2)$$

From eq. (5.2), it is clear that either $C_2 = 0$ or $\sin\left[l\sqrt{\frac{P}{EI}}\right] = 0$. As $C_1 = 0$, then if C_2 is also equal to zero, then from eq. (5.1) we will get $y = 0$. This means that the bending of the column will be zero, which is not true.

$$\therefore \quad \sin\left[l\sqrt{P/EI}\right] = 0 \quad \text{or} \quad l\sqrt{P/EI} = 0, \pi, 2\pi, 3\pi, \ldots$$

Taking the least practical value, $l\sqrt{P/EI} = \pi$, we get,

$$\boxed{P = \frac{\pi^2 EI}{l^2}}$$

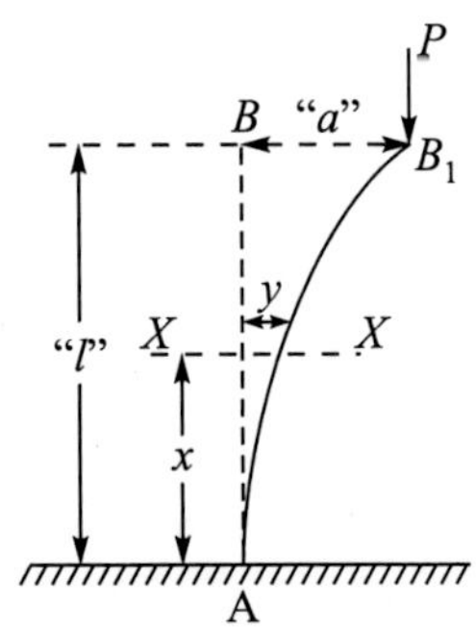

Fig. 5.6

Case II: When one end is fixed and the other end is free: Figure 5.6 shows a column *AB* of length '*l*' whose lower end '*A*' is fixed, the upper end *B* being free. Let '*P*' is the crippling load at which the column has just buckled. Due to the crippling load '*P*', the column will deflect.

Consider any sec. *X-X*, at a distance '*x*' from the fixed end *A*.

Let y = Deflection (Lateral displacement) at the sec.

a = Deflection at the free end *B*.

The moment due to crippling load at the section X-X,

$$= P(a-y) \qquad \text{(+ive sign is taken due to sign conventions)}$$

But moment $= EI\dfrac{d^2y}{dx^2}$

∴ Equating the two moments, we get

$$EI\frac{d^2y}{dx^2} = P(a-y) = Pa - Py$$

or $$EI\frac{d^2y}{dx^2} + P\cdot y = P\cdot a \qquad \text{[Dividing by } EI]$$

The solution of the above differential equation is

$$y = C_1 \cdot \cos\left[x\sqrt{\frac{P}{EI}}\right] + C_2 \cdot \sin\left[x\sqrt{\frac{P}{EI}}\right] + a \qquad ...(5.3)$$

where C_1 and C_2 are the constants of integration.

The values of C_1 and C_2 are obtained as below:

(*i*) For the fixed end, the deflection as well as slope is-zero.

Hence at end A, the deflection $y = 0$ and also slope $\dfrac{dy}{dx} = 0$

Hence at A, $x = 0$ and $y = 0$

Substituting these values in eq. (5.3), we get,

$$C_1 = -a$$

At A, $x = 0$ and $\dfrac{dy}{dx} = 0$.

Differentiating the eq. (5.3), w.r.t. x, we get,

$$\frac{dy}{dx} = C_1(-1)\sin\left[x\sqrt{P/EI}\right] \times \sqrt{P/EI} + C_2\left[x\sqrt{P/EI}\right] \times \sqrt{\frac{P}{EI}} + 0$$

$$= -C_1\sqrt{P/EI}\,\sin\left[x\sqrt{P/EI}\right] + C_2\sqrt{P/EI}\cos\left[x\sqrt{P/EI}\right]$$

But at A, $x = 0$ and $\dfrac{dy}{dx} = 0$.

∴ The above equation becomes,

$$0 = C_2\sqrt{P/EI}$$

From above equation, it is clear that either $C_2 = 0$ or $\sqrt{P/EI} = 0$

But for the crippling load P, the value of $\sqrt{P/EI}$ cannot be equal to zero ∴ $C_2 = 0$

Substituting the values of $C_1 = -a$ and $C_2 = 0$ in eq. (5.3),

we get, $$y = -a\cdot\cos\left[x\sqrt{P/EI}\right] + a \qquad ...(5.4)$$

(*ii*) For free end of the column, $x = l$ and $y = 0$.

Substituting these values in eq. (5.3), we get

$$a = -a \cdot \cos\left[l\sqrt{P/EI}\right] + a \quad \text{or} \quad 0 = -a \cdot \cos\left[l\sqrt{P/EI}\right]$$

or $$a \cdot \cos\left[l\sqrt{P/EI}\right] = 0$$

But 'a' cannot be equal to zero.

$$\therefore \quad \cos\left[l\sqrt{P/EI}\right] = 0$$

$$\therefore \quad \left[l\sqrt{P/EI}\right] = \frac{\pi}{2}, \frac{3\pi}{2}, \frac{5\pi}{2}, \ldots$$

Taking the least practical value,

$$l\sqrt{P/EI} = \frac{\pi}{2}$$

or $$\sqrt{\frac{P}{EI}} = \frac{\pi}{2l}$$

or $$\boxed{P = \frac{\pi^2 EI}{4l^2}}$$

Case III: When both ends of column are fixed: Figure 5.7 shows a column AB of length "l" whose both ends A and B are fixed.

Let 'P' is the crippling load at which the column has buckled. Due to crippling load 'P', the column will deflect. Consider a section X-X at a distance x from the end A. Let the deflection of the column at the section be 'y'. As both ends of the column are fixed and the column carries a crippling load, there will be some fixed end moments at A and B.

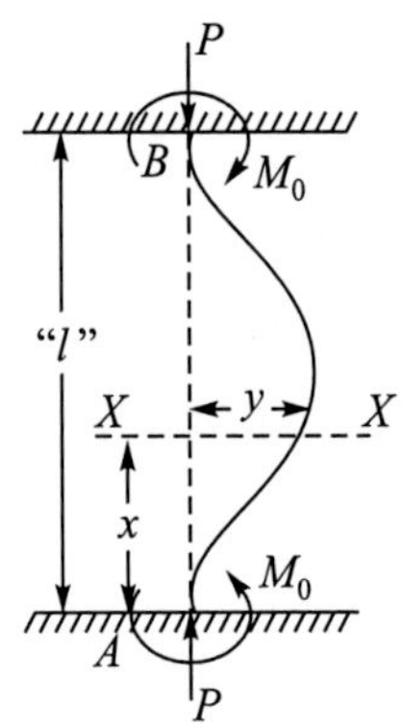

Fig. 5.7

Let M_0 = Fixed end moments at A and B.

Then moment at the section $= M_0 - P \cdot y$

But moment $= EI\dfrac{d^2y}{dx^2}$

$\therefore$ Equating the two moment, we get,

$$EI\frac{d^2y}{dx^2} = M_0 - P \cdot y$$

or $$EI\frac{d^2y}{dx^2} + P \cdot y = M_0$$

or $$\frac{d^2y}{dx^2} + \frac{P}{EI} \cdot y = \frac{M_0}{EI} \qquad \text{(Dividing by } EI\text{)}$$

The solution of the above differential equation is

$$y = C_1 \cdot \cos\left[x \cdot \sqrt{P/EI}\right] + C_2 \cdot \sin\left[x \cdot \sqrt{P/EI}\right] + \frac{M_0}{P} \qquad \text{...(5.5)}$$

where C_1 and C_2 are the constants of integration.
The values of C_1 and C_2 are obtained as below:

(*i*) At A, $x = 0$, $y = 0$ and also $\frac{dy}{dx} = 0$, as A is fixed end.

(*ii*) At B, $x = l$, $y = 0$ and $\frac{dy}{dx} = 0$, as B is also a fixed end.

Substituting the values $x = 0$ and $y = 0$ in eqn. (5.5), we get

$$0 = C_1 + \frac{M_0}{P}$$

$$C_1 = -\frac{M_0}{P}$$

Differentiating eq. (5.5) with respect to x, we get,

$$\begin{aligned}\frac{dy}{dx} &= C_1\,(-1)\sin\left[x \cdot \sqrt{P/EI}\right] \cdot \sqrt{P/EI} \\ &\quad + C_2 \cos\left[x \cdot \sqrt{P/EI}\right] \cdot \sqrt{P/EI} + 0 \\ &\quad - C_1 \sin\left[x\sqrt{P/EI}\right] \cdot \sqrt{P/EI} \\ &\quad + C_2 \cos\left[x \cdot \sqrt{P/EI}\right] \cdot \sqrt{P/EI}\end{aligned}$$

Substituting the values $x = 0$ and $\frac{dy}{dx} = 0$, the above equation becomes,

$$0 = C_2 \sqrt{P/EI}$$

From the above equation, it is clear that either

$C_2 = 0$ or $\sqrt{P/EI} = 0$. But for a given crippling load P, the value of $\sqrt{P/EI}$ cannot be equal to zero.

$$\therefore \qquad C_2 = 0.$$

Now substituting the values of $C_1 = -\frac{M_0}{P}$ and $C_2 = 0$ in eq. (5.5), we get

$$\begin{aligned}y &= -\frac{M_0}{P}\cos\left[x \cdot \sqrt{P/EI}\right] + 0 + \frac{M_0}{P} \\ &= -\frac{M_0}{P}\cos\left[x \cdot \sqrt{P/EI}\right] + \frac{M_0}{P} \qquad \text{...(5.6)}\end{aligned}$$

At the end B, of the column, $x = l$ and $y = 0$

Substituting these values in equation (5.6), above

$$0 = -\frac{M_0}{P}\cos\left[l\cdot\sqrt{P/EI}\right] + \frac{M_0}{P}$$

or $$\frac{M_0}{P}\cos\left[l\cdot\sqrt{P/EI}\right] = \frac{M_0}{P}$$

or $$\cos\left[l\cdot\sqrt{P/EI}\right] = \frac{M_0}{P}\times\frac{P}{M_0} = 1$$

$\therefore$ $$l\sqrt{\frac{P}{EI}} = 0,\ 2\pi,\ 4\pi,\ 6\pi,\ \ldots$$

Taking the least practical value,

$$l\sqrt{P/EI} = 2\pi$$

or $$\boxed{P = \frac{4\pi^2 EI}{l^2}}$$

Case IV: When one end of the column is fixed and the other end is hinged or pinned.

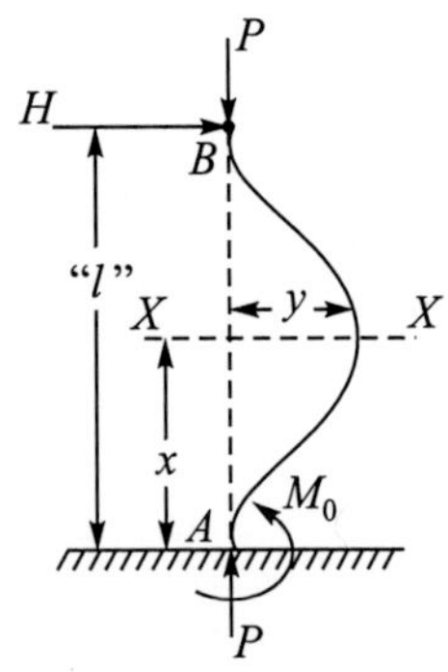

Fig. 5.8

Figure 5.8 shows a column AB of length l, whose upper end B is hinged while its lower end A is fixed.

Let 'P' is the crippling load at which the column has buckled. Due to the crippling load P, the column will deflect.

As the column is fixed at A, there will be some fixed end moment at A. In order to balance the fixing moment at A, there will be a horizontal reaction at B as shown in the Fig. 5.8.

Consider a section X-X at a distance x from A.

Let y = Deflection of the column at the section.

M_0 = Fixed end moment at A.

H = Horizontal reaction at B.

The moment at the section = Moment due to crippling load at B + Moment due to horizontal reaction at B.

$$= -P\cdot y + H(l - x)$$

But moment $$= EI\frac{d^2y}{dx^2}$$

Equating the two moments, we get

$$EI\frac{d^2y}{dx^2} = -P\cdot y + H(l - x)$$

or $$EI\frac{d^2y}{dx^2} + P.y = H(l - x)$$

or $$\frac{d^2y}{dx^2} + \frac{P}{EI}\cdot y = \frac{H}{EI}(l - x)$$ (Dividing by EI)

$$= \frac{H}{EI}(l - x) \times \frac{P}{P} = \frac{P}{EI} \cdot \frac{H(l - x)}{P}$$

The solution of the above differential equation is ,

$$y = C_1 \cos\left[x \cdot \sqrt{P/EI}\right] + C_2 \sin\left[x \cdot \sqrt{P/EI}\right] + \frac{H}{P}(l - x) \qquad \ldots(5.7)$$

where C_1 and C_2 are the constants of integration.

The values of C_1 and C_2 are obtained as below:

(*i*) At the fixed end A, $x = 0$, $y = 0$ and also $\frac{dy}{dx} = 0$.

(*ii*) At the hinged end B, $x = l$ and $y = 0$.

Substituting the value $x =$ and $y = 0$ in eq. (5.7), we get

$$0 = C_1 + \frac{H \cdot l}{P}$$

$$C_1 = -\frac{H}{P} \cdot l$$

Differentiating eq. (5.7) w.r.t. to x, we get,

$$\frac{dy}{dx} = C_1(-1)\sin\left[x \cdot \sqrt{P/EI}\right] \cdot \sqrt{P/EI}$$

$$+ C_2 \cos\left[x \cdot \sqrt{P/EI}\right] \cdot \sqrt{P/EI} - \frac{H}{P}$$

$$= -C_1 \sin\left[x \cdot \sqrt{P/EI}\right] \cdot \sqrt{P/EI}$$

$$+ C_2 \cos\left[x \cdot \sqrt{P/EI}\right] \cdot \sqrt{P/EI} - \frac{H}{P}$$

At A, $x = 0$, and $\frac{dy}{dx} = 0$.

Substituting these values in the above equation, we get

$$0 = C_2 \sqrt{P/EI} - \frac{H}{P}$$

or

$$C_2 = \frac{H}{P}\sqrt{\frac{EI}{P}}$$

Substituting the values of $C_1 = \frac{H}{P} \cdot l$ and $C_2 = \frac{H}{P}\sqrt{\frac{EI}{P}}$ in eq. (5.7), we get,

$$y = -\frac{H}{P} \cdot l\cos\left[x \cdot \sqrt{P/EI}\right] + \frac{H}{P}\sqrt{\frac{EI}{P}} \sin\left[x \cdot \sqrt{P/EI}\right]$$

$$+ \frac{H}{P}(l - x)$$

At the end B, $x = l$ and $y = 0$.

Hence the above equation becomes,

$$0 = -\frac{H}{P}\cdot l\cos\left[l\cdot\sqrt{P/EI}\right] + \frac{H}{P}\sqrt{\frac{EI}{P}}\sin\left[l\cdot\sqrt{P/EI}\right] + \frac{H}{P}(l-l)$$

$$0 = \frac{H}{P}\,l\cos\left[l\cdot\sqrt{P/EI}\right] + \frac{H}{P}\sqrt{\frac{EI}{P}}\sin\left[l\cdot\sqrt{P/EI}\right] + 0$$

or $$\frac{H}{P}\sqrt{\frac{EI}{P}}\sin\left[l\cdot\sqrt{P/EI}\right] = \frac{H}{P}\cdot l\cos\left[l\cdot\sqrt{P/EI}\right]$$

or $$\sin\left[l\cdot\sqrt{P/EI}\right] = \frac{H}{P}\cdot l\times\frac{P}{H}\times\sqrt{\frac{P}{EI}}\cdot\cos\left[l\cdot\sqrt{P/EI}\right]$$

$$= l\cdot\sqrt{P/EI}\,\cos\left[l\cdot\sqrt{P/EI}\right]$$

or $$\tan\left[l\cdot\sqrt{P/EI}\right] = l\cdot\sqrt{\frac{P}{EI}}$$

The solution to the above equation, is

$$l\sqrt{P/EI} = 4.5 \text{ radium.}$$

Squaring both sides, we get

$$l^2\cdot\frac{P}{EI} = (4.5)^2 = 20.25$$

or $$P = 20.25\,\frac{EI}{l^2}$$

But $$20.25\ \Omega\ 2\pi^2$$

$\therefore$ $$\boxed{P = \frac{2\pi^2\,EI}{l^2}}$$

PROBLEMS ON COLUMNS & STRUTS

Example 5.1: *A solid round bar 60 mm ϕ and 2.5 m long is used as a strut, one end of which in fixed, while its other end is hinged. Find the safe compressive load for this strut using Euler's formula.*

Assume E = 200 GN/m² and FOS = 3.

Solution: Given:

$$D = 60 \text{ mm} = 0.06 \text{ m}$$
$$E = 200 \text{ GN/m}^2$$
$$FOS = 3$$
$$l = 2.5 \text{ m}$$

End condition, one end hinged and other fixed.

$$\therefore \qquad le = \frac{l}{\sqrt{2}} = \frac{2.5}{\sqrt{2}} = 1.768 \text{ m.}$$

Safe compressive load,

Crippling load,
$$\text{"}Pe\text{"} = \frac{\pi^2 EI}{le^2} = \frac{\pi^2 \times 200 \times 10^9 \frac{\pi}{64} \times (.06)^4 \times 10^{-3}}{(1.768)^2}$$

$$= 401.7 \text{ kN}$$

$\therefore$ **Safe compressive**

Load
$$= \frac{Pe}{FOS} = \frac{401.7}{3} = 133.9 \text{ kN.}$$

Example 5.2: *A slender pin ended column (Aluminium) 1.8 m long and of circular Cross-section is to have an outside diameter of 50 mm. Calculate the internal diameter to prevent failure by buckling, if the actual load applied is 13.6 KN and the critical load applied is twice the actual load. "E" for Aluminium 70 GN/m².*

Solution: Outside diameter of column,

$$D = 50 \text{ mm} = 0.05 \text{ m.}$$

Inside diameter of column "d"

$$A = \frac{\pi}{4}(D^2 - d^2) = \frac{\pi}{4}(0.05^2 - d^2)$$

$$I = \frac{\pi}{64}(D^4 - d^4) = \frac{\pi}{64}(0.05^4 - d^4)$$

Also, Critical load = 2 × Safe load (given)

$$= 2 \times 13.6 = 27.2 \text{ kN}$$

End conditions: Pin-ended

$$le = l = 1.8 \text{ m.}$$

$Pe = \frac{\pi^2 EI}{le^2}$, we get

$$27.2 \times 10^3 = \frac{\pi^2 \times 70 \times 10^9 \times \frac{\pi}{64}(0.05^4 - d^4)}{(1.8)^2}$$

or
$$(0.05^4 - d^4) = \frac{27.2 \times 10^3 \times 1.8^2 \times 64}{\pi^2 \times 70 \times 10^9 \times \pi} = 2.6 \times 10^{-6}$$

or
$$d^4 = 6.25 \times 10^{-6} - 2.6 \times 10^{-6} = 3.65 \times 10^{-6}$$

$$\text{"}d\text{"} = 0.0437 \text{ m} = 43.7 \text{ mm.}$$

Example 5.3: *A steel rod 5 m long and of 40 mm diameter is used as a column with our end fixed and the other free. Determine the crippling load by Euler's formula.*

Take E as 200 GPa.

Solution: Given:

$$\text{“}l\text{”} = 5\text{ m} = 5 \times 10^3\text{ mm}$$

$$\text{“}d\text{”} = 40\text{ mm},\ E = 200\text{ GPa} = 200 \times 10^3\text{ N/mm}^2$$

$$I = \frac{\pi}{64} \times d^4 = \frac{\pi}{64} \times (40)^4 = 40000\,\pi\text{ mm}^4.$$

In our case,

$$le = 2l = 2 \times (5 \times 10^3) = 10 \times 10^3\text{ mm}.$$

∴ Crippling load

$$Pe = \frac{\pi^2 EI}{le^2} = \frac{\pi^2 \times 200 \times 10^3 \times 40000\pi}{(10 \times 10^3)^2}$$

$$= 2480\text{ N} = 2.48\text{ kN}$$ **Ans.**

Example 5.4: *A bar of length 4 m when used as a simply supported beam and subjected to a u.d.l of 30 kN/m over the whole span, deflects15 mm at the centre. Determine the crippling loads when it is used as a column with following end conditions:* **(KU 2010)**

(*i*) Both ends pin-jointed

(*ii*) One end fixed and other end hinged.

(*iii*) Both ends fixed.

Solution. Length of Bar, $l = 4$ m.

$$\text{u.d.l. “}\omega\text{”} = 30\text{ kN/m}$$

Deflection,

$$\delta = 15\text{ mm} = 0.015\text{ m}$$

we know that,

$$\delta = \frac{5\,\omega l^4}{384\,EI}$$

$$0.015 = \frac{5 \times (30 \times 10^3) \times 4^4}{384\,EI}$$

∴

$$EI = \frac{5 \times (30 \times 10^3) \times 4^4}{0.015 \times 384} = 6.66 \times 10^6\text{ Nm}^2$$

(*i*)

$$Pe = \frac{\pi^2 EI}{le^2}, \qquad (le = l = 4\text{ m})$$

$$= \frac{\pi^2 \times 6.66 \times 10^6}{4^2} \times 10^{-3}\text{ kN} = 4108\text{ kN}$$ **Ans.**

(*ii*)

$$Pe = \frac{\pi^2 EI}{le^2}, \qquad \left[le = \frac{l}{\sqrt{2}} = \frac{4}{\sqrt{2}} = 2.88\text{ m}\right]$$

$$= \frac{\pi^2 \times 6.66 \times 10^6}{2.83^2} \times 10^{-3}\text{ kN} = 8207\text{ kN}$$ **Ans.**

(*iii*)

$$Pe = \frac{\pi^2 EI}{le^2} \qquad \left[le = \frac{l}{2} = \frac{4}{2} = 2\text{ m}\right]$$

$$= \frac{\pi^2 \times 6.66 \times 10^6}{2^2} \times 10^{-3}\text{ kN} = 16432\text{ kN}.$$ **Ans.**

Example 5.5: *A solid round bar 5 m long and 60 mm in diameter was found to extend 6.2 mm under a tensile load of 60 kN. Find buckling load if both ends of col. are fixed.*

Solution: Actual length of bar, "l" = 5 m = 5000 mm

Diameter of column, "d" = 60 mm.

X-sectional area of bar, $A = \frac{\pi}{4} \times d^2 = \frac{\pi}{4} \times 60^2 = 2827.43 \text{ mm}^2$.

Tensile load $W = 60 \text{ kN} = 60000 \text{ N}$.

Extension of bar, $\delta l = 6.2 \text{ mm}$

$$\text{Tensile strain} = \frac{\delta l}{l} = \frac{6.2}{5000}$$

$$\text{Tensile stress} = \frac{W}{A} = \frac{60000}{2827.43} = 21.22 \text{ N/mm}^2.$$

$$\text{Modulus of elasticity, } E = \frac{\text{Tensile stress}}{\text{Tensile strain}} = \frac{21.22 \times 5000}{6.2} = 17112.9 \text{ N/mm}^2$$

$$\text{M.I. of section } I = \frac{\pi}{64} \times d^4 = \frac{\pi}{64} \times (60)^4 = 63.62 \times 10^4 \text{ mm}^4$$

$$\text{Effective length } L = \frac{l}{2} \qquad (\because \text{ Both ends fixed})$$

$$= \frac{5000}{2} = 2500 \text{ mm}$$

$$Pe = \frac{\pi^2 EI}{L^2} = \frac{\pi^2 \times 17112.9^2 \times 63.62 \times 10^4}{(2500)^2} = 17192.4 \text{ N} \qquad \textbf{Ans.}$$

Example 5.6: *Determine the crippling load for a T-sec shown in Fig. 5.9, when it is used as strut of length 5 m with both of its end hinged.*

Take $E = 2 \times 10^5$ N/mm².

Solution:

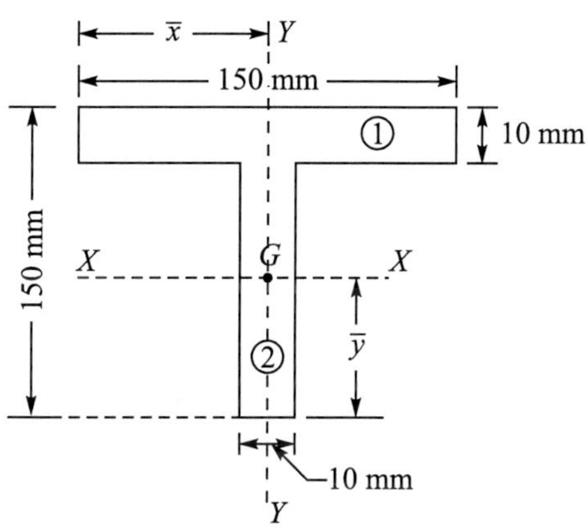

Fig. 5.9

The given sec. is symmetrical about *Y-Y* axis and hence the *C.G.* of the sec will lie on this axis.

Let $\bar{y}$ = Distance of *C.G.* of the see from bottom end.

Divide the given see into two parts.

$$a_1 = 150 \times 10 = 1500 \text{ mm}^2$$

$$y_1 = 150 \frac{-10}{2} = 145 \text{ mm}$$

$$a_2 = 140 \times 10 = 1400 \text{ mm}^2$$

$$y_2 = \frac{140}{2} = 70 \text{ mm.}$$

$$\therefore \quad \bar{y} = \frac{a_1 y_1 + a_2 y_2}{a_1 + a_2} = \frac{1500 \times 145 + 1400 \times 70}{1500 + 1400}$$

$$= \frac{315500}{2900} = 108.8 \text{ mm}$$

M.I. about *X-X* axis,

$$I_{XX} = \left[\frac{150 \times 10^3}{12} + 1500 \times (145 - 108.8)^2\right] + \left[\frac{10 \times 140^3}{12} + 1400 \times (108.8 - 70)^2\right]$$

$$= 12500 + 1965660 + 2286667 + 2107616$$

$$I_{XX} = 6372443 \text{ mm}^4$$

M.I. about *Y-Y* axis,

$$I_{YY} = \frac{10 \times 150^3}{12} + \frac{140 \times 10^3}{12} = 2812500 + 11667 = 2824167 \text{ mm}^4.$$

Least value of M.I. is about *Y-Y* axis.

$$I = 2824167 \text{ mm}^4.$$

Effective length, $L = l =$ 5 m = 5000 mm (∵ Both ends hinged)

Using the relation, $P = \frac{\pi^2 EI}{L^2} = \frac{\pi^2 \times 2 \times 10^5 \times 2824167}{5000^2} = 222987.3 \text{ N} = 223 \text{ kN}$ **Ans.**

Example 5.7: *A built-up beam shown in Fig. 5.10 is simply supported at its ends. Compute its length given that when it is subjected to a load of 40 kN/m, it deflects by 10 mm. Find out the safe load if this beam is used as a column with both ends fixed. Assume a factor of safety as 4. Use Euler's formula. Take $E = 2.1 \times 10^5$ N/mm².*

Solution: M.I. of section about *X-X* axis,

$$I_{XX} = 2\left[\frac{200 \times 10^3}{12} + 200 \times 10 \times (195)^2\right] + \frac{10 \times 380^2}{12}$$

$$= 15213.3 \times 10^4 + 4572.7 \times 10^4 = 19786 \times 10^4 \text{ mm}^4.$$

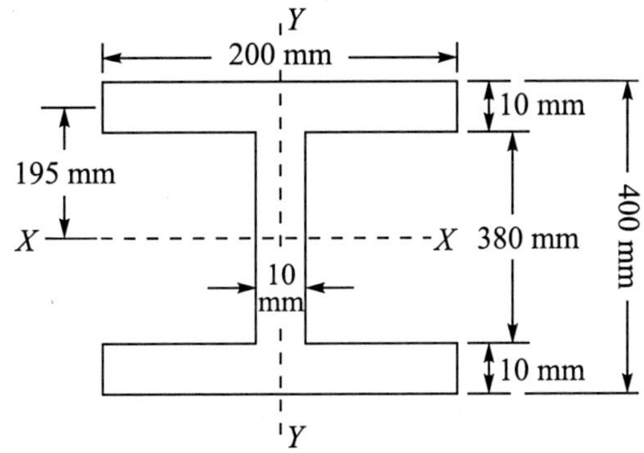

Fig. 5.10

M.I. of section about *Y-Y* aixs,

$$I_{YY} = 2\left[\frac{10\times 200^3}{12}\right] + \frac{380\times 10^3}{12} = 1333.33 \times 10^4 + 3.17 \times 10^4 = 1336.5 \times 10^4 \text{ mm}^4$$

Least value of the M.I. is about *Y-Y* axis.

$$I = 1336.5 \times 10^4 \text{ mm}^4.$$

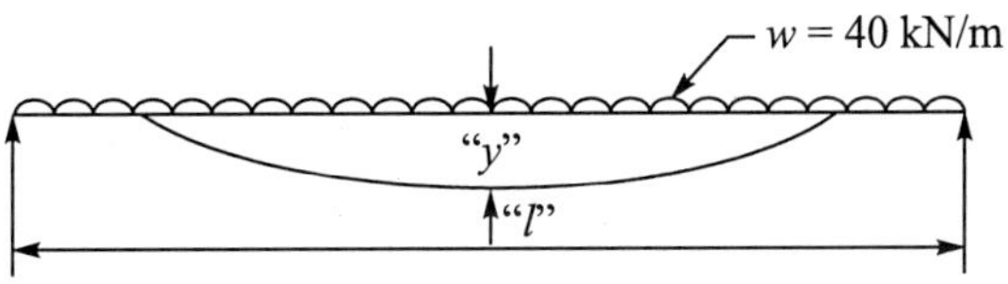

Fig. 5.11

Let l = Length of beam.

Deflection of beam in the centre, $y = 10$ mm

u.d.l. "ω" = 40 kN/m = 40 N/mm.

Using the relation,

$$y = \frac{5\,\omega\, l^4}{384\, EI}$$

$$10 = \frac{5\times 40\times l^4}{384\times 2.1\times 10^5\times 1336.5\times 10^4}$$

$$l^4 = \frac{10\times 384\times 2.1\times 10^5\times 1336.5\times 10^4}{5\times 40} = 538876.8 \times 10^8$$

or "l" = 2710 mm = 2.71 m

Safe load, the beam can carry as a column.

Actual length of column, $l = 2710$ mm

Effective length of column, $L = \frac{l}{2} = \frac{2710}{2} = 1355$ mm (∵ Both ends fixed)

Using the relation,

$$Pe = \frac{\pi^2 EI}{L^2} = \frac{\pi^2 \times 2.1 \times 10^5 \times 1336.5 \times 10^4}{(1355)^2} = 15087227 \text{ N.}$$

$$\therefore \quad \text{Safe load} = \frac{\text{Cripping load}}{FOS}$$

$$= \frac{15087227}{4} = 3771807 \text{ N} = 3771.81 \text{ kN.} \qquad \textbf{Ans.}$$

Example 5.8: *Determine the ratio of buckling strength of two circular columns, one hollow and other solid. Both are made of same material and have same length, X-sec area and end conditions. The internal diameter of the hollow column is half of its external diameter.*

Solution: Hollow Circular Column

Let "d" = Internal diameter of hollow column

D = External diameter of hollow column

$d = 0.5\, D$ (given)

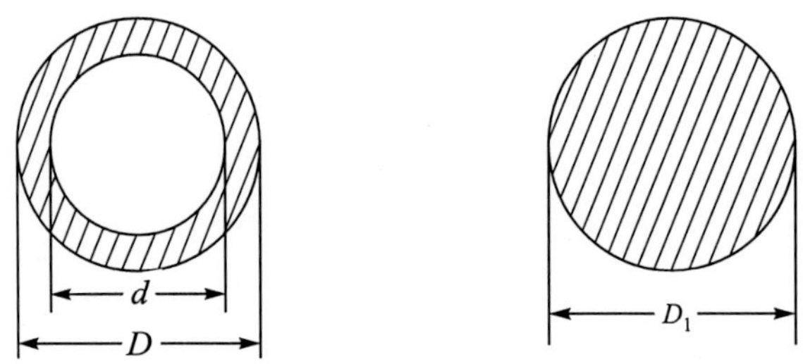

"Hollow circular col" "Solid circular column"

Fig. 5.12

Cross-sectional area of hollow column,

$$A_H = \frac{\pi}{4}(D^2 - d^2) = \frac{\pi}{4}[D^2 - (0.5\,D)^2]$$

$$= \frac{\pi}{4}[D^2 - 0.25\,D^2]$$

$$A_H = \frac{\pi}{4} \times 0.75\,D^2.$$

M.I. of hollow column,

$$I_H = \frac{\pi}{64}[D^4 - (0.5\,D)^4] = \frac{\pi}{64}[D^4 - 0.0625\,D^4]$$

$$= \frac{\pi}{64} \times 0.9375^4$$

Buckling strength of the hollow column,

$$P_H = \frac{\pi^2\, EIH}{L^2} = \frac{\pi^2 E \times \pi \times 0.9375\,D^2}{64\,L^2} \qquad \ldots(i)$$

Solid Circular Column: Let D_1 = Diameter of solid column
Cross-sectional area of solid column,

$$A_S = \frac{\pi}{64} \times D_1^2 \quad \text{(given)}$$

$$A_H = A_S$$

$$\therefore \quad \frac{\pi}{4} \times 0.75\, D^2 = \frac{\pi}{4} \times D_1^2$$

or $$D_1 = 0.866\, D$$

M.I. of solid column, $$I_S = \frac{\pi}{64} \times D_1^4 = \frac{\pi}{64} \times (0.866D)^4 = \frac{\pi \times 0.562\, D^4}{64}$$

Buckling strength of solid column, Ps

$$= \frac{\pi^2\, EI_S}{L^2} = \frac{\pi^2 E \times \pi \times 0.562\, D^4}{64\, L^2} \quad \ldots(ii)$$

Dividing eq. (*i*) by (*ii*), we get,

$$\frac{P_H}{P_S} = \frac{\pi^2 E \times \pi \times 0.9375\, D^4}{64\, L^2} \times \frac{64\, L^2}{\pi^2 E \times \pi \times 0.9375\, D^4}$$

or $$\frac{P_H}{P_S} = \frac{0.9375}{0.562} = 1.67 \qquad \textbf{Ans.}$$

Example 5.9: *An I-section joist 400 mm × 200 mm × 20 mm and 6 m long is used as a strut with both ends fixed. What is Euler's crippling load for the column? Take Young's modulu for the joist as 200 GPa.*
(MDU 2010)

Solution: Given:

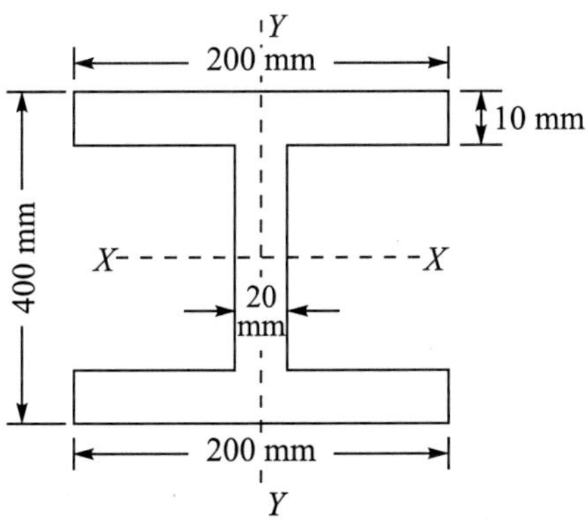

Fig. 5.13

Outer depth "D" = 400 mm,
Outer width "B" = 200 mm
Length "l" = 6 m = 6×10^3 mm
E = 200 GPa = 200×10^3 N/mm^2.

From the geometry of the Fig. 5.13, we find that inner depth,

$$d = 400 - (2 \times 20) = 360 \text{ mm}$$

and inner width, $$b = 200 - 20 = 180 \text{ mm}.$$

We know that M.I. of the joist section about X-X axis

$$I_{XX} = \frac{1}{12}\,[BD^2 - ba^3]$$

$$= \frac{1}{12}\,[200 \times (400)^3 - 180 \times (360)^3]\ \text{mm}^4.$$

$$= 366.8 \times 10^6\ \text{mm}^3 \qquad \ldots(i)$$

Similarly,

$$I_{YY} = \left[2 \times \frac{2 \times (200)^3}{12}\right] + \frac{360 \times (20)^3}{12}\ \text{mm}^4$$

$$= 2.91 \times 10^6\ \text{mm}^4 \qquad \ldots(ii)$$

Since $I_{YY} < I_{XX}$, the joist will tend to buckle in Y-Y direction.

Thus we will take the value of I as I_{YY}

$$= 2.91 \times 10^6\ \text{mm}^4$$

$$Le = \frac{l}{2} \qquad \text{[Column fixed at both the ends]}$$

$$= \frac{6 \times 10^3}{2} = 3 \times 10^3\ \text{mm}.$$

∴ Euler's crippling load for the column

$$Pe = \frac{\pi^2 EI}{Le^2} = \frac{\pi^2 \times (200 \times 10^3) \times (2.91 \times 10^6)}{(3 \times 10^3)^2}$$

$$= 638.2 \times 10^3\ \text{N} = 638.2\ \text{kN}$$ **Ans.**

Example 5.10: *A 3 m length of a tube has a crippling load of 2000 N when used as strut with pin-jointed ends. Determine the crippling load for a length 4 m of the same tube when used as a strut with both ends fixed.*

Solution: Case 1: Both ends pin-jointed

Actual length of strut, $l = 3\ \text{m} = 3000\ \text{mm}$

Effective length of strut, $L = l = 3000\ \text{mm}.$

Crippling load, $P = 2000\ \text{N}$

$$P = \frac{\pi^2 EI}{L^2}$$

or $\pi^2 EI = PL^2 = 2000 \times 3000^2.$

Case 2: Both ends fixed

Actual length of strut, $l = 4\ \text{m} = 4000\ \text{mm}$

Effective length of strut, $L = \frac{l}{2} = \frac{4000}{2} = 2000\ \text{mm}.$

Crippling load,

$$P = \frac{\pi^2 EI}{L^2} = \frac{2000 \times 3000^2}{2000^2}$$

$$= 4500\ \text{N}. \qquad [\because p^2 EI = 2000 \times 3000^2]$$ **Ans.**

SUMMARY

- A structured member, subjected to an axial compressive force is called a strut. A vertical strut used in buildings or frames is called a column.
- All short columns of the air due to crushing whereas long columns fail due to buckling and crushing.
- The load at which the column just buckles is known as buckling load or crippling load.
- Slenderness ratio is the ratio of the length of column to the least radius of gyration.

  Mathematically, slenderness ratio = $\frac{l}{k}$
- Depending upon the slenderness ratio or length to diameter ratio, columns are divided into three classes :
  (*i*) Short columns : Length < 8 times diameter or slenderness ratio < 32
  (*ii*) Medium columns : Length 8 to 30 times diameter or slenderness ratio between 32 and 120
  (*iii*) Long columns : Length > 30 times diameter or slenderness ratio > 120
- The crippling load for a column by Euler's formula for any type of end condition

$$P_c = \frac{\pi^2 EI}{l_e^2}$$

where P_c = crippling load
E = Youngs modulus of the material of the column,
I = Least M.I. of the column
l_e = Effective length of column
= l, when both ends hinged
= Zl, when one end is fixed and other is free
= $\frac{l}{2}$, when both ends are fixed
= $\frac{l}{\sqrt{2}}$, when one end is fixed and other is hinged.

where l = Actual length of the column.

- Buckling factor is the ratio between the equivalent length of the column to the minimum radius of gyration.
- Strength of a column depends upon the slenderness ratio and end conditions of the column.
- The compressive strength of a column decreases as the slenderness ratio of column increases.
- Column has four types of ends conditions :
  (*i*) Both ends of the column are hinged.
  (*ii*) Both ends of the column are fixed.
  (*iii*) One end is fixed and the other end is hinged.
  (*iv*) One end is fixed and the other end is free.

OBJECTIVE TYPE QUESTIONS

1. A member of structure or bar which carries an axial compressive load is called :
   (*a*) Strut (*b*) Tie (*c*) Shaft (*d*) None of these
2. When the strut is vertical *i.e.*, inclined at 90° to the horizontal is known as
   (*a*) Column (*b*) Pillar (*c*) Stanchion (*d*) Any of above

3. Which of the following is an example of the strut?
   (*a*) Piston rods (*b*) Connecting rods
   (*c*) Side links in forging machines (*d*) All of the above
4. The slenderness ratio of a long column is
   (*a*) 10-20 (*b*) 20-30 (*c*) 50-60 (*d*) above 30
5. Euler's formula is applicable to
   (*a*) Short columns (*b*) Medium columns (*c*) Long columns
6. A beam column is one which carries
   (*a*) Axial load (*b*) Transverse loads
   (*c*) Eccentric loads (*d*) Axial as well as transverse loads.
7. The ratio of equivalent length of the column to the minimum radius of gyration is called
   (*a*) Poisson's ratio (*b*) Buckling factor (*c*) Factor of safety
8. A column that fails due to direct stress is called
   (*a*) Short column (*b*) Long column
   (*c*) Weak column (*d*) Medium column

Answers: 1. (*a*), 2. (*d*), 3. (*d*), 4. (*d*), 5. (*c*), 6. (*d*), 7. (*b*), 8. (*a*)

EXERCISE FOR PRACTICE

1. A strut 2.50 m long is 60 mm in diameter. Find the safe compressive load for the member using Euler's formula if: (*i*) one end is fixed and other is higed, (*ii*) One end is fixed and the other is free. Take $E = 2 \times 10^5$ N/mm^2, and FOS = 3.5. **(MDU 2011)**
2. A column of circular section made of cast iron 200 mm external diameter and 20 mm thick is used as a column. The length of the column is 4 m. Both ends of the column are fixed. The column carries a load of 200 kN at an eccentricity of 25 mm from the axis of the column. Find the stresses developed in the extreme fibres. Take $E = 94000$ N/mm^2. **(MDU 2011)**
3. An I-section R.S.J. 300 × 150 × 12.5 mm is used as column 6 m long. Determine the Euler's crippling load if:
   (*i*) Both ends fixed
   (*ii*) One end fixed, other end free
   Take $E = 2 \times 10^5$ N/mm^2. **(MDU 2009)**
4. A hollow circular column of length 4.5 m, external dia. 200 mm and internal dia. 150 mm is fixed at both ends. It carries a load of 180 kN at an eccentricity of 17 mm from the axis of column. Find the max stress developed. What should be the value of limitng eccentricity. If no tension develops. Take $E = 0.96 \times 10^5$ N/mm^2. **(MDU 2009)**
5. Determine the ratio of the buckting strengths of two circular columns–one hollow and hte other solid. Both the columns are made of the same material and have the same length, cross sectional area and end conditions. The internal diameter of the hollow column is half of its external diameter.
6. A round steel rod of diameter 15 mm and length 2 m is subjected to a gradually increasing axial compressive load. Using Euler's formula find the buckling load. Find also the max. lateral deflection corresponding to the buckling condition. Both ends of the rod may be taken as hinged.
   Take $E = 2.1 \times 10^5$ N/mm^2 and the yield stress of steel = 250 N/mm^2.

❑❑❑

CHAPTER

6 Deflection of Beams

Relation between Slope, Deflection and Radius of Curvature: Consider a small portion PQ of a beam, bent into an arc as shown in Fig. 6.1.

Let ds = Length of the beam PQ.

R = Radius of the arc, into which the beam has been bent.

C = Centre of the arc.

ψ = Angle which the tangent at P makes with X-X axis.

$\psi + d\psi$ = Angle which the tangent at Q makes with the X-X axis.

From the geometry of the Fig. 6.1.

We find that,

$$\angle PCQ = d\psi$$

and

$$ds = R \cdot d\psi$$

$$\therefore \quad R = \frac{ds}{d\psi} = \frac{dx}{d\psi}$$

(Considering $ds = dx$)

or

$$\frac{1}{R} = \frac{d\psi}{dx}$$

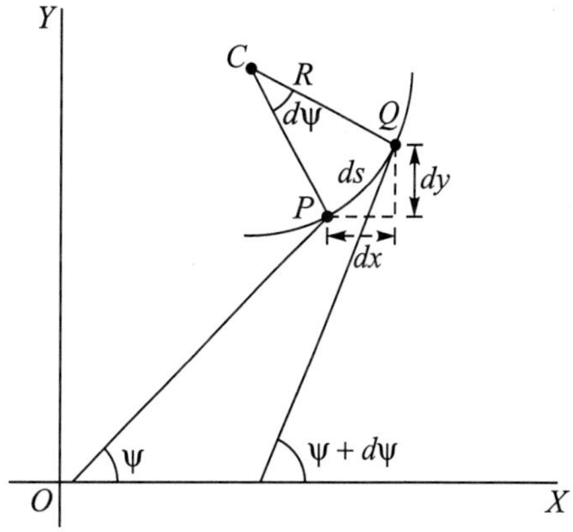

Fig. 6.1

We know that if dx and dy are the co-ordinates of point P, then

$$\tan \psi = \frac{dy}{dx}$$

Since ψ is very small angle, therefore taking $\tan \psi = \psi$

$$\frac{d\psi}{dx} = \frac{d^2 y}{dx^2} \qquad \left[\because \frac{1}{R} = \frac{d\psi}{dx}\right]$$

We also know that, (from the theory of bending stress)

$$\frac{M}{I} = \frac{E}{R} \quad \text{or} \quad M = EI \times \frac{1}{R}$$

$$\therefore \quad M = EI \times \frac{d^2 y}{dx^2} \qquad \left(\text{Substituting value of } \frac{1}{R}\right)$$

Methods for Slope and Deflection at a Section: The important methods used for finding out the slope and deflection at a section in a loaded beam are:

1. Double Integration Method.
2. Moment Area Method.
3. Macaulay's Method.

The first two methods are suitable for a single load, where as the last one is suitable for several load.

1. Double Integration Method: We know from the simple theory of bending in beams,

$$M = EI\frac{d^2y}{dx^2}$$

Integrating the above = *n*,

$$EI\frac{dy}{dx} = \int M$$

and integrating the above = *n*, one again

$$EI \cdot y = \iint M$$

It is thus obvious that after first integration of the original differential equation, we get the value of slope at any point and on further integration we get the value of deflection.

We will now discuss various cases with different load conditions.

Case I: Simply Supported Beam with a Central Point Load: Consider a simply supported beam *AB* of length "*l*" and carrying a point load *W* at the centre of beam '*C*' as shown in Fig. 6.2.

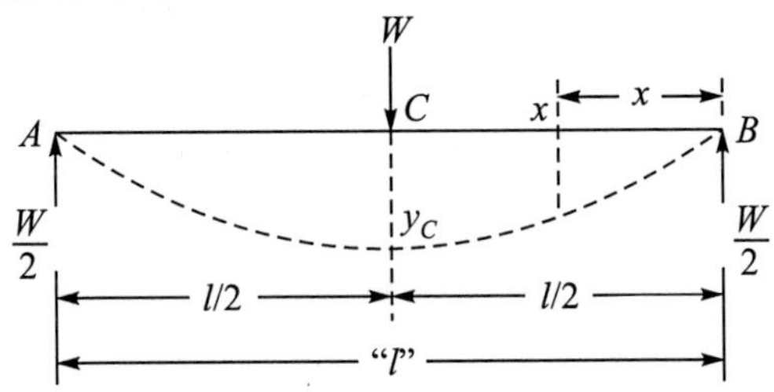

Fig. 6.2

From the geometry of the Fig. 6.2, we find that the reaction at *A*,

$$R_A = R_B = \frac{W}{2}$$

Consider a section '*X*' at a distance *x* from *B*.

We know that the *B.M.* at this section,

$$M_X = R_B \times x = \frac{W}{2}x = \frac{Wx}{2} \qquad \text{(Plus sign due to sagging)}$$

$$\therefore \qquad EI\frac{d^2y}{dx^2} = \frac{Wx}{2} \qquad \ldots(6.1)$$

Intigrating the above equation,

$$EI\frac{dy}{dx} = \frac{Wx^4}{4} + C_1 \qquad \ldots(6.2)$$

where C_1 is the first constant of integration.

We know that when $x = \frac{l}{2}$, then $\frac{dy}{dx} = 0$.

Substituting these values in eq. (6.2), we get

$$0 = \frac{Wl^2}{16} + C_1 \quad \text{or} \quad C_1 = -\frac{Wl^2}{16}$$

Substituting this value of C_1 in eq. (6.2),

$$EI\frac{dy}{dx} = \frac{Wx^2}{4} - \frac{Wl^2}{16} \qquad \ldots(6.3)$$

This is the required equation for the slope, at any section. It will be interesting to know that the max. slope occurs at A and B. Thus for max. slope at B, substituting $x = 0$ in eq. (6.3),

$$EI \cdot i_B = -\frac{Wl^2}{16}$$

or $$i_B = -\frac{Wl^2}{16\,EI}$$ (Minus sign means that the tangent at B makes an angle with AB in anticlock wise direction)

$\therefore$ $$i_B = \frac{Wl^2}{16\,EI} \text{ radians.}$$

By symmetry, $$i_A = \frac{Wl^2}{16\,EI} \text{ radians.}$$

Integrating the eq. (6.3) once again,

$$EI \cdot y = \frac{Wx^3}{12} - \frac{Wl^2 x}{16} + C_2 \qquad \ldots(6.4)$$

where C_2 is the second constant of integration. We know that when $x = 0$, then $y = 0_1$ substituting these values in eq. (6.4), we get $C_2 = 0$.

$\therefore$ $$EI \cdot y = \frac{Wx^3}{12} - \frac{Wl^2 x}{16} \qquad \ldots(6.5)$$

This is the required equation for deflection, at any section. A little consideration will show that max. deflection occurs at the mid point C. Thus for max. deflection, substituting $x = \frac{l}{2}$ in eq. (6.5),

$$EI\,y_C = \frac{W}{12}\left(\frac{l}{2}\right)^3 - \frac{Wl^2}{16}\left(\frac{l}{2}\right)$$

$$= \frac{Wl^3}{96} - \frac{Wl^3}{32} = -\frac{Wl^3}{48}$$

or $$\text{“}y_C\text{”} = -\frac{Wl^3}{48\,EI}$$ (Minus sign means that the deflection is downwards)

$$= \frac{Wl^3}{48\,EI}$$

Case II: Simply Supported Beam with a Uniformaly Distributed Load: Consider a simply supported beam of length "*l*" and carrying a uniformaly distributed load of *w* per unit length as shown in Fig. 6.3. From the geometry of the Fig. 6.3.

We know that the reaction at *A*.

$$R_A = R_B = \frac{wl}{2} \quad \ldots(6.6)$$

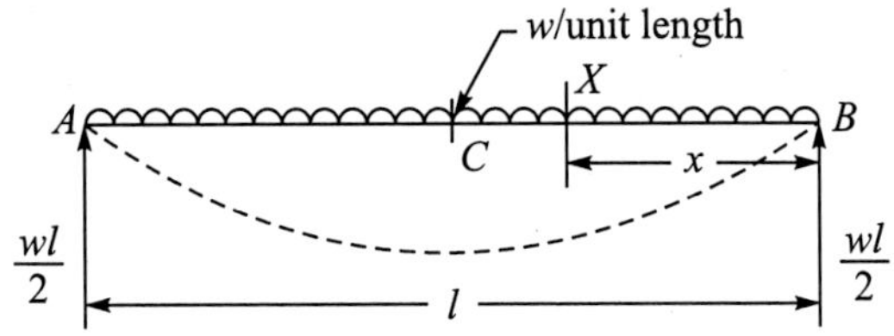

Fig. 6.3

Consider a section *X* at a distance *x* from *B*, we know that the *B.M.* at this section,

$$M_X = \frac{wlx}{2} - \frac{wx^2}{2} \quad \text{(+ sign due to sagging)}$$

$$EI\frac{d^2y}{dx^2} = \frac{wlx}{2} - \frac{wx^2}{2} \quad \ldots(6.7)$$

Integrating the above equation,

$$EI \cdot \frac{dy}{dx} = \frac{wlx^2}{4} - \frac{wx^2}{6} + C_1 \quad \ldots(6.8)$$

where C_1 is the first constant of integration.

We know that when $x = \frac{l}{2}$, then $\frac{dy}{dx} = 0$

Substituting these values in the above equation,

$$0 = \frac{wl}{4}\left(\frac{l}{2}\right)^2 - \frac{w}{6}\left(\frac{l}{2}\right)^3 + C_1 = \frac{wl^3}{16} - \frac{wl^3}{48} + C_1$$

or $$C_1 = -\frac{wl^3}{24}$$

Substituting this value of C_1 in equation (6.8),

$$\therefore \quad EI\frac{dy}{dx} = \frac{wlx^2}{4} - \frac{wx^3}{6} - \frac{wl^3}{24} + C_1 \quad \ldots(6.9)$$

This is the required equation for the slope at any section. We know that the max. slope occurs at *A* and *B*.

Thus for max. slope, substituting $x = 0$ in equation (6.9),

$$EI \cdot i_B = -\frac{wl^3}{24},$$ (Minus sign means that the tangent at *A* makes an angle with *AB* in the anticlock wise direction)

$$\therefore \quad i_B = -\frac{wl^3}{24\,EI}$$

or $$i_B = \frac{wl^3}{24\,EI}$$

By symmetry, $$i_A = \frac{wl^3}{24\,EI}$$

Integrating the eq. (6.9), once again,

$$EI \cdot y = \frac{wlx^3}{12} - \frac{wx^4}{24} - \frac{wl^3 x}{24} + C_2 \qquad \ldots(6.10)$$

where C_2 is the second constant of integration, we know that when $x = 0$, then $y = 0$. Substituting these values in eq. (6.9), we get, $C_2 = 0$

$$\therefore \qquad EI \cdot y = \frac{wlx^3}{12} - \frac{wx^4}{24} - \frac{wl^3 x}{24} \qquad \ldots(6.11)$$

This is the required equation for the deflection at any point. We know that max. deflection occurs at the mid point '*C*'. Thus max. deflection, substituting $x = \frac{l}{2}$ in equation (6.11), we get

$$EI \cdot y_C = \frac{wl}{12}\left(\frac{l}{2}\right)^3 - \frac{w}{24}\left(\frac{l}{2}\right)^4 - \frac{wl^3}{24}\left(\frac{l}{2}\right)$$

$$= \frac{wl^4}{96} - \frac{wl^4}{384} - \frac{wl^4}{48} = -\frac{5\,wl^4}{384}$$

or $$\text{“}y_C\text{”} = -\frac{5\,wl^4}{384\,EI}$$ (Minus sign means that the deflection is downwards)

or $$\text{“}y_C\text{”} = \frac{5\,wl^4}{384\,EI}$$

Case III: Simply Supported Beam with a Gradually Varring Load: Consider a simply supported beam *AB* of length '*l*' and carrying a gradually varying load from zero at '*B*' to *w* per unit length at *A* as shown in the Fig. 6.4. From the geometry of the Fig. 6.4, we find that the reaction at *A*,

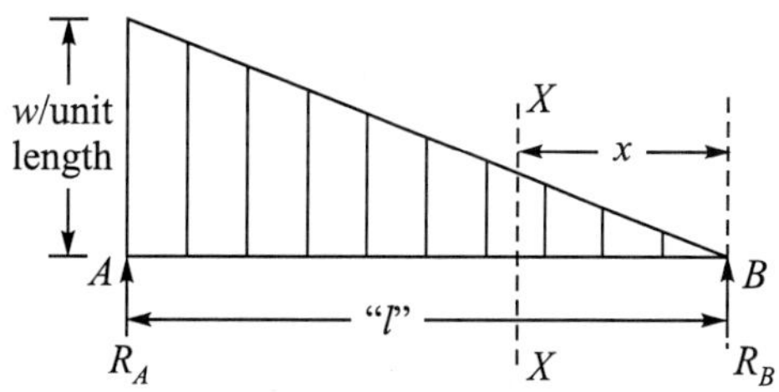

Fig. 6.4

$$R_A = \frac{wl}{3} \quad \text{and} \quad R_B = \frac{wl}{6}$$

Now consider a section *X-X* at a distance '*x*' from *B*. We know that the *B.M.* at this section,

$$M_X = R_B \cdot x - \left(\frac{wx}{l} \times \frac{x}{2} \times \frac{x}{3}\right) = \frac{wlx}{6} - \frac{wx^3}{6l}$$

$$\therefore \qquad EI \cdot \frac{d^2 y}{dx^2} = \frac{wlx}{6} - \frac{wx^3}{6l} \qquad \ldots(6.12)$$

Integrating the above equation,

$$EI\,\frac{dy}{dx} = \frac{wlx^2}{12} - \frac{wx^4}{24\,l} + C_1 \qquad \ldots(6.13)$$

where C_1 is the first constant of integration. Integrating the equation (6.13) once again,

$$EI \cdot y = \frac{wlx^3}{36} - \frac{wx^5}{120\,l} + C_1 x + C_2 \qquad \text{...(6.14)}$$

where C_2 is the second constant of integration.

We know that when $x = 0$, then $y = 0$. Therefore $C_2 = 0$. We also know that when $x = l$, then $y = 0$. Substituting these values in eq. (6.14),

$$0 = \frac{wl}{36} \times l^3 - \frac{w}{120\,l} \times l^5 + C_1\, l = \frac{wl^4}{36} - \frac{wl^4}{120} + C_1\, l$$

$$\therefore \qquad C_1 = \frac{wl^3}{36} + \frac{wl^3}{120} = -\frac{7\,wl^3}{360}$$

Now substituting this value of C_1 in eq. (6.13),

$$EI \cdot \frac{dy}{dx} = \frac{wlx^2}{12} - \frac{wx^4}{23\,l} - \frac{7\,wl^3}{360} \qquad (6.15)$$

This is the required equation for slope at any section, by which we can get the slope at any section on the beam.

A little consideration will show that the max. slope will be either at the support *A* or *B*. Thus for the slope at *A*, substituting $x = l$ in eq. (6.15),

$$EI \cdot i_A = \frac{wl}{12} \times l^2 - \frac{w}{24\,l} \times l^4 - \frac{7\,wl^3}{360} = \frac{wl^3}{45}$$

$$\therefore \qquad i_A = \frac{wl^3}{45\,EI}$$

Now, for slope at *B*, substituting $x = 0$ in eq. (6.15)

$$EI \cdot i_B = -\frac{7\,wl^3}{360}$$

$$\therefore \qquad i_B = -\frac{7\,wl^3}{360\,EI},$$ (Minus sign means that the tangent at *B* makes an angle with *AB* in the anticlockwise direction.)

$$= -\frac{7\,wl^3}{360\,EI} \text{ radians.}$$

Now substituting the value of C_1 in eq. (6.14),

$$EI \cdot y = \frac{wlx^3}{36} - \frac{wx^5}{120\,l} - \frac{7\,wl^3 x}{360}$$

$$\therefore \qquad y = \frac{1}{EI}\left(\frac{wlx^3}{36} - \frac{wx^5}{120\,l} - \frac{7\,wl^3 x}{360}\right)$$

This is the required equation for the deflection at any section, by which we can get deflection at any section on the beam. For the deflection at the centre of the beam, substituting $x = \frac{l}{2}$ in eq. (6.16),

$$y_C = \frac{1}{EI}\left[\frac{wl}{36}\left(\frac{l}{2}\right)^3 - \frac{w}{120\,l}\left(\frac{l}{2}\right) - \frac{7\,wl^3}{360}\left(\frac{l}{2}\right)\right]$$

$$= -\frac{0.00651\,wl^4}{EI} \qquad \text{(Minus sign means that the deflection is down words)}$$

$$= \frac{0.00651\,wl^4}{EI}$$

We know that the max. deflection will occur, where slope of the beam is zero. Therefore, equating the eq. (6.15) to zero.

$$\frac{wlx^2}{12} - \frac{wx^4}{24\,l} - \frac{7\,wl^3}{360} = .0.$$

$$\therefore \qquad x = 0.519\,l$$

Now substituting this value of x in eq. (6.16),

$$y_{max} = \frac{1}{EI}\left[\frac{wl}{36}(0.519\,l)^3 - \frac{w}{120\,l}(0.519\,l)^5 - \frac{7\,wl^3}{360}(0.519\,l)\right]$$

$$= \frac{0.00651\,wl^4}{EI}, \text{ (Minus sign means that deflection is down wards)}$$

Macaulay's Method

Following rules are observed while using Macaulay's method:

1. Always take origin on the extreme left of beam.
2. Take left clockwise moments as –ve, and left anticlockwise moments at +ve.
3. While calculating the slopes and deflection it is convenient to use the values first in terms of kN and meters.

Case I: Simply Supported Beam with a Central Point Load: Consider a simply supported beam AB of length 'l' and carrying a point load 'W' at the centre of the beam 'C' as shown in the Fig. 6.5.

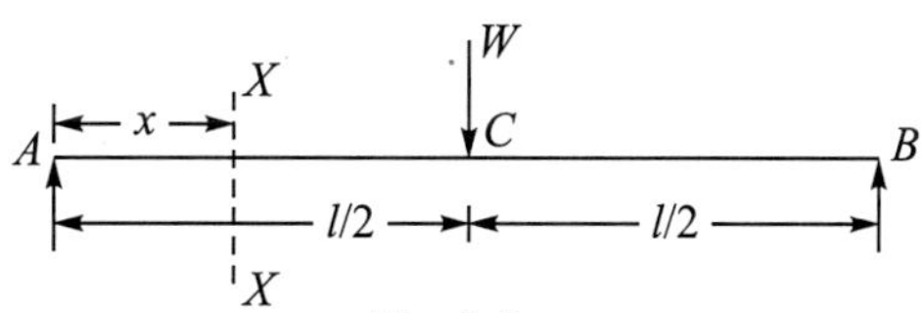

Fig. 6.5

Take 'A' as the origin.

We know that $B.M.$ at any point, in section AC at a distance 'x' from A.

$$M_X = -\frac{W}{2}x, \qquad \text{(Minus sign due to left clockwise)}$$

and the $B.M.$ at any point in section CB at a distance 'x' from A,

$$M_X = -\frac{W}{2}x + W\left(x - \frac{l}{2}\right) \qquad \ldots(6.16)$$

Thus, we can express the *B.M.* for all the sections of the beam in a single equation,

$$M_X = -\frac{W}{2}x \;\vdots\; + W\left(x - \frac{l}{2}\right)$$

For any point in section *AC*, stop at the dotted line, and for any point in section *CB* add the expression beyond the dotted line also.

Now rewriting the above equation,

$$EI\frac{d^2y}{dx^2} = -\frac{Wx}{2} \;\vdots\; + W\left(x - \frac{l}{2}\right) \qquad \ldots(6.17)$$

Integrating the above equation,

$$EI \cdot \frac{dy}{dx} = -\frac{Wx^2}{4} + C_1 \;\vdots\; + \frac{W}{2}\left(x - \frac{l}{2}\right) \qquad \ldots(6.18)$$

It may be noted that the integration of $\left(x - \frac{l}{2}\right)$ has been made as a whole and not for individual terms for the expression. This is only due to this simple integration that Macaulay's method is more effective. This type of integration is also justified as the constant of integration C_1 is not only valid for the section *AC*, but also for section *CB*.

Integrating the eq. (6.18) once again,

$$EI \cdot y = -\frac{Wx^3}{12} + C_1 x + C_2 \;\vdots\; + \frac{W}{6}\left(x - \frac{l}{2}\right)^3 \qquad \ldots(6.19)$$

It may again be noted that the integration of $\left(x - \frac{l}{2}\right)^2$ has again been made as a whole and not for individual terms.

We know that when $x = 0$, then $y = 0$. Substituting these values in equation (6.19), we find $C_2 = 0$.

We also know that when $x = l$, then $y = 0$.

Substituting these values of x and y and $C_2 = 0$ in eq. (6.19),

$$0 = -\frac{Wl^3}{12} + C_1 l + \frac{W}{6}\left(\frac{l}{2}\right)^3$$

$$\therefore \qquad C_1 l = -\frac{Wl^3}{12} - \frac{Wl^3}{48} = \frac{3Wl^3}{48} = \frac{Wl^3}{16}$$

$$\text{or} \qquad C_1 = \frac{Wl^2}{16}$$

Now substituting this value of C_1 in eq. (6.18),

$$\therefore \qquad EI\frac{dy}{dx} = \frac{Wx^2}{4} + \frac{Wl^2}{16} \;\vdots\; + \frac{W}{2}\left(x - \frac{l}{2}\right)^2 \qquad \ldots(6.20)$$

This is the required equation for slope at any section. We know that max. slope occurs at A and B. Thus for max. slope at A, substituting $x = 0$ in eq. (6.20) upto the dotted line only.

$$EI \cdot i_A = \frac{Wl^2}{16}$$

$$\therefore \qquad i_A = \frac{Wl^2}{16\,EI}$$

By symmetry,

$$i_B = \frac{Wl^2}{16\,EI}$$

Substituting the value of C_1 again in eq. (6.19) and

$$C_2 = 0,$$

$$EI \cdot y = -\frac{Wx^3}{12} + \frac{Wl^2 x}{16} \;\vdots\; + \frac{W}{6} + \left(x - \frac{l}{2}\right)^3 \qquad (6.21)$$

This is the required equation for deflection at any section. We know that max. deflection occurs at C. Thus for max. deflection, substituting $x = \frac{l}{2}$ in equation (6.21) for the portion AC only (remembering that C lies in AC)

$$EI \cdot y_C = -\frac{W}{12}\left(\frac{l}{2}\right)^3 \;\vdots\; + \frac{Wl^2}{16}\left(\frac{l}{2}\right) = \frac{Wl^3}{48}$$

or

$$\text{“}y_C\text{”} = \frac{Wl^3}{48\,EI}$$

Case II: Simply Supported Beam with an Eccentric Point Load: Consider a simply supported beam AB of length 'l' and carring an eccentric point load 'W' at C, such that $AC = a$ and $CB = b$ as shown in Fig. 6.6.

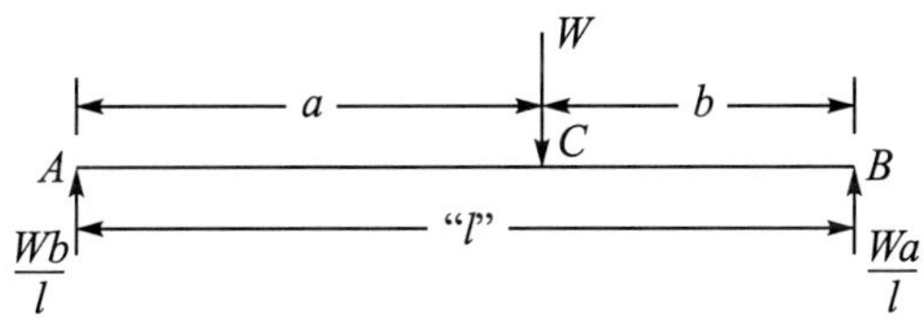

Fig. 6.6

Take A as the origin. The $B.M.$ at any point in section AC at a distance 'x' from A,

$$M_X = -\frac{Wb}{l}x \qquad \text{(Minus sign due to left clockwise)}$$

$B.M.$ at any point in section CB at a distance 'x' from A.

$$M_X = -\frac{Wbx}{l} \;\vdots\; + W(x-a) \qquad \ldots(6.22)$$

or

$$EI\frac{d^2y}{dx^2} = \frac{Wbx}{l} \;\vdots\; + W(x-a) \qquad \ldots(6.23)$$

Integrating the above equation,

$$EI\frac{dy}{dx} = -C\frac{Wbx^2}{2l} + C_1 \;\vdots\; + \frac{W(x-a)^2}{2} \qquad \ldots(6.24)$$

Integrating the above equation, once again,

$$EI\,y = -\frac{Wbx^3}{6l} + C_1 x + C_2 \;\vdots\; + \frac{W(x-a)^3}{6} \qquad \ldots(6.25)$$

we know that when $x = 0$, then $y = 0$.

Substituting these values in equation (6.25) upto the dotted line only.

Therefore, $C_2 = 0$. We also know that when $x = l$, then $y = 0$. Substituting these values again in equation (6.25) and $C_2 = 0$.

$$EI \cdot y = -\frac{Wb}{6l}(l)^3 + C_1 l \;\vdots\; + \frac{W(l-a)^3}{6}$$

$$= -\frac{Wbl^2}{6} + C_1 l \;\vdots\; + \frac{Wb^3}{6}, \qquad [\because (l-a) = b]$$

$$\therefore \quad C_1 l = \frac{Wbl^2}{6} + \frac{Wb^3}{6} = \frac{Wb}{6}(l^2 - b^2)$$

or

$$C_1 = \frac{Wb}{6l}(l^2 - b^2)$$

Now substituting this value of C_1 in eq. (6.24),

$$EI\frac{dy}{dx} = -\frac{Wbx^2}{2l} + \frac{Wb}{6l}(l^2 - b^2) \;\vdots\; + \frac{w(x-a)^2}{2} \qquad \ldots(6.26)$$

This is the required equation for slope at any point. We know that slope is max. at *A* or *B*. Substituting $x = 0$, upto dotted line only (remembering that *C* lies in *AC*).

$$EI \cdot i_A = \frac{Wb}{6l}(l^2 - b^2)$$

or

$$i_A = \frac{Wb}{6\,EI\,l}(l^2 - b^2)$$

Similarly,

$$i_B = \frac{Wa}{6\,EI\,l}(l^2 - b^2)$$

Substituting the value of C_1 again in eq. (6.25) and $C_2 = 0$

$$EI \cdot y = -\frac{Wbx^3}{6l} + \frac{Wbx}{6l}(l^2 - b^2) \;\vdots\; + \frac{W(x-a)^3}{6}$$

This is the required equation for deflection at any point. For deflection in *AC*, consider the equation upto the dotted line only,

$$EI \cdot y = -\frac{Wbx^3}{6l} + \frac{Wbx}{6l}(l^2 - b^2)$$

$$= \frac{Wbx}{6l}(l^2 - b^2 - x^2)$$

$$\therefore \quad \text{Deflection "}y\text{"} = \frac{Wbx}{6\,EI\,l}(l^2 - b^2 - x^2)$$

6.1. CANTILEVER WITH A POINT LOAD AT ITS FREE END

Consider a cantilever AB of length 'l' and carrying a point load W at the free end as shown in Fig. 6.7. Consider a section X-X at a distance 'x' from the free end B.

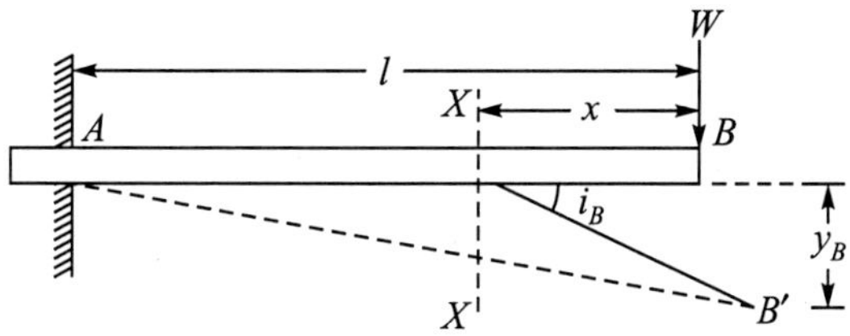

Fig. 6.7

We know that the $B.M.$ at this section,

$$M_X = -W \cdot x \qquad \text{(Minus sign due to hogging)}$$

$$\therefore \qquad EI\frac{d^2y}{dx^2} = -W \cdot x \qquad \text{...(6.27)}$$

Integrating the above equation,

$$EI \cdot \frac{dy}{dx} = -\frac{Wx^2}{2} + C_1 \qquad \text{...(6.28)}$$

where C_1 is the first constant of integration.

We know that when $x = l, \dfrac{dy}{dx} = 0.$

Substituting these values in the above equation,

$$0 = -\frac{Wl^2}{2} + C_1 \quad \text{or} \quad C_1 = -\frac{Wl^2}{2}$$

Now substituting this value of C_1 in eq. (6.28),

$$EI\frac{dy}{dx} = -\frac{Wx^2}{2} + \frac{Wl^2}{2} \qquad \text{...(6.29)}$$

This is the required $= n$ for the slope, at any section by which we can get the slope at any point on the cantilever. We know that max. slope occurs at the free end. Now let us see the abbreviation 'i' for the angle of inclination (in radian) and considering $i = \tan i$, for very small angles. Thus for max. slope,

Substituting $x = 0$ in eq. (6.29),

$$EI \cdot i_B = \frac{Wl^2}{2}$$

$$i_B = \frac{Wl^2}{2\,EI} \text{ radians.}$$

Plus sign means that the tanget at B makes an angle with AB in the clockwise direction.

Integrating the eq. (6.29), once again,

$$EI \cdot y = -\frac{Wx^3}{6} + \frac{Wl^2 x}{2} + C_2 \qquad \text{...(6.30)}$$

where C_2 is the second constant of integration.

We know that when $x = l, y = 0.$

Substituting these values in the above equation,

$$0 = -\frac{Wl^3}{6} + \frac{Wl^3}{2} + C_2 = \frac{Wl^3}{3} + C_2$$

or $$C_2 = -\frac{Wl^3}{3}$$ (Minus sign means that the deflection is downwards)

Substituting this value of C_2 in eq. (6.30),

$$EI \cdot y = -\frac{Wx^3}{6} + \frac{Wl^2 x}{2} - \frac{Wl^3}{3}$$

$$= \frac{Wl^2 x}{2} - \frac{Wx^3}{6} = \frac{Wl^3}{3} \quad \text{...(6.31)}$$

This is the required equation to the deflection at any section. We know that max. deflection occurs at the free end. Therefore, for max. deflection, substituting $x = 0$ in eq. (6.31)

$$EI \cdot y_B = -\frac{Wl^3}{3}$$

or $$y_B = -\frac{Wl^3}{3\,EI}$$

$$'y_B' = \frac{Wl^3}{3\,EI}.$$

6.2. CANTILEVER WITH A UNIFORMLY DISTRIBUTED LOAD (UDL)

Consider a cantilever AB of length 'l' and carrying uniformly distributed load (*udl*) of 'w' per unit length as shown in the Fig. 6.8.

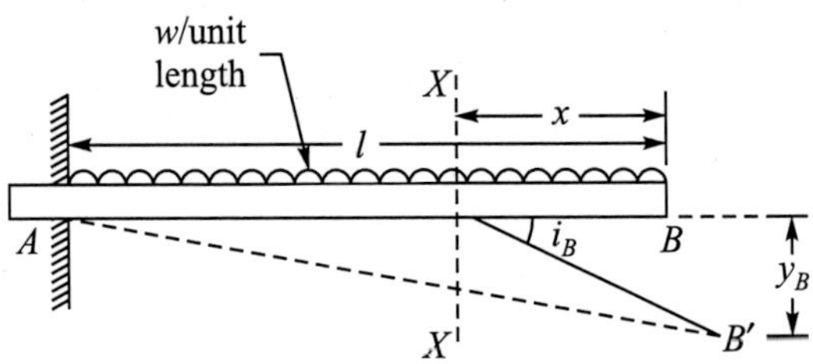

Fig. 6.8

Consider a section X-X at a distance 'x' from the free end B.

We know that the $B.M.$ at the section,

$$M_X = (-)\frac{wx^2}{2},$$ (Minus sign due to hogging)

$$\therefore \quad EI\frac{d^2 y}{dx^2} = -\frac{wx^2}{2} \quad \text{...(6.32)}$$

Integrating the above equation,

$$EI\frac{dy}{dx} = -\frac{wx^3}{6} + C_1, \qquad \ldots(6.33)$$

where C_1 is the first constant of integration.

We know that when $x = l$, then $\frac{dy}{dx} = 0$.

Substituting these values in eq. (6.33),

$$0 = -\frac{wl^3}{6} + C_1 \quad \text{or} \quad C_1 = \frac{wl^3}{6}$$

Substituting this value of C_1 in eq. (6.33),

$$EI\frac{dy}{dx} = -\frac{wx^3}{6} + \frac{wl^3}{6} \qquad \ldots(6.34)$$

This is the required equation for the slope at any section. We know that max. slope occurs at the free end *B*. Therefore, for the max. slope, substituting $x = 0$ in eq. (6.34)

$$EI \cdot i_B = \frac{wl^3}{6}$$

or $$i_B = \frac{wl^3}{6\,EI} \text{ radians}$$ (Plus sign means that the tangent at *B* makes an angle with *AB* in clockwise direction)

Integrating the eq. (3), once again,

$$EI \cdot y = -\frac{wx^4}{24} + \frac{wl^3 x}{6} + C_2 \qquad \ldots(6.35)$$

where C_2 is the second constant of integration.

We know that when $x = l$, then $y = 0$.

Substituting these values in the above equation,

$$0 = -\frac{wl^4}{24} + \frac{wl^4}{6} + C_2$$

or $$C_2 = -\frac{wl^4}{8}$$

Substituting this value of C_2 in eq. (6.35),

$$EI \cdot y = -\frac{wx^4}{24} + \frac{wl^3 x}{6} - \frac{wl^4}{8} = \frac{wl^3 x}{6} - \frac{wx^4}{24} - \frac{wl^4}{8} \qquad \ldots(6.36)$$

This is the required equation of deflection at any section we know that max. deflection occurs at free end.

∴ For max. slope, substituting $x = 0$ in eq. (6.36),

$$EI \cdot y_B = \frac{wl^4}{8}$$

or $$y_B = -\frac{wl^4}{8\,EI}$$ (Minus sign means that deflection is downwards)

or $$\text{“}y_B\text{”} = \frac{wl^4}{8\,EI}.$$

6.3. CANTILEVER WITH A GRADUALLY VARYING LOAD

Consider a cantilever AB of length 'l' and carrying a gradually varying load from zero at B to 'w' per unit length at 'A' as shown in Fig. 6.9.

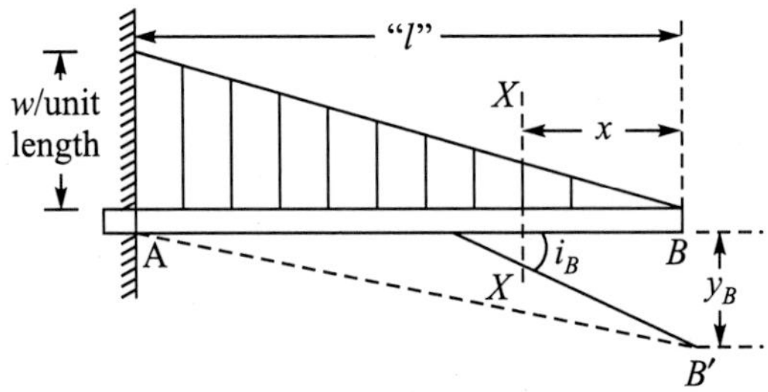

Fig. 6.9

Now consider a section X-X, at a distance 'x' from the free end B. We know that the $B.M.$ at the section,

$$M_X = -\frac{1}{2} \times \frac{wx}{l} \times x \times \frac{x}{3} = -\frac{wx^3}{6\,l},$$ (Minus sign due to hogging)

$$\therefore \quad EI\,\frac{d^2y}{dx^2} = -\frac{wx^3}{6\,l} \qquad \text{...(6.37)}$$

Integrating the above equation,

$$EI \cdot \frac{dy}{dx} = -\frac{wx^4}{24\,l} + C_1, \qquad \text{...(6.38)}$$

where 'C_1' is the constant of integration.

We know that when $x = l$, then $\frac{dy}{dx} - 0.$

Substituting these values in eq. (6.38),

$$0 = -\frac{wx^4}{24\,l} + C_1 \quad \text{or} \quad C_1 = +\frac{wl^3}{24}$$

$$\therefore \quad EI \cdot \frac{dy}{dx} = -\frac{wx^4}{24\,l} + \frac{wl^3}{24}, \qquad \text{...(6.39)}$$

This is the required equation for slope at any section, by which we can get the slope at any section on the cantilever, we know that the max. slope occurs at the free end. Therefore, for max. slope, substituting $x = 0$ in eq. (6.39),

$$EI \cdot i_B = \frac{wl^3}{24}$$

or $$i_B = \frac{wl^3}{24\ EI} \text{ radians} \quad \ldots(6.40)$$

Integrating eq. (6.39), again,

$$EI \cdot y = -\frac{wx^5}{120l} + \frac{wl^3}{24} + C_2 \quad \ldots(6.41)$$

where C_2 is the constant of integration.

We know that when $x = l$, then $y = 0$

Substituting these values in above equation,

$$0 = -\frac{wl^4}{120} + \frac{wl^4}{24} + C_2 \quad \text{or } C_2 = -\frac{wl^4}{30}$$

$$\therefore \quad EI \cdot y = -\frac{wx^5}{120\,l} + \frac{wl^3 x}{24} - \frac{wl^4}{30} \quad \ldots(6.42)$$

This is the required equation for deflection at any section. We know for max. deflection occurs at the free end. Therefore, for max. slope, substituting $x = 0$ in eq. (6.42),

$$EI \cdot y_B = -\frac{wl^4}{30}$$

or $$y_B = -\frac{wl^4}{30\ EI},$$ (Minus sign mean that deflection is downwards)

or $$\text{“}y_B\text{”} = \frac{wl^4}{30\ EI}$$

PROBLEMS ON DEFLECTION OF BEAMS

Example 6.1: *A simply supported beam of span 3 m is subjected to a central load of 10 kN. Find the max. slope and deflection of the beam.*

Take $I = 12 \times 10^6$ mm^4 and $E = 200$ Gpa.

Solution: Given, span “l” = 3 m = 3×10^3 mm,

$W = 10$ kN $= 10 \times 10^3$ N.

$I = 12 \times 10^6$ mm^4

$E = 200$ Gpa $= 200 \times 10^3$ N/mm^3.

Max. slope of the beam,

$$i_A = \frac{Wl^2}{16\ EI} = \frac{(10\times10^3)\times(3\times10^3)^2}{16\times(200\times10^3)\times(12\times10^6)}$$

$$= 0.0023 \text{ rad.}$$

Max. deflection of the beam,

$$y_C = \frac{Wl^3}{48\ EI} = \frac{(10\times10^3)\times(3\times10^3)^3}{48\times(200\times10^3)\times(12\times10^6)}$$

$$= \mathbf{2.3\ mm.}$$

Example 6.2: *A wooden beam 140 mm wide and 240 mm deep has a span of 4 m. Determine the load, that can be placed at its centre to cause the beam a deflection of 10 mm. Take E = 6 Gpa.*

Solution: Given:

Width (b) = 140 mm, Depth (d) = 240 mm

Span (l) = 4 m = 4×10^3 mm

Central deflection (y_C) = 10 mm and

$E = 6$ Gpa $= 6 \times 10^3$ N/mm^2.

Let W = Magnitude of load.

$$\text{M.I., } I = \frac{bd^3}{12} = \frac{140 \times (240)^3}{12}$$

$$= 161.3 \times 10^6 \text{ mm}^4$$

and deflection
$$y_C = \frac{Wl^3}{48\,EI}$$

or
$$10 = \frac{W \times (4 \times 10^3)^3}{48 \times (6 \times 10^3) \times (161.3 \times 10^6)}$$

or
$$\text{“}W\text{”} = \frac{10}{1.38 \times 10^{-3}} = 7.25 \times 10^3 \text{ N.}$$

Example 6.3: *A steel joist, simply supported over a span of 6 m carries a point load of 50 kN at 1.2 m from the left hand support. Find the position and magnitude of the max deflection.*

Take EI = 14 × 10^{12} N-mm^2.

Solution: Given, span (l) = 6 m = 6×10^3 mm.

Point load W = 50 kN = 50×10^3 N.

Distance between point load and left end (a)

= 1.2 m = 1.2×10^3 mm and flexural rigidity

$EI = 14 \times 10^{12}$ N-mm^2.

Position of max. deflection: We know that the position of the max. deflection,

$$x = \sqrt{\frac{l^2 - a^2}{3}} = \sqrt{\frac{(6 \times 10^3)^2 - (1.2 \times 10^3)^2}{3}}$$

$$= 3.39 \times 10^3 \text{ mm} = 3.39 \text{ m.}$$

Magnitude of max. deflection

$$y_{max} = \frac{Wa}{9\sqrt{3} \cdot EI\, l}\,(l^2 - a^2)^{3/2}$$

$$= \frac{(50 \times 10^3) \times (1.2 \times 10^3)}{9\sqrt{3} \times (14 \times 10^{12}) \times (6 \times 10^3)} \times \left[(6 \times 10^3)^2 - (1.2 \times 10^3)^2\right]^{3/2}$$

$$= (0.0458 \times 10^{-9}) \times (2.052 \times 10^{11})$$

$$= \mathbf{9.4\ mm}$$

Example 6.4: *A simply supported beam AB of span 5 meters is carrying a point load of 30 kN at a distance 3.75 m from left end A. Calculate the slopes at A and B and deflection under the load.*

Take $EI = 26 \times 10^{12}$ N-mm^2.

Solution: Given, span 'l' = 5 m = 5×10^3 mm

Point load 'W' = 30 kN = 30×10^3 N

Distance between point load and left end (a) = 3.75 m = 3.75×10^3 mm.

Distance between point load and right end (b) = 5 – 3.75 = 1.25 m = 1.25×10^3 mm.

$EI = 26 \times 10^{12}$ N – mm^2.

Slope at A,

We know that slope at 'A',

$$i_A = \frac{Wb}{6\,EI\,l}(l^2 - b^2).$$

$$= \frac{(30\times10^3)\times(1.25\times10^3)}{6\times(26\times10^{12})\times(5\times10^3)} \times \left[(5\times10^3)^2 - (1.25\times10^3)^2\right]$$

$$= (0.0481 \times 10^{-9}) \times (23.4375 \times 10^6)$$

$$= 0.00113 \text{ rad.}$$

Slope at B,

$$i_B = \frac{Wa}{6\,EI\,l}(l^2 - a^2)$$

$$= \frac{(30\times10^3)\times(3.75\times10^3)}{6\times(26\times10^{12})\times(5\times10^3)} \times \left[(5\times10^3)^2 - (3.75\times10^3)^2\right]$$

$$= (0.1442 \times 10^{-9}) \times (10.9375 \times 10^6)$$

$$= 0.00158 \text{ rad.}$$

Deflection under the load,

$$'y_C' = \frac{Wb}{6\,EI\,l}(l^2 - a^2 - b^2)$$

$$= \frac{(30\times10^3)\times(3.75\times10^3)\times(1.25\times10^3)}{6\times(26\times10^{12})\times(5\times10^3)} \times \left[(5\times10^3)^2 - (3.75\times10^3)^2 - (1.25\times10^3)^2\right]$$

$$= (0.18 \times 10^{-6}) \times (9.375 \times 10^6)$$

$$= \mathbf{1.69\ mm}$$ **Ans.**

Example 6.5: *A simply supported beam of span 4 m is carrying a udl of 2 kN/m over the entire span. Find the max. slope and deflection of beam.*

Take $EI = 80 \times 10^9$ N-mm^2.

Solution: Given: span 'l' = 4 m = 4×10^3 mm.

'w' = 2 kN/m = 2 N/mm.

$E = 80 \times 10^9$ N – mm^2.

Max. slope of the beam,

$$i_A = \frac{wl^3}{24\ EI} = \frac{2\times(4\times10^3)^3}{24\times(80\times10^9)} = 0.067 \text{ rad.}$$

Max. deflection of the beam,

$$'y_C' = \frac{5\ wl^4}{384\ EI} = \frac{5\times2\times(4\times10^3)^4}{384\times(80\times10^9)} = 83.3 \text{ mm.}$$ **Ans.**

Example 6.6: *A simply supported beam AB of span 4 m is carring a triangular load varying from zero at A to 5 kN/m at B. Determine the max. deflection of the beam. Take EI = 1.25 × 10^{12} N-mm^2.*

Solution: Given $l = 4 \text{ m} = 4 \times 10^3$ mm.

$w = 5$ kN/m $= 5$ N/mm, $EI = 1.25 \times 10^{12}$ N-mm^2.

$$y_{max} = \frac{0.00652\ wl^2}{EI} = \frac{0.00652\times5\times(4\times10^3)^4}{1.25\times10^{12}} = 6.68 \text{ mm.}$$

Example 6.7: *A horizontal steel girder having uniform X-sec is 14 m long and is simply supported at its ends. It carries two concentrated loads as shown in Fig. 6.10. Calculate the deflections of the beam under loads C&D. Take E = 200 GPa and I = 120 × 10^6 mm^4.*

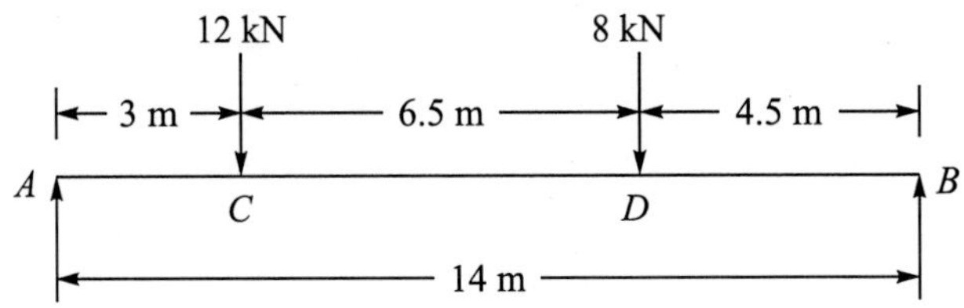

Fig. 6.10

Solution: Given, span "l" = 14 m = 14×10^3 mm,

Load at C, $W_1 = 12$ kN $= 12 \times 10^3$ N.

Load at D, $W_2 = 8$ kN $= 8 \times 10^3$ N

$E = 200$ Gpa $= 200 \times 10^3$ N/mm^2

$I = 160 \times 10^6$ mm^4.

Taking moments about A,

$$R_B \times 14 = (12 \times 3) + (8 \times 9.5) = 112$$

$$\therefore \quad R_B = \frac{112}{14} = 8 \text{ kN} = 8 \times 10^3 \text{ N}$$

and $R_A = (12 + 8) - 8 = 8$ kN $= 12 \times 10^3$ N

Now taking A as the origin and using Macaulay's method, the $B.M.$ at any sec X-X at a distance x from A,

$$EI\frac{d^2y}{dx^2} = -(12 \times 10^3)\ x + \vdots\ (12 \times 10^3) \times [x - (3 \times 10^3)]$$
$$+ (8 \times 10^3) \times [x - (9.5 \times 10^3)]$$

Integrating the above equation,

$$EI\frac{dy}{dx} = -(12 \times 10^3)\frac{x^2}{2} + \vdots\, C_1 + (12 \times 10^3) \times \left[\frac{x-(3\times10^3)}{2}\right]^2$$

$$+ (8 \times 10^3) \times \left[\frac{x-(9.5\times10^3)}{2}\right]^2$$

$$= -(6 \times 10^3)\,x^2 + C_1 \,\vdots\, + (6 \times 10^3) \times [x - (3 \times 10^3)]^2 + \,\vdots\, (4 \times 10^3)$$

$$\times [x - (9.5 \times 10^3)] \quad \ldots(i)$$

Integrating the above equation, once again.

$$EI \cdot y = -(6 \times 10^3) \times \frac{x^3}{3} + C_1\,x + C_2 + \,\vdots\, (6 \times 10^3)$$

$$\times \left[\frac{x-(3\times10^3)}{3}\right]^3 \,\vdots\, + (4 \times 10^3) \times \left[\frac{x-(9.5\times10^3)}{3}\right]^3$$

$$= (2 \times 10^3)\,x^3 + C_1\,x + C_2 \,\vdots\, + (2 \times 10^3)\,[x - (3 \times 10^3)]^3 \,\vdots$$

$$+ \frac{4\times10^3}{3} \times [x - (9.5 \times 10^3)]^3 \quad \ldots(ii)$$

We know that when $x = 0$, then $y = 0$

Therefore, $C_2 = 0$, and when $x = (14 \times 10^3)$ mm,

then $y = 0$.

$$0 = -(2 \times 10^3) \times (14 \times 10^3) + C_1 \times (14 \times 10^3) + (2 \times 10^3)$$

$$\times [(14 \times 10^3) - (3 \times 10^3)]^3 + \frac{4\times10^3}{3} \times [(14 \times 10^3) - (9.5 \times 10^3)]^3$$

$$= -(5488 \times 10^{12}) + (14 \times 10^3)\,C_1 + (2662 \times 10^{12}) + 121.5 \times 10^{12}$$

$$= -(2704.5 \times 10^{12}) + (14 \times 10^3)\,C_1$$

$$\therefore \qquad C_1 = \frac{2704.5\times10^{12}}{14\times10^3} = 193.2 \times 10^9$$

Substituting the value of $C_1 = 193.2 \times 10^9$ and $C_2 = 0$ in eq. (*ii*),

$$EIy = -2 \times 10^3 x^3 + 193.2 \times 10^9\,x \,\vdots\, + 2 \times 10^3\,[x - (3 \times 10^3)]^3 \,\vdots$$

$$+ \frac{4\times10^3}{3} \times [x - (9.5 \times 10^3)]^3 \quad \ldots(iii)$$

Now for deflection under 12 kN load, substituting $x = 3$ (or 3×10^3 mm) in eq. (*iii*) upto first dotted line only,

$$EIy_C = -2 \times 10^3 \times (3 \times 10^3)^3 + 193.2 \times 10^9 \times (3 \times 10^3)$$
$$= -(54 \times 10^{12}) + (579.6 \times 10^{12})$$
$$= 525.6 \times 10^{12}$$

$$\therefore \quad y_C = \frac{525.6 \times 10^{12}}{EI} = \frac{525.6 \times 10^{12}}{(200 \times 10^3) \times (160 \times 10^6)}$$
$$= 16.4 \text{ mm.}$$

Similarly, deflection under 8 kN load,

Substituting $x = 9.5$ m (9.5×10^3 mm) in eq. (6.45) upto second dotted line only,

$$EIy_D = -2 \times 10^3 \times (9.5 \times 10^3)^3 + 193.2 \times 10^9 \times (9.5 \times 10^3)$$
$$+ 2 \times 10^3 \times [(9.5 \times 10^3) - (3 \times 10^3)]$$
$$= -(1714.75 \times 10^{12}) + (1835.4 \times 10^{12}) + (549.25 \times 10^{12})$$
$$= 669.9 \times 10^{12}$$

$$\therefore \quad y_D = \frac{66.99 \times 10^{12}}{EI} = \frac{669.9 \times 10^{12}}{(200 \times 10^3) \times (160 \times 10^6)}$$
$$= \mathbf{20.9 \text{ mm.}}$$

6.4. DEFLECTION OF BEAMS (MOMENT AREA METHOD)

Whenever a beam is loaded, it deflects from its original position. The amount by which a beam deflects, depends upon its cross-section and the bending moment.

The two design criteria for the deflection of a beam are:

Strength	Stiffness
It should be strong enough to resist the *B.M* and *S.F.*	It should be stiff enough to resist the deflection of beam.

Moment area method is convenient in case of beams acted upon with point loads in which case *B.M.* area consists of triangles and rectangles. This is a graphical method for the slopes and deflections of beams and cantilever. The method is based on Mohr's theorems which are stated below:

Mohr's Theorem I: "The change of slope between any two points, on an elastic curve is equal to the net area of *B.M.* diagram between these points divided by *EI.*"

Mohr's Theorem II: "The intercept taken on a vertical reference line of tangents at any two points on an elastic curve, is equal to the moment of the *B.M.* diagram between these points about the reference line divided by *EI.*"

Area and Centroid of Diagrams

To find the slope and deflection of beam by moment area method, area and centroid of the diagram should be known.

So, some propeties of area and centroid is given below:

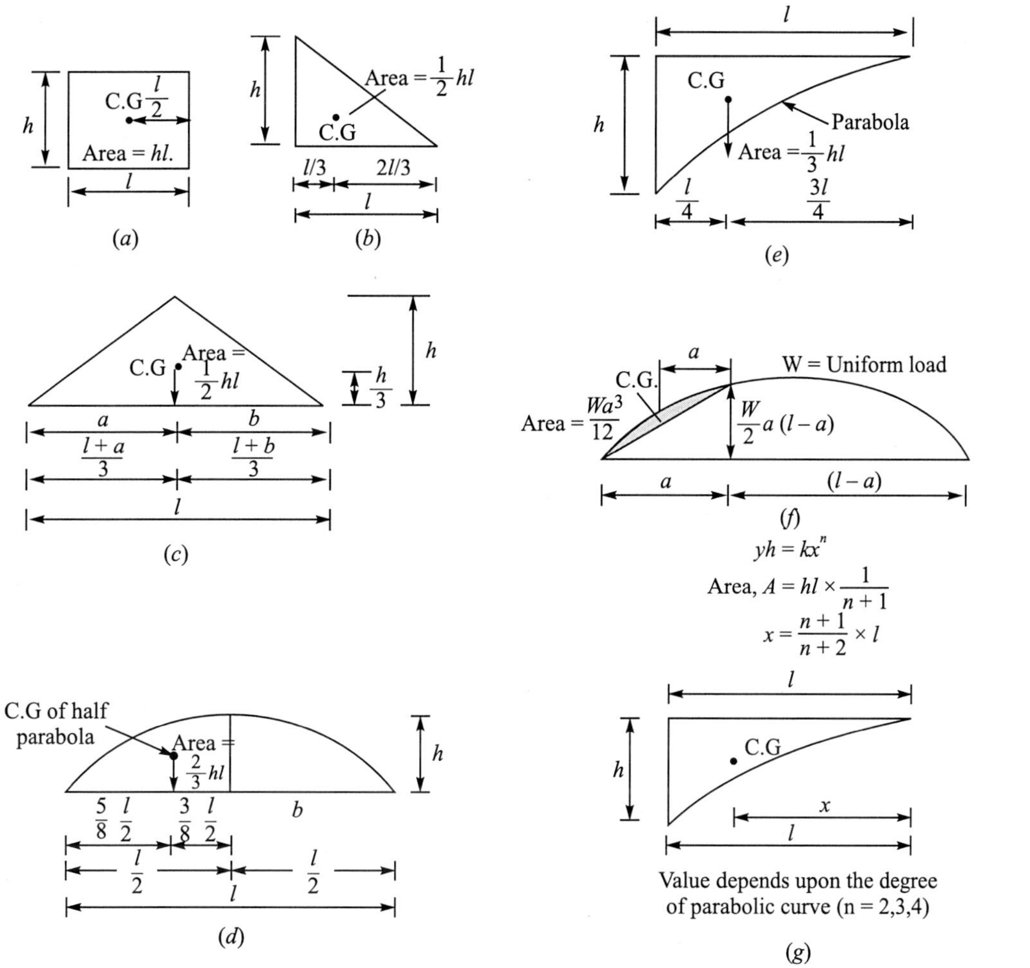

Fig. 6.11

To understand the method, few cases are illustrated below:

Case I: Simply Supported Beam with a Concentrated Point Load: *B.M.* at *A*, due to R_B say $M_1 = (+)\ \dfrac{W}{2} \times l$ similarly, *B.M.* at *A*, due to load "*W*" say $M_2 = (-)\ W \times \dfrac{l}{2}$. Now draw the *B.M.* diagram.

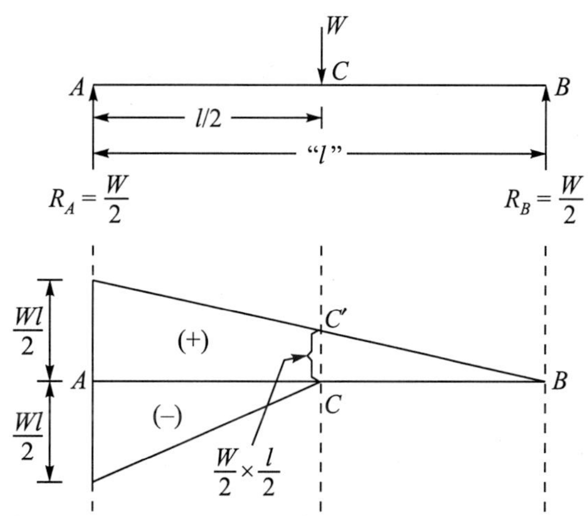

Fig. 6.12

In order to find slope at 'B', let us consider the $B.M.$ diagram from C to B.

$\therefore$ Area of $B.M.$ diagram. from C to B.

$$\text{"}A\text{"} = \text{Area of } \Delta\, BC'C$$

$$= \frac{1}{2} \times \frac{Wl}{4} \times \frac{l}{2} = \frac{Wl^2}{16}$$

and the distance of $C.G.$ of $B.M.$ diagram from 'B'.

$$\bar{x} = \frac{2}{3} \times \text{span}$$

$$= \frac{2}{3} \times \frac{l}{2} = \frac{l}{3}$$

$$\therefore \quad \text{Slope at } B, \textit{ viz, } i_B = \frac{A}{EI} = \frac{\frac{Wl^2}{16}}{EI} = \frac{Wl^2}{16\, EI}$$

By symmetry, slope at A,

$$\textit{viz } i_A = \frac{Wl^2}{16\, EI}$$

and deflection at "C", viz. $y_C = \dfrac{A\bar{x}}{EI}$

$$= \frac{\frac{Wl^2}{16} \times \frac{l}{3}}{EI} = \frac{Wl^3}{48\, EI}.$$

Case II: Simply Supported Beam with an Eccentric Point Load:

$$R_A = \frac{Wb}{l} \text{ and } R_B = \frac{Wa}{l}$$

$B.M.$ at "A" due to reaction R_B,

say $\quad M_1 = (+)\ \dfrac{Wa}{l} \times l = (+)\ Wa$

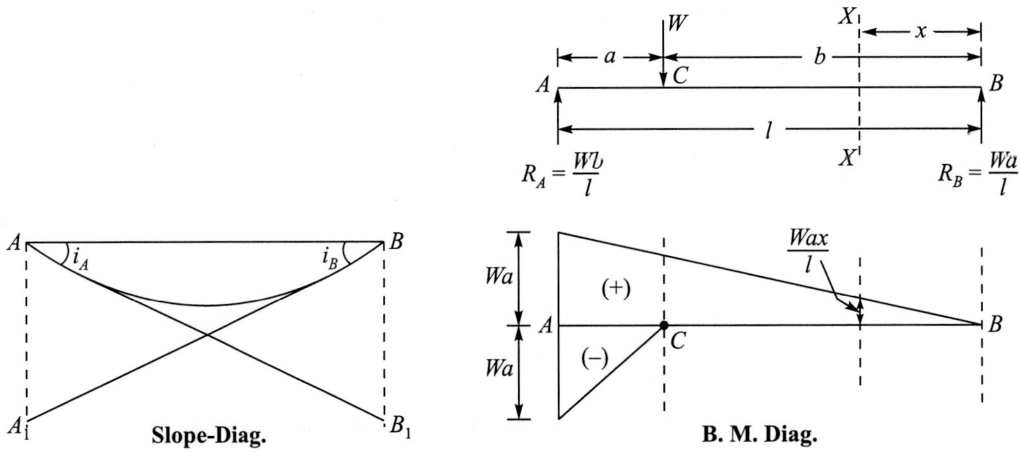

Fig. 6.13

Similarly, *B.M.* at "*A*" due to load "*W*",

say $$M_2 = (-)\, Wa$$

Now draw the *B.M.* diagram.

Area $$\text{“}A_1\text{”} = \frac{1}{2} \times Wa \times l = \frac{Wal}{2} \qquad \text{(+ve } B.M. \text{ Diagram)}$$

and Area $$\text{“}A_2\text{”} = \frac{1}{2} \times Wa \times a = \frac{Wa^2}{2} \qquad \text{(–ve } B.M. \text{ Diagram)}$$

From the geometry of the loading, we see that slope at any section is not known. Hence draw vertical lines through *A* and *B*. Let AA_1 and BB_1 be equal intercepts.

$$AA_1 = i_B \times l \quad \text{or} \quad i_B = \frac{AA_1}{l}.$$

But $$AA_1 = \frac{A_1x_1 - A_2x_2}{EI} = \frac{I}{EI}\left[\left(\frac{Wal}{2} \times \frac{l}{3}\right) - \left(\frac{Wa^2}{2} \times \frac{a}{3}\right)\right]$$

$$\therefore \quad AA_1 = \frac{Wa}{6\,EI}(l^2 - a^2) \qquad \text{...(6.43)}$$

$$\therefore \quad i_B = \frac{AA_1}{l} = \frac{Wa}{6\,EI \times l}(l^2 - a^2)$$

Similarly, $$i_A = \frac{Wb}{6\,EI \times l}(l^2 - a^2).$$

Now consider any section *X-X* at a distance '*x*' from '*B*'. We find that the area of *B.M.* diagram between '*X*' and '*B*',

$$\text{“}A\text{”} = \frac{1}{2} \times \frac{Wax}{l} \times x = \frac{Wax^2}{2\,l} \qquad \text{...(6.44)}$$

∴ Change of slope between "*X*" and "*B*",

$$= \frac{A}{EI} = \frac{Wax^2}{2\,l \times EI} \qquad \text{...(6.45)}$$

Now for the max. deflection, the slope at *X-X* should be zero, or the change of slope between '*B*' and '*X*' should be = Slope at "*B*".

i.e., $$\frac{Wa}{6\,l\,EI}(l^2 - a^2) = \frac{Wax^2}{2\,l\,EI}$$

$$\therefore \quad x^2 = \frac{l^2 - a^2}{3} \quad \text{or} \quad x = \sqrt{\frac{(l^2 - a^2)}{3}}$$

from eq. (6.44) above, we have seen,

$$A = \frac{Wax^2}{2\,l}$$

Distance of *C.G.* of *B.M.* diagram from *B*,

$$\bar{x} = \frac{2x}{3}$$

$$\therefore \qquad y_x = \frac{A\bar{x}}{EI} = \frac{\frac{Wax^2}{2\,l} \times \frac{2}{3}x}{EI} = \frac{Wax^3}{3\,EI\,l}$$

Now for max. deflection,

Substituting the value of $\quad 'x' = \sqrt{\frac{(l^2 - a^2)}{3}}$

$$\therefore \qquad y_{max} = \frac{Wa}{3\,EI\,l}\left(\sqrt{\frac{(l^2 - a^2)}{3}}\right)^3$$

$$= \frac{Wa}{9\sqrt{3}\,l\,EI}\,(l^2 - a^2)^{3/2}.$$

Case-III: Simply Supported Beam with udl

B.M. at '*A*' due to R_B,

say $\qquad M_1 = \frac{wl}{2} \times l = \frac{wl^2}{2}$

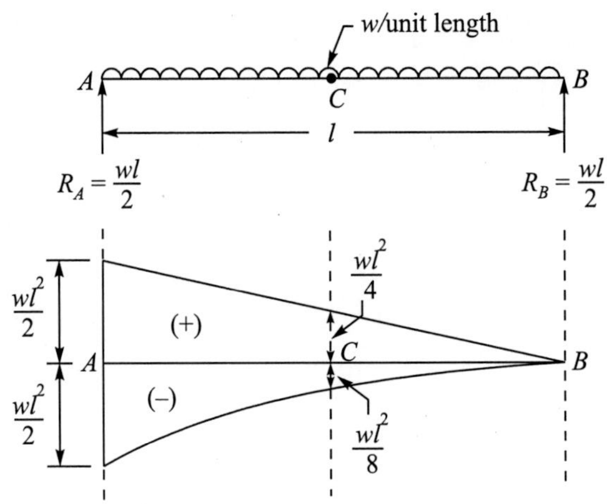

Fig. 6.14

B.M. at '*A*' due to load "*w*",

say $\qquad M_2 = (-)\ w \times l \times \frac{l}{2} = (-)\ \frac{wl^2}{2}$

Area of (+) *B.M.* diagram between *C* and *B*,

$$"A_1" = \frac{1}{2} \times \frac{wl^2}{4} \times \frac{l}{2} = \frac{wl^3}{16}$$

Area of (–) *B.M.* diagram between *C* and *B*,

$$"A_2" = \frac{1}{3} \times \frac{wl^3}{8} \times \frac{l}{2} = \frac{wl^3}{48}$$

∴ Net area of *B.M.* diagram from *C* and *B*,

$$A = A_1 - A_2 = \frac{wl^3}{16} - \frac{wl^3}{48} = \frac{wl^3}{24}$$

Distance of *C.G.* of the (+) *B.M.* diagram on *CB* from '*B*',

$$\bar{x} = \frac{3}{4} \times \frac{l}{2} = \frac{3l}{8}$$

Now, slope at '*B*', *viz.*, $i_B = \dfrac{A}{EI} = \dfrac{wl^3}{24\,EI}$

By symmetry,

Slope at '*A*', *viz.*, $\text{“}i_A\text{”} = \dfrac{wl^3}{24\,EI}$

Deflection at C, $\text{“}y_C\text{”} = \dfrac{A\bar{x}}{EI} = \dfrac{A_1\bar{x}_1 - A_2\bar{x}_2}{EI}$

$$= \frac{\left[\dfrac{wl^3}{16} \times \dfrac{l}{3}\right] - \left[\dfrac{wl^3}{48} \times \dfrac{3l}{8}\right]}{EI}$$

∴ $\text{“}y_C\text{”} = \dfrac{5\,wl^4}{384\,EI}$.

Case IV: To Find the Central Deflection with the Given Loading in the *S.S.* Beam.

B.M. at '*A*' due to reaction at *B*, say, $M_1 = (+)\ w \times a \times l$

B.M. at '*A*' due to load *w*/unit

say, $M_2 = (-)\ 2\ wa \times \dfrac{l}{2} = (-)\ wal$

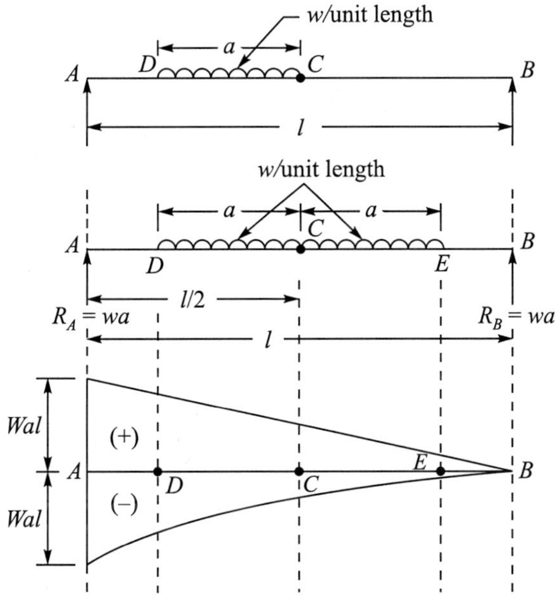

Fig. 6.15

Draw the *B.M.* diagram,

(+) *B.M.* diagram area between *C* and *B*,

$$A_1 = \frac{1}{2} \times \frac{wal}{2} \times \frac{l}{2} = \frac{wal^2}{8}$$

(–) *B.M.* diagram area between *C* and *B*,

$$A_2 = -\frac{1}{3} \times \frac{wa^2}{2} \times a = -\frac{wa^3}{6}$$

$$\bar{x}_1 = \frac{2}{3} \times \frac{l}{2} = \frac{l}{3}$$

also,
$$\bar{x}_2 = \frac{l}{2} - a + \frac{3a}{4} = \left(\frac{l}{2} - \frac{a}{4}\right)$$

$$2y_C = \frac{A\bar{x}}{EI} = \frac{A_1\bar{x}_1 + A_2\bar{x}_2}{EI}$$

$$= \frac{\left[\frac{wal^2}{8} \times \frac{l}{3}\right] + \left[(-)\frac{wa^3}{6} \times \left(\frac{l}{2} - \frac{a}{4}\right)\right]}{EI}$$

$$= \frac{\frac{wal^3}{24} - \frac{wa^3}{24}(2l - a)}{EI}$$

$$= \frac{wa}{24\,EI}(l^3 - 2la^2 + a^3)$$

$\therefore$
$$\text{“}y_C\text{”} = \frac{wa}{48\,EI}(l^3 - 2la^2 + a^3)$$

Case V: Cantilever with Point Load at Free End: We know that the *B.M.* will be zero, at '*B*' and will increase by a straight line law to $W \times l$ at *A*.

Area
$$\text{‘}A\text{’} = \frac{1}{2} \times Wl \times l = \frac{W \times l^2}{2}$$

$$\bar{x} = \frac{2}{3} \times l$$

$\therefore$ Slope at *B*, *viz.*,
$$i_B = \frac{A}{EI} = \frac{Wl^2}{2\,EI} \text{ radians}$$

Slope at *B*,
$$y_B = \frac{A\bar{x}}{EI} = \frac{\frac{Wl^2}{2} = \frac{2l}{3}}{EI}$$

$$\text{“}y_B\text{”} = \frac{Wl^3}{3\,EI}.$$

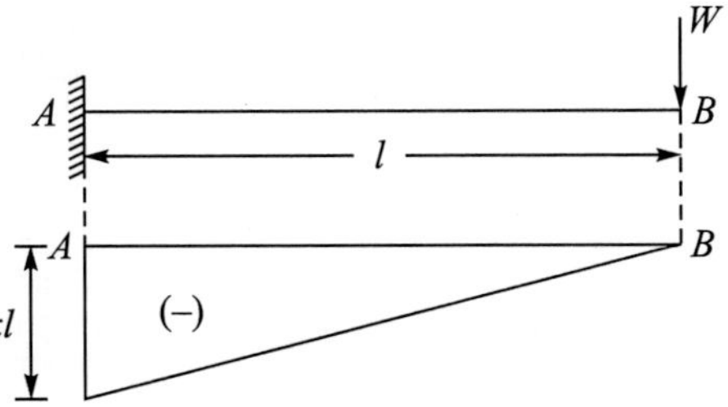

Fig. 6.16

Case VI: Cantilever with a Point Load at any Point: *B.M.* will be zero at *B* and *C* and will increase by a straight line law to Wl_1 upto '*A*'.

Area of *B.M.* diagram,

$$\text{"}A\text{"} = \frac{1}{2} \times Wl_1 \times l_1$$

$$= \frac{Wl_1^2}{2}$$

$$\bar{x} = \frac{2}{3} l_1 + (l - l_1)$$

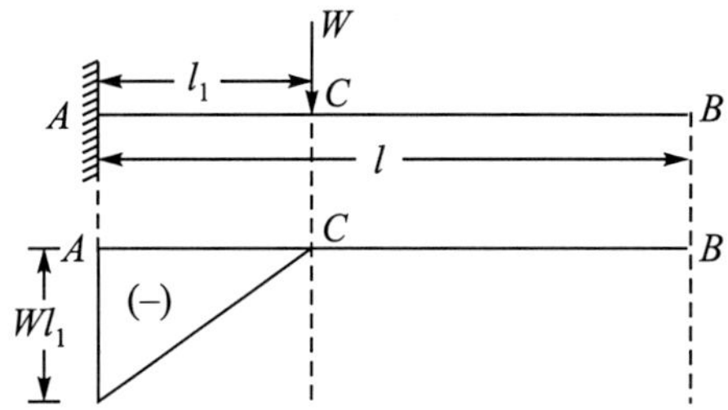

Fig. 6.17

Slope at '*B*', *viz.*, $$i_B = \frac{A}{EI} = \frac{Wl_1^2}{2\,EI} \text{ radians}$$

Deflection at '*B*', *viz.*, $$y_B = \frac{A\bar{x}}{EI} = Wl_1^2 \times \frac{\left[\frac{2\,l_1}{3} + (l - l_1)\right]}{EI}$$

$$\therefore \qquad \text{"}y_B\text{"} = \frac{Wl_1^3}{3\,EI} + \frac{Wl_1^2}{2\,EI}(l - l_1)$$

Case VII: Cantilever with udl: *B.M.* will be zero at '*B*', and will increase in the form of a parabola to $\frac{wl^2}{2}$ at '*A*'.

$$\text{"}A\text{"} = \frac{wl^2}{2} \times l \times \frac{1}{3} = \frac{wl^3}{6}$$

$$\bar{x} = \frac{3}{4} \times l$$

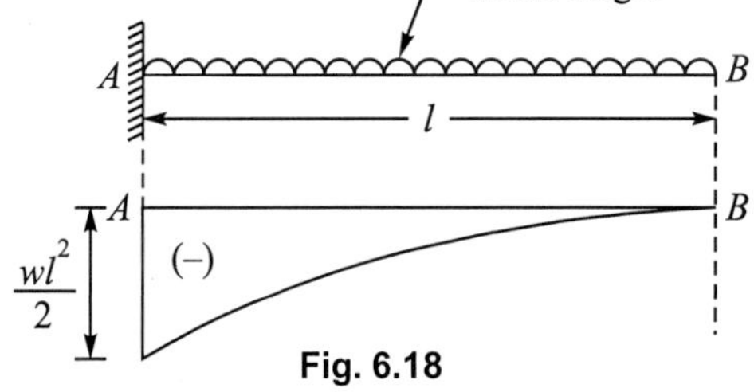

Fig. 6.18

Slope at *B*, *viz.*, $$i_B = \frac{A}{EI} = \frac{wl^3}{6\,EI} \text{ radians}$$

Deflection at '*B*', *viz.*,

$$\text{'}y_B\text{'} = \frac{A\bar{x}}{EI} = \frac{\frac{wl^3}{6} \times \frac{3}{4} l}{EI} \quad \text{or} \quad \text{"}y_B\text{"} = \frac{wl^4}{8\,EI}.$$

Example 6.8: *A simply supported beam of 2 m span carries a pt. load of 20 kN at its mid. pt. Determine the max. slope and deflection of the beam. Take flexural rigidity (EI) of the beam as 500 × 10⁹ N-mm².*

Solution: Given:

Span $\text{"}l\text{"} = 2 \text{ m} = 2 \times 10^3 \text{ mm}$

$W = 20 \text{ kN} = 20 \times 10^3 \text{ N}$

$EI = 500 \times 10^9 \text{ N-mm}^2$

Max. slope
$$\text{“}i_B\text{”} = \frac{wl^2}{16\,EI} = \frac{(20\times10^3)\times(2\times10^3)^2}{16\times500\times10^9}$$
$$= 0.01 \text{ radians.}$$

Max. deflection
$$\text{“}y_C\text{”} = \frac{wl^3}{48\,EI} = \frac{(20\times10^3)\times(2\times10^3)^3}{48\times(500\times10^9)}$$
$$= 6.67 \text{ mm.}$$

Example 6.9: *A simply supported beam of 2.4 m span is subjected to a udl of 6 kN/m over the entire span. Calculate the max. slope and deflection of the beam.*

Given $EI = 8 \times 10^{12}$ *N-mm²*.

Solution: Given:

$$\text{Span “}l\text{”} = 2.4 \text{ m} = 2.4 \times 10^3 \text{ mm}$$
$$\text{“}w\text{”} = 6 \text{ kN/mm} = 6 \text{ N/mm}$$
$$EI = 8 \times 10^{12} \text{ N-mm}^2$$
$$\text{Slope “}i_A\text{”} = \frac{wl^3}{24\,EI} = \frac{6\times(2.4\times10^3)^3}{24\times8\times10^{12}} = 0.00043 \text{ radians.}$$
$$\text{Deflection ‘}y_C\text{’} = \frac{5\,wl^4}{384\,EI} = \frac{5\times6\times(2.4\times10^3)^4}{384\times(8\times10^{12})} = 0.324 \text{ mm.}$$

Example 6.10: *A cantilever beam of span 2.4 m carries a point load of 15 kN at a distance of 1.8 m from the fixed end. What are the values of slope and deflection at the free end of the cantilever, if the flexural rigidity for the beam sec. is* 9×10^{12} *N-mm²*.

Solution: Given:

$$\text{Span “}l\text{”} = 2.4 \text{ m} = 2.4 \times 10^3 \text{ mm.}$$
$$\text{Point load “}W\text{”} = 15 \text{ kN} = 15 \times 10^3 \text{ N}$$
$$EI = 9 \times 10^{12} \text{ N-mm}$$

Distance of load from fixed end $l_1 = 1.8$ m
$$= 1.8 \times 10^3 \text{ mm.}$$

Slope is given by,
$$\text{“}i_B\text{”} = \frac{Wl_1^2}{2\,EI} = \frac{(15\times10^3)\times(1.8\times10^3)^3}{2\times(9\times10^{12})} = 0.0027 \text{ radians}$$

Deflection is given by,
$$\text{“}y_B\text{”} = \frac{Wl_1^3}{3\,EI} + \frac{Wl_1^2}{2\,EI}(l - l_1)$$
$$= \frac{(15\times10^3)\times(1.8\times10^3)^3}{3\times(9\times10^{12})} + \frac{(15\times10^3)\times(1.8\times10^3)^2}{2\times(9\times10^{12})} \times [(2.4\times10^3) - (1.8\times10^3)]$$
$$= 3.2 + 1.6 = 4.8 \text{ mm.}$$

Example 6.11: *A cantilever of length 2 a is carring a load of W at the free end, and another load of W at its centre. Determine, by moment Area method the slope and deflection of the cantilever at the free end.*

Solution: Given:

$$\text{Span } (l) = 2\,a, \quad \text{Load} = W$$

The cantilever *AB* of length 2 *a* and carring point loads of *W* each at *C* and *B* is shown in Fig. 6.19.

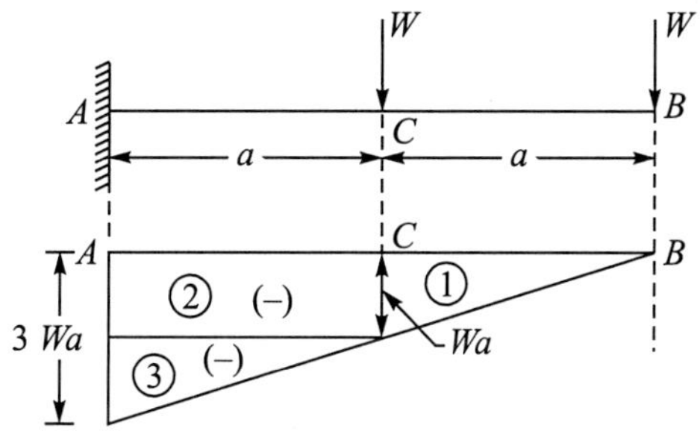

Fig. 6.19

Slope at free end:

$$B.M. \text{ at } B,\ M_B = 0$$

$$B.M. \text{ at } C,\ M_C = -\ Wa$$

$$B.M. \text{ at } A,\ M_A = -\ (W \times 2\,a) - (Wa) = (-)\ 3\ Wa$$

Now draw the *B.M.* diagram (1)

Area of *B.M.* diagram

$$A_1 = \frac{1}{2} \times Wa \times a = \frac{wa^2}{2}$$

Similarly, area of *B.M.* diagram (2)

$$A_2 = Wa \times a = Wa^2$$

and area of *B.M.* diagram

$$A_3 = \frac{1}{2} \times 2\ Wa \times a = Wa^2.$$

∴ Total area of *B.M.* diagram (3)

$$\text{“}A\text{”} = A_1 + A_2 + A_3 = \frac{Wa^2}{2} + Wa^2 + Wa^2$$

$$= \frac{5\ Wa^2}{2}.$$

Slope at *B*, $i_B = \dfrac{A}{EI} = \dfrac{5\ Wa^2}{2\ EI}.$

Deflection at the free end: We know that total moment of the *B.M.* diagram about *B*,

$$A\,\bar{x} = A_1\,\bar{x}_1 + A_2\,\bar{x}_2 + A_3\,\bar{x}_3$$

$$= \left(\frac{Wa^2}{2} \times \frac{2\,a}{3}\right) + \left(Wa^2\ \frac{3\,a}{2}\right) + \left(Wa^2 \times \frac{5\,a}{3}\right) = \frac{7\ Wa^3}{2}$$

Deflection at *B*,

$$\therefore \qquad \text{“}y_B\text{”} = \frac{A\bar{x}}{EI} = \frac{7\,wa^3}{2\,EI}$$

Example 6.12: *Cantilever with a Gradually varying load:*

We know that the B.M. will be zero at B and will increase in the form of a cubic parabola to $\frac{wl^2}{6}$ *at "A".*

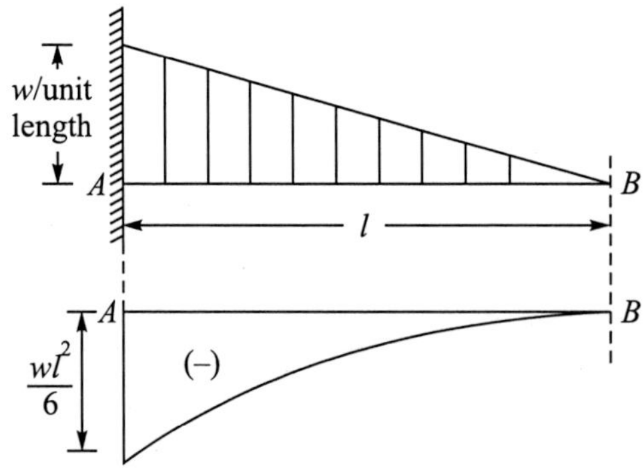

Fig. 6.20

Solution: ∴ Area of *B.M.* diagram,

$$\text{“}A\text{”} = \frac{wl^2}{6} \times l \times \frac{1}{4} = \frac{wl^3}{24}$$

and distance between *C.G.* of the *B.M.* diagram and '*B*'

$$\bar{x} = l \times \frac{4}{5}$$

Slope at *B*,
$$i_B = \frac{A}{EI} = \frac{wl^3}{24\,EI}$$

and Deflection
$$\text{“}y_B\text{”} = \frac{A\bar{x}}{EI} = \frac{\frac{wl^3}{24} \times l \times \frac{4}{5}}{EI} = \frac{wl^4}{30\,EI}$$

6.5. DEFLECTION OF BEAMS (CONJUGATE BEAM METHOD)

The conjugate beam method is a modified form of moment area method and can be conveniently used for finding slope and deflection of cantilevers and simply supported beams.

This method is specially useful for beams with different *EI* (Flexural rigidities).

Conjugate beam: Is an imaginary beam of length equal to that of the original beam with width equal to $\frac{1}{EI}$ and loaded with *B.M.* diagram.

The slope and deflection is found out by following two theorems known as Mohr's theorems.

Mohr's First Theorem: It states, "The *S.F.* at any section of the conjugate beam is equal to the slope of the elastic curve at the corresponding section of the actual beam".

Mohr's Second Theorem: It states,"The *B.M.* at any section of the conjugate beam is equal to the deflection of the elastic curve at the corresponding section of the actual beam".

"Relation between an actual beam and the conjugate beam"

S. No	Actual beam	Conjugate beam	Remarks
1.	Fixed end	Free end	Slope and deflection at fixed end of the actual beam is zero. *S.F.* and *B.M.* at the free end of the conjugate beam is also zero.
2.	Free end	Fixed end	Slope and deflection at the free end of the actual beam exists. *S.F.* and *B.M.* at the fixed end of the conjugate beam also exists.
3.	Simply supported or roller supported ends	Simply supported end	Slope at the free end of the actual beam exists. But deflection is zero *S.F.* at the simply supported end of the conjugate beam exists. But the *B.M.* is zero.

We shall now discuss the applications of the conjugate beam with various types of loadings on cantilevers and beams.

Note: When the deflection equation is differentiated successively, we get the relations as given below:

$$EI \cdot y = \text{Deflection} \quad \ldots(6.46)$$

$$EI \cdot \frac{dy}{dx} = \text{Slope} \quad \ldots(6.47)$$

$$EI \cdot \frac{d^2y}{dx^2} = \text{Moment} = M \quad \ldots(6.48)$$

$$EI \cdot \frac{d^3y}{dx^3} = \text{Shear} = S = \frac{dM}{dx} \quad \ldots(6.49)$$

$$EI \cdot \frac{d^4y}{dx^4} = \text{Load} = \frac{ds}{dx} = \frac{d^2M^2}{dx^2} \quad \ldots(6.50)$$

From the above we find that a similarity of relations exists among load, shear and moment [eqns. (6.48) (6.49) and (6.50)] and among moment, slope and deflection [eqns. (6.46), (6.47) and (6.48)]. This gives us an idea of a method by which slope and deflection can be calculated. In conjugate beam method, the beam (conjugate beam) is loaded with not the actual loads, but with the elastic weight $\frac{M}{EI}$ corresponding to the actual load.

Case I: Cantilever with a Point Load at the Free End: Consider a cantilever *AB* of length "*l*" and carrying a point load *W* at the free end '*B*' as shown in Fig. 6.20. We know that the *B.M.* will be zero at '*B*' and will increase by a straight line law to *Wl* at *A*.

Now draw the conjugate beam *AB* (free at *A*, fixed at *B* and loaded with *B.M.* diagram).

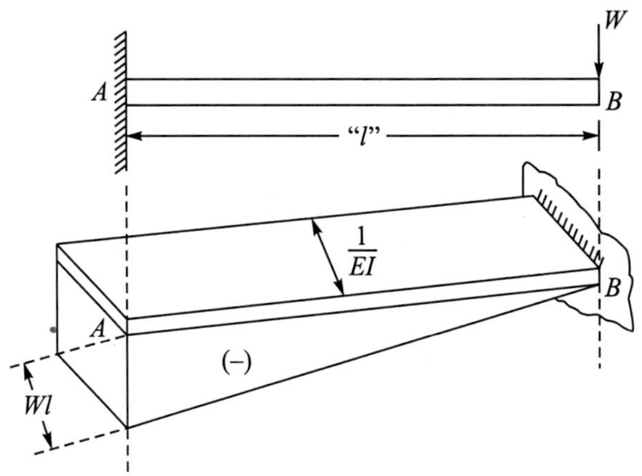

Fig. 6.21

From the geometry of the Fig. 6.21, we find that the total load of the conjugate beam,

$$\text{“}P\text{”} = \frac{1}{2}\,(-\,Wl) \times l \times \frac{1}{EI} = (-)\,\frac{Wl^2}{2\,EI}$$

Now *S.F.* at *B* (on the conjugate beam)

$$F_B = (-)\,P = (-)\left(-\frac{Wl^2}{2\,EI}\right) \qquad \text{(–ive sign due to left downwards)}$$

$$\text{“}F_B\text{”} = \frac{Wl^2}{2\,EI}$$

$\therefore$ Slope at $B,\ i_B = F_B = \dfrac{Wl^2}{2\,EI}$ radians.

and the *B.M.* at '*B*' (on the conjugate beam *AB*),

$$M_B = -\,P \times \frac{2}{3}\,l = -\left(-\frac{Wl^2}{2\,EI} \times \frac{2}{3}\,l\right) \qquad \text{(–ive sign due to hogging)}$$

$$\text{‘}M_B\text{’} = \frac{Wl^3}{3\,EI}$$

$\therefore$ Deflection at '*B*', $y_B = M_B = \dfrac{Wl^3}{3\,EI}$.

Case II: Cantilever with Uniformaly Distributed Load: Consider a cantilever *AB* of length (*l*) and carring a *udl* "*w*" per unit length as shown in the Fig 6.21. We know that the *B.M.* will be zero at *B*, and will increase in the form of a parabola to $\left(-\dfrac{wl^2}{2}\right)$ at *A*.

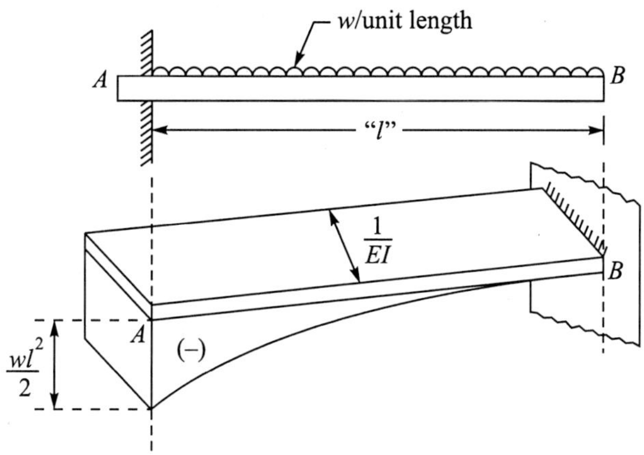

Fig. 6.22

Now draw the conjugate beam *AB* (free at *A*, fixed at *B* and loaded with the *B.M.* diagram) as shown.

From the geometry of the Fig. 6.22, we find that the total load of the conjugate beam,

$$“P” = \frac{1}{3}\left(-\frac{wl^2}{2}\right) \times l\,\frac{1}{EI} = (-)\,\frac{wl^3}{6\,EI}$$

Now, *S.F.* at '*B*' (on conjugate beam)

$$F_B = -P = \left(-\frac{wl^3}{6\,EI}\right)$$

$$= \frac{wl^3}{6\,EI} \qquad \text{(ive sign due to left downward)}$$

$\therefore$ Slope at '*B*'
$$i_B = F_B = \frac{wl^3}{6\,EI}$$

and the *B.M.* at '*B*' (on the conjugate beam)

$$M_B = -P \times \frac{3}{4}\,l = -\left(-\frac{wl^3}{6\,EI} \times \frac{3}{4}\,l\right) \qquad \text{(–ive sign due to hogging)}$$

$$= \frac{wl^4}{8\,EI}$$

$\therefore$ Deflection at *B*,
$$y_B = M_B = \frac{wl^4}{8\,EI}$$

Case III: Cantilever with a Gradually Varying Load: Consider a cantilever *AB* of length "*l*" and carring a gradually varring load from zero at *B* to '*w*' per unit length at *A* as shown in Fig. 6.22. We know that the *B.M.* will be zero at *A* and will increase in the form of cubic parabola to $\left(-\frac{wl^2}{6}\right)$ at *A*.

Now draw the conjugate beam (free at *A*, fixed at *B* and loaded with *B.M.* diagram) as shown.

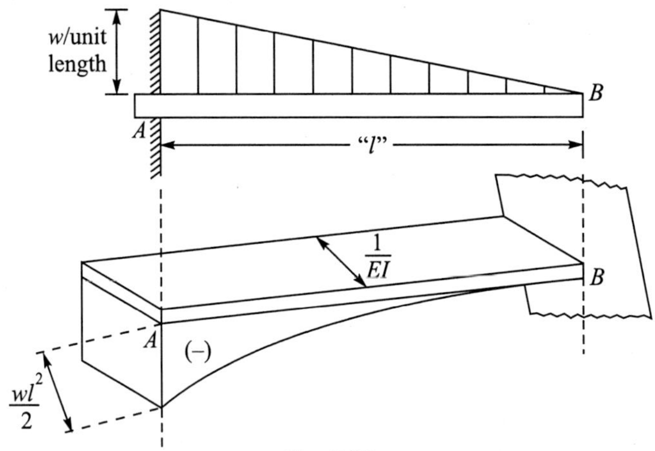

Fig. 6.23

From the geometry of the Fig. 6.23, we find that the total load of the conjugate beam,

$$P = \frac{1}{4}\left(-\frac{wl^2}{6}\right) \times l \times \frac{1}{EI} = (-)\ \frac{wl^3}{24\ EI}$$

Now, *S.F.* at '*B*' (on the conjugate beam)

$$F_B = -P = (-)\left(-\frac{wl^3}{24\ EI}\right) \qquad \text{(–ve sign due to left downwards)}$$

or $$F_B = P = \frac{wl^3}{24\ EI}.$$

$\therefore$ Slope at *B*, $$i_B = F_B = \frac{wl^3}{24\ EI}$$

and the *B.M.* at '*B*' (on the conjugate beam)

$$M_B = -P \times \frac{4}{5}\ l = (-)\left(-\frac{wl^3}{24\ EI} \times \frac{4}{5}\ l\right) \qquad \text{(–ve sign due to hogging)}$$

$$= \frac{wl^4}{30\ EI}$$

$\therefore$ Deflection at "*B*", $$y_B = M_B = \frac{wl^4}{30\ EI}$$

Case IV: Simply Supported Beam with a Central Point Load: Consider a simply supported beam *AB* of length "*l*" and carrying a point load '*W*' at *C* (*i.e.*, the centre of beam) as shown in Fig. 6.23.

We know that the *B.M.* will be zero at *A* and *B* and will increase by a straight line law to (+) $\frac{Wl}{4}$ at *C*.

Now draw the conjugate beam *AB* (Simply supported at *A* and *B* and loaded with the *B.M.* diagram) as shown in the Fig. 6.24.

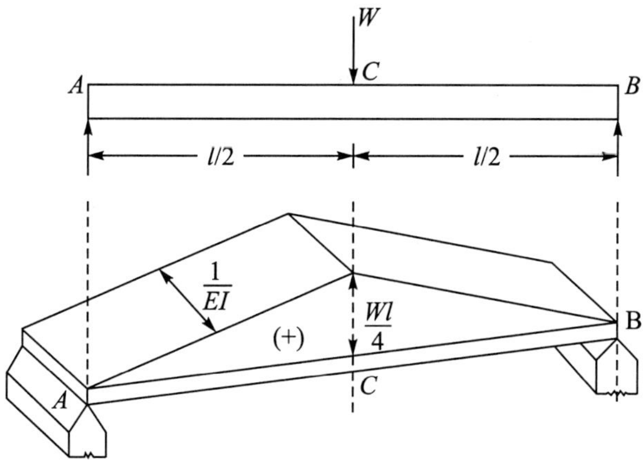

Fig. 6.24

From the geometry of Fig. 6.24, we find that the total load of the conjugate beam,

$$\text{“}P\text{”} = \frac{1}{2}\left(+\frac{Wl}{4}\right) \times l \times \frac{1}{EI} = \frac{Wl^2}{8\,EI}$$

Now *S.F.* at *AB* on the conjugate beam,

$$F_A = +R_A = +\frac{P}{2} = \frac{1}{2} \times \frac{Wl^2}{8\,EI}$$

$$= \frac{Wl^2}{16\,EI} \qquad \text{(Plus sign due to left upwards)}$$

∴ Slope at *A*, $i_A = F_A = \dfrac{Wl^2}{16\,EI}$

Similarly, $i_B = \dfrac{Wl^2}{16\,EI}$

and *B.M.* at *C* (on conjugate beam *AB*)

$$M_C = +\left[\left(\frac{Wl^2}{16\,EI} \times \frac{l}{2}\right) - \left(\frac{Wl^2}{16\,EI} \times \frac{l}{6}\right)\right]$$

$$= \frac{Wl^3}{48\,EI} \qquad \text{(Plus sign due to sagging)}$$

∴ Deflection at *C*, $y_c = M_C = \dfrac{Wl^3}{48\,EI}$

Case V: Simply Supported Beam with an Eccentric Point Load: Consider a simply supported beam *AB* of length "*l*" and carrying a point load *W* at *C*, such that *AC* = *a* and *CB* = *b* as shown in Fig. 6.25. We know that the *B.M.* will be zero at *A* and *B* and will increase by a straight line law to $\left(+\dfrac{Wab}{l}\right)$ at "*C*".

Now draw the conjugate beam *AB* (simply supported at *A* and *B*, and loaded with *B.M.* diagram as shown in Fig. 6.25). The reaction at *B* of the conjugate beam may be found out by taking moments about *A*,

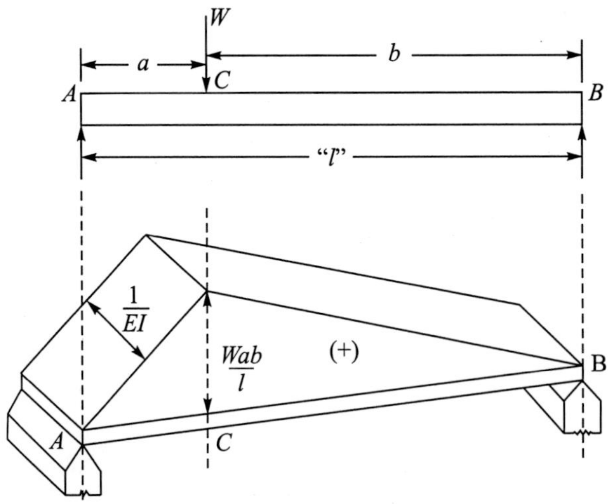

Fig. 6.25

$$R_B \times l = \left[\frac{1}{2}\times\frac{Wab}{l}\times a\times\frac{1}{EI}\times\frac{2a}{3}\right]+\left[\frac{1}{2}\times\frac{Wab}{l}\times b\times\frac{1}{EI}\times\left(a+\frac{b}{3}\right)\right]$$

$$= \frac{Wab}{3\,EI\,l}\times a^2+\frac{Wab}{2\,EI\,l}\left(ab+\frac{b^2}{3}\right)$$

$$= \frac{Wab}{6\,EI\,l}[2a^2+3ab+b^2]$$

$$= \frac{Wab}{6\,EI\,l}[a^2+b^2+2ab+a^2+ab]$$

$$= \frac{Wab}{6\,EI\,l}[(a+b)^2+a(a+b)]$$

$$= \frac{Wa\,(l-a)}{6\,EI\,l}\,[l^2+al] \qquad (\because a+b=l)$$

$$\therefore \qquad R_B \times l = \frac{Wal\,(l+a)}{6\,EI\,l}$$

or
$$\text{"}R_B\text{"} = \frac{Wa}{6\,EI\,l}\,(l^2-a^2)$$

Now *S.F.* at '*B*' (on the conjugate beam *AB*).

$$F_B = -R_B \qquad \text{(–ve sign due to right upwards)}$$

$\therefore$ Slope at 'B', $\quad i_B = F_B = (-)\ \dfrac{Wa}{6\,EI\,l}\,(l^2 - a^2)$ (minus sign means that the tangent at B makes an angle with AB in the anticlockwise direction)

$$= \frac{Wa}{6\,EI\,l}$$

Similarly, $\quad i_A = \dfrac{Wb}{6\,EI\,l}\,(l^2 - b^2)$ (Substituting a for b)

and $B.M.$ at 'C' (on the conjugate beam AB),

$$M_C = R_B \times b - \frac{1}{2} \times \frac{Wab}{l} \times b \times \frac{1}{EI} \times \frac{b}{3}$$

$$= \frac{Wa}{6\,EI\,l}\,(l^2 - a^2)\,b - \frac{Wab}{6\,EI\,l} \times b^2$$

$$= \frac{Wab}{6\,EI\,l}\,(l^2 - a^2 - b^2)$$

$\therefore$ Deflection at 'C'

$$y_C = M_C = \frac{Wab}{6\,EI\,l}\,(l^2 - a^2 - b^2)$$

From the geometry of the figure, we find that the max. deflection will take place in CB because $b > a$. We know the $S.F.$ at any section X, at a distance 'x' from B,

$$F_X = R_B - \frac{Wab}{EI\,l} \times \frac{x}{b} \times \frac{x}{2} = \frac{Wa}{6\,EI\,l}\,(l^2 - a^2) - \frac{Wax^2}{2\,EI\,l}$$

$$= \frac{Wa}{6\,EI\,l}\,(l^2 - a^2 - 3x^2).$$

We know that the max. deflection will take place at a section, where the slope of the elastic curve is zero (In other words max. $B.M.$ takes place at a section where $S.F.$ is zero), therefore equating the above equation to zero,

$$\frac{Wa}{6\,EI\,l}\,(l^2 - a^2 - 3x^2) = 0$$

$$x = \sqrt{\frac{l^2 - a^2}{3}}$$

We also know that the $B.M.$ at any section X, from at a distance 'x' from B.

$$M_X = R_B \cdot x - \frac{Wab}{EI\,l} \times \frac{x}{b} \times \frac{x}{2} \times \frac{x}{3} = \frac{Wa}{6\,EI\,l}\,(l^2 - a^2)\,x - \frac{Wax^3}{6\,EI\,l}$$

$$= \frac{Wax}{6\,EI\,l}\,(l^2 - a^2 - x^2)$$

Now, for max. *B.M.*, substituting the value of x in the above eq., we get,

$$M_{max} = \frac{Wax}{6\,EI\,l}\sqrt{\frac{l^2 - a^2}{3}} \times \left[l^2 - a^2 - \left(\frac{l^2 - a^2}{3}\right)\right]$$

$$M_{max} = \frac{Wax}{6\,EI\,l}\sqrt{\frac{l^2 - a^2}{3}} \times \left[\frac{2}{3}(l^2 - a^2)\right]$$

$$= \frac{Wa}{9\sqrt{3}\cdot EI\,l}\,(l^2 - a^2)^{3/2}.$$

∴ Max. deflection,

$$y_{max} = M_{max} = \frac{Wa}{9\sqrt{3}\cdot EI\,l}\,(l^2 - a^2)^{3/2}$$

Case VI: Simply Supported Beam with a Uniformally Distributed Load: Consider a simply supported beam *AB* of length "*l*" and carrying a unifromally distributed load of "*w*" per unit length as shown in the Fig. 6.26. We know that the *B.M.* at *A* and *B* will be zero and will increase in the form of parabola to $\left(+\frac{wl^2}{8}\right)$ at '*C*' (*i.e.,* the centre of the beam).

Now draw the conjugate beam *AB* (simply supported at *A* and *B* and loaded with B.M. diagram) as shown in the Fig. 6.25.

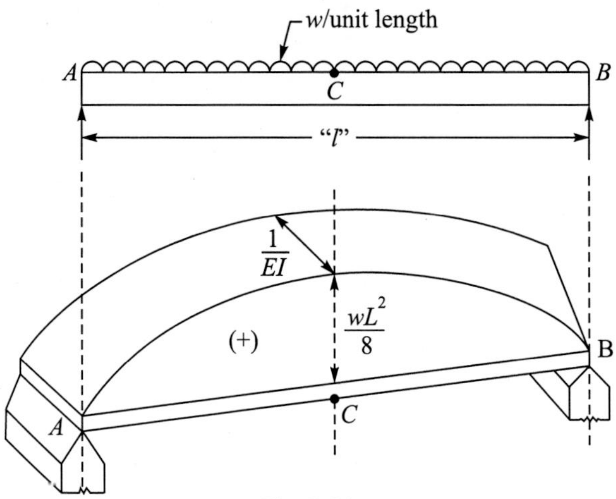

Fig. 6.26

From the geometry of the Fig. 6.26, we find that the total load of the conjugate beam,

$$\text{"}P\text{"} = \frac{2}{3}\left(+\frac{wl^2}{8}\right)\times l \times \frac{1}{EI} = \frac{wl^3}{12\,EI}$$

Now, *S.F.* at *A*, (on the conjugate beam *AB*)

$$F_A = +R_A = +\frac{P}{2} = \frac{1}{2}\times\frac{wl^3}{12\,EI} \qquad \text{(Plus sign due to left upwards)}$$

$$= \frac{wl^3}{24\,EI}$$

$\therefore$ Slope at A, $$i_A = F_A = \frac{wl^3}{24\,EI}$$

Similarly, $$i_B = \frac{wl^3}{24\,EI},$$

and $B.M.$ at 'C' (on conjugate beam AB),

$$M_C = \left(\frac{wl^3}{24\,EI} \times \frac{l}{2}\right) - \left(\frac{wl^3}{24\,EI} \times \frac{3}{16}\,l\right) = \frac{5\,wl^4}{384\,EI}$$

$\therefore$ Deflection at 'C',

$$y_C = M_C = \frac{5\,wl^4}{384\,EI}$$

Case VII: Simply Supported Beam with a Gradually Varing Load: Consider a simply supported beam AB of length 'l' and carrying a gradually varying load from zero at B to 'w' per unit length at A, as shown in the Fig. 6.27.

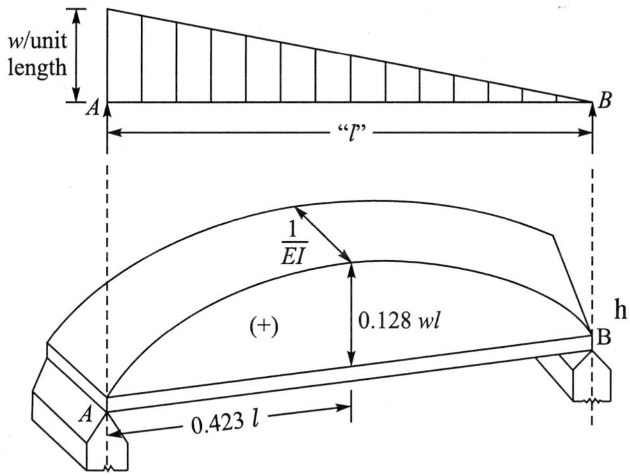

Fig. 6.27

We know that the $B.M.$ will be zero at A and B and will increase in the form of a parabola to max. equal to 0.128 wl at a distance of 0.423 l from A.

Now draw the conjugate beam AB (simply supported at A and B and loaded with the $B.M.$ diagram) as shown in the Fig. 6.27.

The reaction at B of the conjugate beam may be found out by taking moments about A.

$$R_B \times l = \frac{wl^3}{24\,EI} \times \frac{8\,l}{15} = \frac{wl^4}{45\,EI}$$

$$R_B = \frac{wl^4}{45\,EI}$$

Now *S.F.* at *B* (on the conjugate beam *AB*),

$$F_B = -R_B \qquad \text{(Minus sign due to right upwards)}$$

$\therefore$ Slope at *B*, $i_B = F_B = -\dfrac{wl^3}{45\,EI}$ (Minus sign means that the tangent at *B* makes an angle with *AB* in anticlockwise direction)

$$= \frac{wl^3}{45\,EI}$$

Similarly, $R_A = \dfrac{wl^3}{24\,EI} - \dfrac{wl^3}{45\,EI} = \dfrac{7\,wl^3}{360\,EI}$

$\therefore$ Slope at *A*, $i_A = F_A = R_A = \dfrac{7\,wl^3}{360\,EI}$.

SUMMARY

- Whenever a beam is loaded, it deflects from its original position. The amount by which a beam deflects, depends upon its cross-sectional area and the bending moment.
- The two design criterian for the deflection of a cantilever or a beam are 'strength' and 'stiffness'. As per the strength criterian, the beam should be strong enough to resist the bending moment and shear force. Stiffness criterian calls for the stiffness of the beam to resist its deflection.
- Deflection plays a significant role in the design of structures. If floor beams or joists deflect too far, the plaster on the ceiling under them may crack. Although no damage to the structure may result, the appearance on the ceiling may be ruined.
- Moment area method for finding deflection of beams is a graphical method and is based on Mohr's two theorems. This method is partially convenient is case of beams acted upon with point loads in which case B.M. area consists of triangles and rectangles.
- Conjugate beam method for finding deflection of beams is a modified form of moment area method and may be conveniently used for finding out slope and deflection of cantilevers and simply supported beams. This method is specially useful for beams and cantilevers with varying flexural rigidities.
- In this method the slope and deflection is found out by two theorems known as Mohr's theorems.
- The width of a conjugate beam is 1/EI.

SHORT ANSWER QUESTIONS

1. What is the relation between slope, deflection and radius of curvature of a simply supported beam ?
2. What is a conjugate beam ?
3. State Mohr's theorem.
4. Give the relation between an actual beam and a conjugate beam when the former has fixed ends.
5. What is moment area method for the slope and deflection of a simply supported beam ?

EXERCISE FOR PRACTICE

1. Using conjugate beam method, find the mid span deflection of the beam shown in the Fig. 6.28. Take $E = 200 \times 10^6$ kN/m² and $I = 200 \times 10^{-4}$ m⁴.

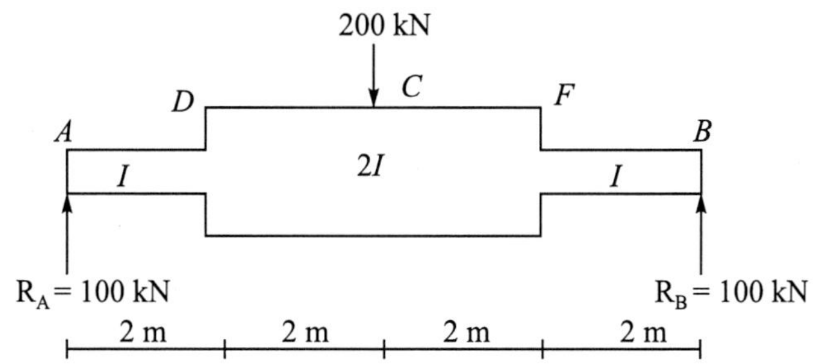

Fig. 6.28.

2. A beam *ABCD* is simply supported at its ends *A* and *D* over a span of 30 m. It is madeup of 3 portions *AB*, *BC* and *CD* each 10 m in length. *M.I.* of the section over each of these individual portion is uniform and the *I* values for them are *I*, 3*I* and 2*I* respectively, where $I = 20 \times 10^9$ mm⁴. The beam carries a point load of 150 kN at *B* and another load of 300 kN at *C* as shown in Fig. 6.29.

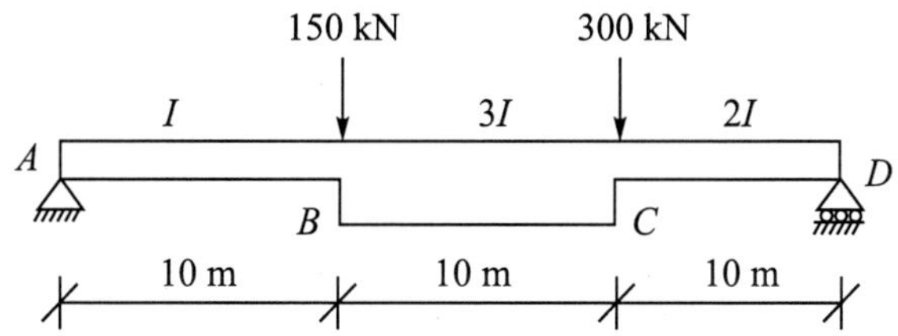

Fig. 6.29.

Neglect the self weight of beam, calculate the deflection at *B* and the slope at *C*. Take *E* = 200 Gpa using conjugate beam method.

3. A simply supported beam *AB* of 2.8 m span carries a point load of 60 kN a distance of 1 m from the left hand support *A*. What is the position of the maximum deflection of the beam ? Also find the magnitude of the deflection under the load. Take *EI* for the beam section as 4×10^{12} N-mm². Use moment area method.

4. A cantilever beam of span 2.4 m carries a point load of 15 kN at a distance of 1.8 m from the fixed end. What are the vlaues of slope and deflection at the free end of the cantilever, if $EI = 9 \times 10^2$ N – mm², using moment area method.

❑❑❑

CHAPTER 7

Principle of Virtual Work Done

Let a member be subjects to a force W (Applied Gradually)

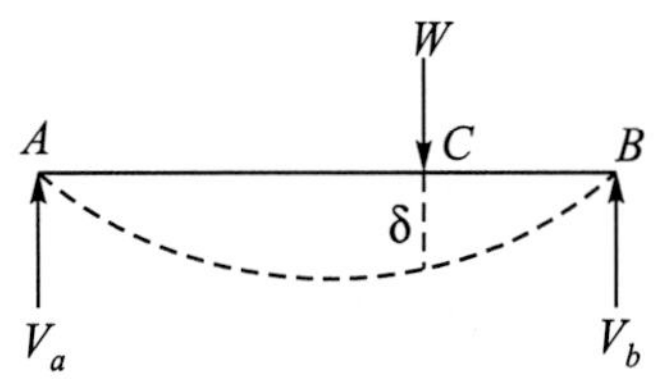

Fig. 7.1. (i)

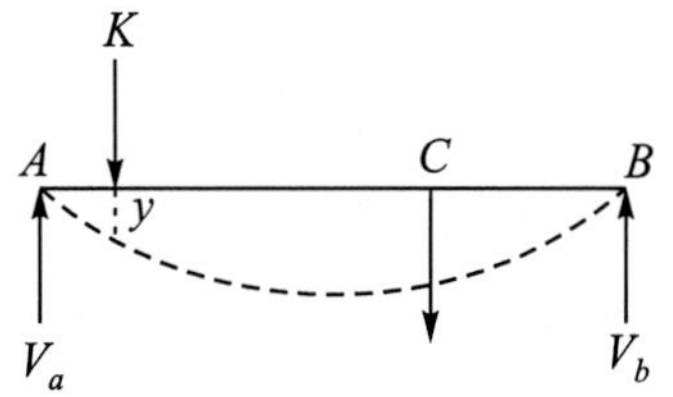

Fig. 7.1. (ii)

Let δ be the deflection of the member in the line of action of force. In this case the deflection δ is caused by the force W Fig. 7.1(*i*).

$$\text{Work done by the force} = \frac{1}{2}\, W \times \delta$$

Figure 7.1(*ii*) shows a member carrying a load k. This load by virtue of its own direct action will produce a deflection.

Let us for a moment consider the load 'k' as a very small load. Hence the deflection produced by 'k' is very small and the work done by 'k' may be ignored.

Suppose now due to the action of some external agency, the member is subjected to a deformation. Let 'y' be the deflection produced in the line of action of the load k.

Now the work done by the load 'k' due to the above deformation = $k \times y$.

Thus we find that:

(*i*) If a load acts on a member and produces a deflection δ in its line of action by virtue of its own direct action, the work done by the load

$$\text{"}W\text{"} = \frac{1}{2} \times W \times \delta$$

(*ii*) If a member subjected to a load 'k' is given a deformation 'y' in the line of action of 'k' by virtue of some other external agency, the work done by the load 'k' due to the above deformation = $k \times y$.

7.1. MAXWELL'S RECIPROCAL DEFLECTION THEOREM

"In any beam or truss, the deflection at any point '*D*' due to load '*W*' at any other point '*C*' is the same as the deflection at '*C*' due to the same load '*W*' applied at '*D*'."

James Clerk Maxwell, a canvendish professor at Cambridge university offered this law in 1864.

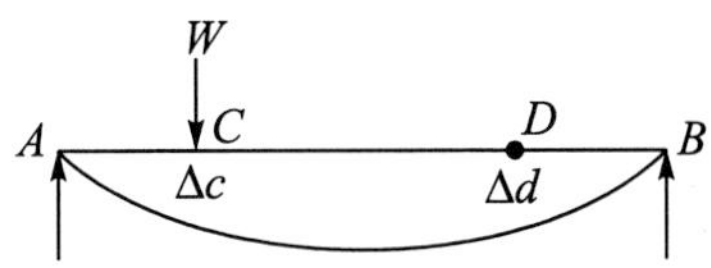

Fig. 7.2. (i)

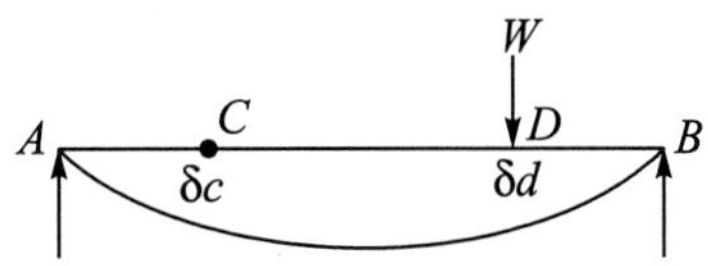

Fig. 7.2. (ii)

The Fig. 7.2 (*i*) shows a structure *AB* carring a load '*W*' applied at any point, '*C*'. Let the deflection at '*C*' be Δ_c. Let the deflection at any other point '*D*' be Δ_d.

Figure 7.2 (*ii*) above shows the same structure *AB* carring the same load '*W*' at *D*. Let the deflection at *C* and *D* be δ_c and δ_d respectively.

$$\text{Work done on the structure} = \frac{1}{2}\, W\Delta_c$$

As the structure is loaded with the load '*W*' at *C*, let another equal load '*W*' be applied at *D*. There will be further deflection of δ_c and δ_d at *C* and *D* as shown in Fig. 7.3.

$$\text{Total work done at this stage} = \frac{1}{2}\, W\Delta_c + \frac{1}{2}\, W\delta_d + W\delta_c$$

Let now the order of loading be changed.

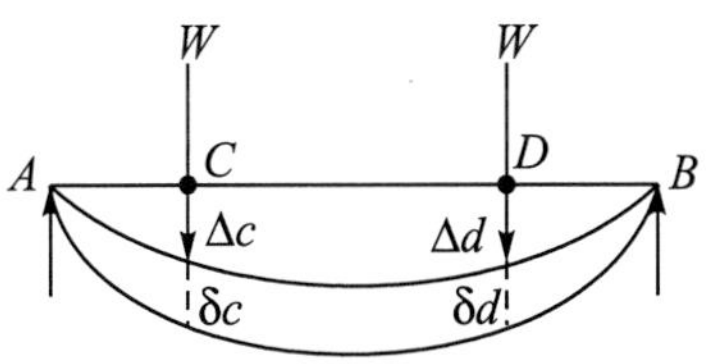

Fig. 7.3.

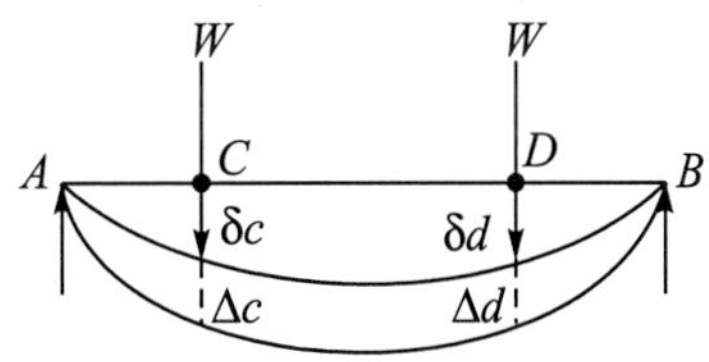

Fig. 7.4.

Let the structure be first loaded as shown in Fig. 7.4, with the load '*W*' at *D*.

For this condition the work done $= \frac{1}{2}\, W\delta d$. As the structure is loaded with load '*W*' at *D*, let an equal load '*W*' be applied at *C*. Further deflections of Δ_c and Δ_d will occur at *C* and *D* respectively Fig. 7.4.

$$\therefore \quad \text{Total work done at this stage} = \frac{1}{2}\, W\delta_d + \frac{1}{2}\, W\Delta_c + W\Delta_d$$

Equating the two expressions obtained for the total work done, when both the loads are present on the structure, we have

$$\frac{1}{2}W\Delta_c + \frac{1}{2}W\delta_d + W\delta_c = \frac{1}{2}W\delta_d + \frac{1}{2}W\Delta_c + W\Delta_d$$

$\therefore \qquad \delta_c = \Delta_d$

i.e., the deflection at '*C*' due to load '*W*' at '*D*'

= Deflection at '*D*' due to same load '*W*' at '*C*'.

7.2. WILLIOT-MOHR'S DIAGRAM

For deflection of Truss-Joints, by Graphical Method, Williot-Mohr's method is adopted.

In the case of a loaded truss, its members are subjected to axial forces (Compressive or Tensile). Hence the members undergo changes in their lengths. Due to these changes in the lengths of the members deflections of the joints take place.

Williot, a French engineer developed a graphical method of determining the deflection of joints of trusses, but this method had limited application and was improved by the German prof. Mohr by introducing corrections to make up for the short commings of the method given by Willot.

Let us first discuss the Williot diagram.

ABC is a triangular frame consisting of members *AC* and *BC*, meeting at joint '*C*'. The members are connected to a wall at '*A*' and '*B*'. There is no members as *AB*. *A* and *B* are fixed points whose positions do not change. Let due to the load carried by structure, the member *AC* be subjected to an increase in length say δl_1.

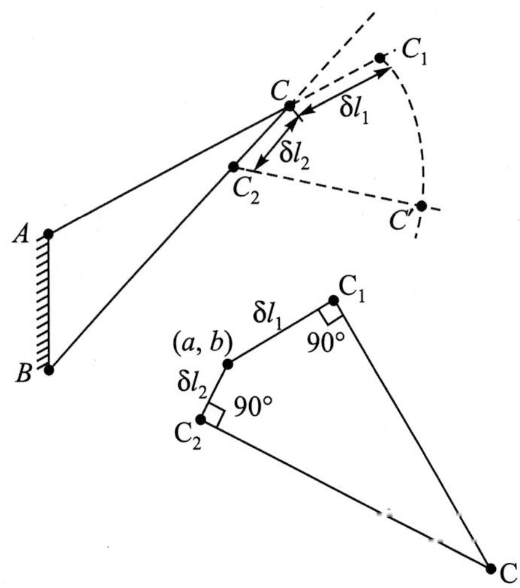

Fig. 7.5

Let δl_2 be the decrease in length of the member *BC*. Let due to these changes in the lengths of the members of the joint '*C*' move to position *C*'.

The position of *C*' can be obtained as follows: Since *AC* is subjected to any extension, produce *AC* to C_1, making $CC_1 = \delta l_1$. Since *BC* is subjected to a contraction, set off CC_2 along *CB* making $CC_2 = \delta l_2$. *A* as centre, AC_1 as radius draw and arc. *B* as centre, BC_2 as radius draw another arc. Let the two arcs intersect at *C*'.

Now C' represents the new position of C. In the actual cases, we come across the deformations δl_1 and δl_2 which are extremely small. Hence, the arcs C_1C' and C_2C' may be replaced by straight lines perpendicular to C_1C and CC_2 respectively. Compared with the length of truss members, the deformation δl_1 and δl_2 are very small and hence the Fig. 7.5. CC_1 $C'C_2$ will be extremely small to scale off the displacement CC'. It is therefore convenient to draw this Fig. 7.5 to a bigger scale so that the displacements CC' may be scaled off easily.

This is drawn as under. Mark a point and designate it as (*a* , *b*). This point represents the vertex '*C*' of the truss. Now set off ac_1 to represent δl_1 to a suitable scale, parallel to *AC*. Set off bc_2 to represent δl_2 in the direction of *CB* (Since this member is compressed). Set off C_1C perpendicular to ac_1 and C_2C perpendicular bc_2. Let these perpendiculars meet at '*C*'. Now the length from (*a, b*) to '*C*' represents the displacement of vertex '*C*'. This Fig. 7.5 drawn in this manner shows again the Fig. 7.5 CC_1 $C'C_2$ to a magnified scale.

The Mohr's diagram has got its own importance, as it gives us the ratio and not the absolute values, of the displacements of joints. The use of Williot-Mohr's diagrams may be best understood from the following example.

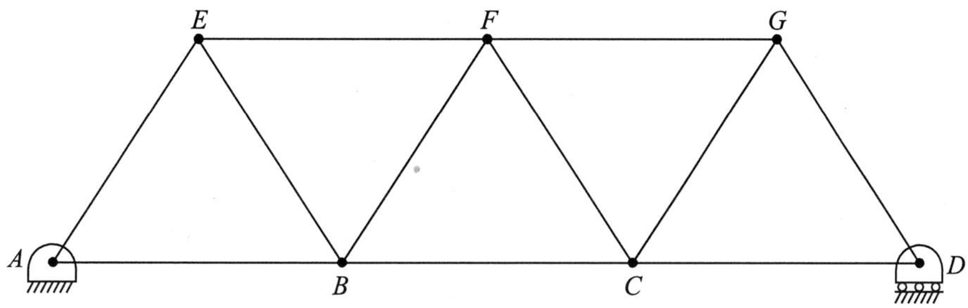

Fig. 7.6

The Fig. 7.6 shown above is a Warren Truss subjected to some external loads. The members *AB, BC, CD, BE* and *CG* are in tension, where as the remaining members are in compression. As a result of tensile force the extension of each member is 2.5 mm and due to compressive force the contraction of each member is 2 mm. Draw the Williot-Mohr's diagram and find out the vertical deflection of joint '*C*'.

Solution: Given: Increase in the length of members due to tensile force $\delta l_1 = 2.5$ mm and decrease in the length of members due to compressive force, $\delta l_2 = 2$ mm.

First draw the Williot diagram as described below:

1. First of all draw the space diagram of the frame as shown in Fig. 7.6.
2. Take some suitable point *a*, to represent the fixed position of '*A*' and assume the direction of the joint '*B*' with respect to joint '*A*' to be fixed. Through '*a*' draw a line parallel to *AB.* Now mark the position of '*b*' such that *ab* is equal to $\delta l_1 = 2.5$ mm to some scale (the direction of '*b*' with respect to '*a*', *i.e.*, whether towards left or right should be decided from the arrow head near the joint *A*) be zero. This discrepancy has crept in, due to the fact, that we assumed the direction of the joint *B* to be fixed with respect to the joint *A*, while drawing the vector diagram. But we know that the direction of the joint *B,* with respect to the joint *A*, is not fixed, but rotates about the joint *A*.

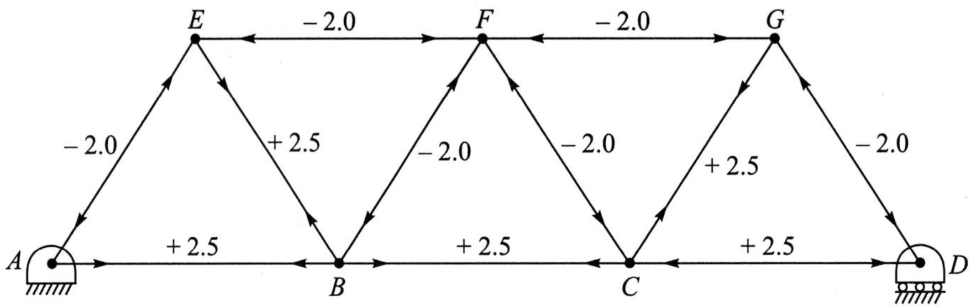

"Space-Diag"

Fig. 7.7

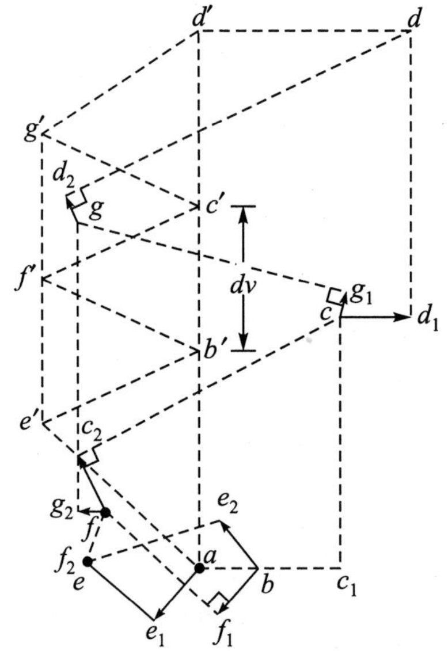

"Williot-Mohr's diag."

Fig. 7.8

Williot-Mohr's Diagram

This discrepancy may be easily corrected, while the help of Mohr's diagram. The whole truss should be given a slight clockwise rotation about the joint *A*, in such a way that the joint '*D*' is restored to the same level as that of joint *A*.

3. Through '*a*' draw ae_1 equal to δl_2 = 2 mm and parallel to *AE* (the direction should be decided from the arrow head near *A*). Similarly, through *b* draw be_2 equal to δl_1 = 2.5 mm and parallel to *BE* (the direction should be decided from the arrow head near *B*).
4. Now through e_1 and e_2, draw e_1e and e_2e at right angles to ae_1 and be_2 respectively meeting at *e*. The point *e* gives the final position of the joint *E* with respect to the fixed joints *A* and *B*.
5. Similarly, locate the position '*f*' of the joint *F* from the joint *B* and *E*.
6. Now locate the position *C* of the joint *C* from joints *B* and *F*.
7. Now locate the position of *g* of the joint *G* from the joints *F* and *C*.
8. Now locate the position of *d* of the joint *D* from the joints *C* and *G*.

From the Williot diagram as drawn, we see that final position of the joint *D* is *d*, with respect to joints *A* and *B*. Thus the vertical distance between *a* and *d*, gives the vertical deflection of the joint *D* with respect to the joint *A*. But from the geometry of the Fig. 7.8, we find that the joint *D* is supported on rollers and both the joints *A* and *D* are at the same level. It is thus obvious, that the vertical deflection of the joint *D* with respect to joint *A* must show that as a result of this rotation, the vertical displacement given to the joint *D* is equal to the height of '*d*' above '*a*', where as the horizontal component of displacement will be given by the horizontal component of '*a*' and '*d*'. It may be noted that the rotation of the truss will proportionately displace all the joints.

In order to draw the Mohr's diagram, draw a st. line *dd'* parallel to *BA*, to meet the vertical line through '*a*' at *d'*. The line *ad'* represents the lower chord *AD* of the truss. Now with *AD* as base, draw a figure *a'b'c'd'e'f'g'* similar to the truss. This figure should be drawn in such a way that after rotating it through 90°, it becomes parallel and similar to the given truss. The figure *a'b'c'd'e'f'g'* is called the Williot diagram and combined diagram is called "Williot-Mohr's" diagram. Now the vertical and horizontal deflection of a joint is given by the respective component of the joint in the Williot diagram and Mohr's diagram (*e.g.*, the vertical and horizontal deflection of a joint *C* is given by the respective components of *C* and *C'*).

By measurement, we find that the vertical deflection of joint *C*, (given by the vertical component of vector *CC'*) "δv" = **10.5 mm.**

❑❑❑

PART II

CHAPTER

8 Strain Energy Concept: Indeterminate Structures

8.1. CASTIGLIANO'S THEOREMS

These theorems are very useful in analysis of indeterminate structures.

8.1.1 First Theorem

"The first theorem of Castigliano states that the partial derivative of the total strain energy in any structure with respect to applied force or moment gives the displacement or rotation respectively at the point of application of the force or moment in the direction of applied force or moment."

Consider a body subjected to forces P_1, P_2 and P_3 as shown in the Fig. 8.1.

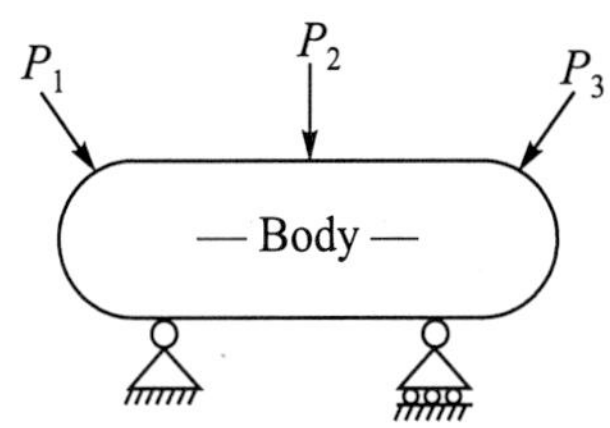

Fig. 8.1

Let the displacements be Δ_1, Δ_2 and Δ_3 in the directions of P_1, P_2 and P_3 respectively. Strain energy stored will be equal to,

$$U = \frac{P_1 \Delta_1}{2} + \frac{P_2 \Delta_2}{2} + \frac{P_3 \Delta_3}{2}$$

or

$$2U = P_1 \Delta_1 + P_2 \Delta_2 + P_3 \Delta_3 \qquad \ldots(8.1)$$

Now let the force P_1 be increased by an amount δP_1. This increment of force will cause additional displacements in the direction of P_1, P_2 and P_3.

These displacements in direction of Δ_1, Δ_2 and Δ_3 will be,

$$\delta\Delta_1 = \frac{\partial \Delta_1}{\partial P_1} \times \delta P_1,$$

$$\delta\Delta_2 = \frac{\partial \Delta_2}{\partial P_1} \times \delta P_1,$$

and

$$\delta\Delta_3 = \frac{\partial \Delta_3}{\partial P_1} \times \delta P_1,$$

The extra energy stored will be,

$$\frac{\partial U}{\partial P_1} \times \delta P_1 = P_1\, \delta\Delta_1 + P_2\, \delta\Delta_2 + P_3\, \delta\Delta_3$$

$$= P_1 \times \frac{\partial \Delta_1}{\partial P_1} \times \delta P_1 + P_2 \times \frac{\partial \Delta_2}{\partial P_1} \times \delta P_1 + P_3 \times \frac{\partial \Delta_3}{\partial P_1} \times \delta P_1$$

$$\therefore \qquad \frac{\partial U}{\partial P_1} = P_1 \frac{\partial \Delta_1}{\partial P_1} + P_2 \frac{\partial \Delta_2}{\partial P_1} + P_3 \frac{\partial \Delta_3}{\partial P_1} \qquad \ldots(8.2)$$

Differentiating eq. (8.1) with respect to P_1,

$$2 \times \frac{\partial U}{\partial P_1} = \Delta_1 + P_1 \frac{\partial \Delta_1}{\partial P_1} + P_2 \frac{\partial \Delta_2}{\partial P_1} + P_3 \frac{\partial \Delta_3}{\partial P_1} \qquad \ldots(8.3)$$

Substituting eq. (8.2) from eq. (8.3), we get,

$$\frac{\partial U}{\partial P_1} = \Delta_1$$

"Thus, the partial derivative of strain energy with respect to P_1 gives displacement in the direction of P_1."

8.1.2. Second Theorem of Castigliano

This is also known as the principle of least work or second theorem of castigliano. If states— "In any and every case of statically indeterminate structure where in an indefinite number of different values of the redundant forces satisfy the conditions of statical equilibrium, their actual values are given by those that render the total strain energy stored to a minimum."

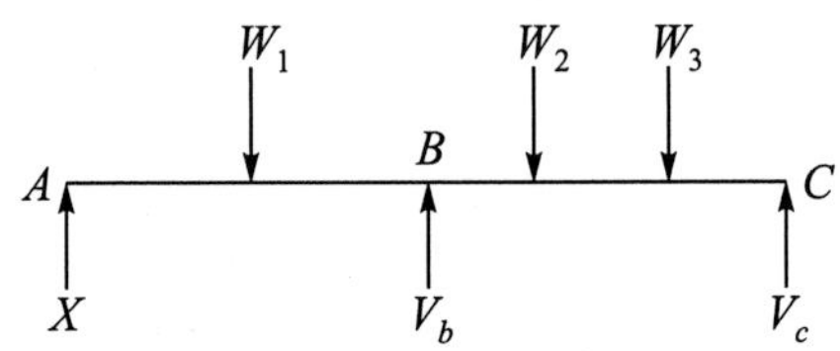

Fig. 8.2 (a)

Figure 8.2 (*a*) shows a contineous beam.

If the reaction at '*A*' is *X*, then the conditions of equilibrium are not sufficient to find the value of *X*. For instance assuming any value of *X*, the reactions V_b and V_c can be determined. Hence, infinite values of *X* with corresponding values of V_b and V_c along with the given loads, satisfy the conditions of equilibrium. Hence our problem is to find out the actual value of *X*. This can be determined from the condition that the deflection at *A* is zero. But by first theorem of castigliano the deflection at *A* is given by $\frac{\partial W_1}{\partial X}$, where W_1 is the strain energy stored by the structure. Hence the value of *X* is given by the condition, $\frac{\partial W_1}{\partial X} = 0$.

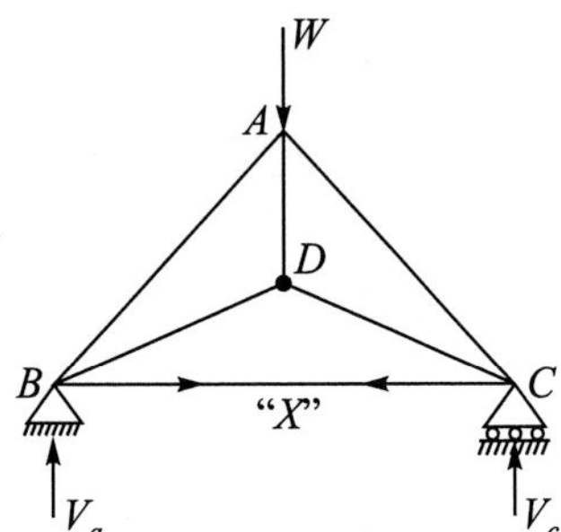

Fig. 8.2 (b)

Similarly, the tension '*X*' in the member *AC* of the redundant frame in Fig. 8.2(*b*) can not be found by using the conditions of equilibrium. But *X* can be determined from the condition that the horizontal movement of *C* with respect to '*A*' is zero *i.e.*, by the condition, $\frac{\partial W_1}{\partial X} = 0$.

Similarly, in Fig. 8.2(*c*), the horizontal thrust '*X*' for the two hinged arch cannot be determined by using the conditions of equilibrium. But '*X*' can be determined from the condition that the horizontal movement of end '*B*' with respect to *A*, equal to zero,

i.e., by the condition, $\frac{\partial W_1}{\partial X} = 0$.

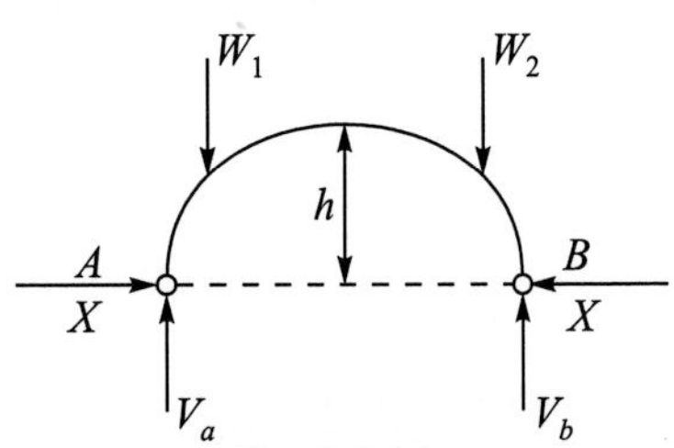

Fig. 8.2 (c)

Hence, if W_i be the strain energy stored by a redundant structure which is a function of a redundant quantity X, then the redundant quantity X is given by the condition, $\frac{\partial W_i}{\partial X} = 0$.

The strain energy stored W_i, being a function of redundant quantity X, the condition $\frac{\partial W_i}{\partial X} = 0$, is a condition for maximum or minimum value of the function W_i.

In general, the strain energy stored by a structure subjected to bending or axial loading is given by,

$$W_i = \sum \frac{M^2\, ds}{2\,EI} \Bigg| + \sum \frac{S^2\, L}{2\,AE}$$

where, M = $B.\ M.$ at any section and

S = Axial force in any member.

The first part of the above expression will hold good for the case in which energy is stored due to bending. The second part of the above expression will hold good for the case in which energy is stored due to axial loading.

For the condition of maximum or minimum value of W_i, we have $\frac{\partial W_i}{\partial X} = 0$

$$\frac{\partial W_i}{\partial X} = \sum \int \frac{ds}{EI} \times M \times \frac{\partial M}{\partial X} \Bigg| + \sum \frac{L}{AE} \times S \times \frac{\partial S}{\partial X} = 0$$

To test for the condition for maximum or minimum values of W_i, differentiating again,

$$\frac{\partial^2 W_i}{\partial X^2} = \sum \int \frac{ds}{EI}\left[M \frac{\partial^2 M}{\partial X^2} + \left(\frac{\partial M}{\partial X}\right)^2\right] + \sum \frac{L}{AE}\left[S \frac{\partial^2 S}{\partial X^2} + \left(\frac{\partial S}{\partial X}\right)^2\right]$$

The $B.M.$ 'M' at any section, or the force 'S' in any member of the redundant truss, is a linear function of X.

$\frac{\partial M}{\partial X}$ and $\frac{\partial S}{\partial X}$ are constants.

$$\therefore \qquad \frac{\partial^2 M}{\partial X^2} = \frac{\partial^2 S}{\partial X^2} = 0$$

Further, the terms $\left(\frac{\partial M}{\partial X}\right)^2$ and $\left(\frac{\partial S}{\partial X}\right)^2$ are positive.

Hence, $\frac{\partial^2 W_i}{\partial X^2}$ is positive.

i.e., the condition $\frac{\partial W_i}{\partial X} = 0$ is a condition for minimum value of W_i.

Hence of all the infinite values of 'X' satisfying the conditions of equilibrium, the actual value of 'X' is given by that value for which the strain energy stored is a minimum.

Points to Remember while Applying Castigliano's Theorem

1. (*a*) Treat all the loads and couples/moments as variables and carry out the partial differentiation.
   (*b*) Substitute the numerical values of different loads and couples in the above equation.
2. To find out the deflection or rotation at a point of the structure where there is no load or couple acting, then it may be assumed that a dummy load *W* or dummy moment/couple is acting at that point and give a value zero at the end.

i.e., $$x = \left(\frac{\partial U}{\partial W}\right)_{W=0}, \quad \text{and} \quad \phi = \left(\frac{\partial U}{\partial M}\right)_{M=0}$$

(*i*) **Deflection under axial load:**

Strain energy under axial load *W*,

$$U = \int \frac{1}{2}\frac{W^2 dx}{AE}$$

$$\therefore \quad \delta = \frac{\partial U}{\partial W} = \int \frac{W\,dx}{AE}$$

(*ii*) **Deflection under bending:**

Strain energy under bending moment *M*,

$$U = \int \frac{M^2 dx}{2EI}$$

$$\therefore \quad \delta = \frac{\partial U}{\partial W} = \frac{\partial U}{\partial M} \times \frac{\partial M}{\partial W} = \int \frac{M}{EI} dx \times \frac{\partial M}{\partial W} = \int \frac{M}{EI}\frac{\partial M}{\partial W} dx$$

(*iii*) **Deflection under torsion:**

Strain energy under torque *T*,

$$U = \int \frac{T^2 dx}{2CI_p}$$

$$\therefore \quad \delta = \frac{\partial U}{\partial W} = \frac{\partial U}{\partial T} \times \frac{\partial T}{\partial W} = \int \frac{T\,dx}{CI_p}\frac{\partial T}{\partial W} = \int \frac{T}{CI_p}\frac{\partial T}{\partial W} dx$$

(*iv*) **Deflection under shear:**

Strain energy under shear free *F*,

$$U = \int \frac{F^2 dx}{2AC}$$

$$\therefore \quad \delta = \frac{\partial U}{\partial F} = \int \frac{F\,dx}{AC}$$

(*v*) **Deflection under horizontal shear:**

$$U = \left(\frac{1 + \frac{1}{m}}{EI^2}\right) \int F^2 dx \iint \frac{(A\bar{y})^2}{b^2} dA$$

where, F is a function of W.

$$\delta = \frac{\partial U}{\partial W}$$

(*vi*) Rotation under bending:

$$U = \int \frac{M^2}{2EI}$$

$$\phi = \frac{\partial U}{\partial M} = \int \frac{M\,dx}{EI}$$

(*vii*) Rotation under torsion:

$$U = \int \frac{T^2 dx}{2CI_p}$$

$$\theta = \frac{\partial U}{\partial T} = \int \frac{T\,dx}{CI_p}$$

8.1.3. Strain Energy Concept

When a structure is loaded, work is done on the structure. The work done is transformed either partially or completely in the form of potential energy referred as "Strain Energy". If the strain produced remains within elastic limit, the potential energy of strain due to loading can be recovered on unloading of the structure. The strain energy in a structure may be due to direct stresses, shear stresses and bending stresses. In case of beams and frames, the strain energy due to direct shear is very small in comparison to strain energy due to bending stresses, hence is generally neglected while finding the indeterminate reactions and moments in beams and frames. However, in final checks the effect of direct and shear stresses may be taken into account.

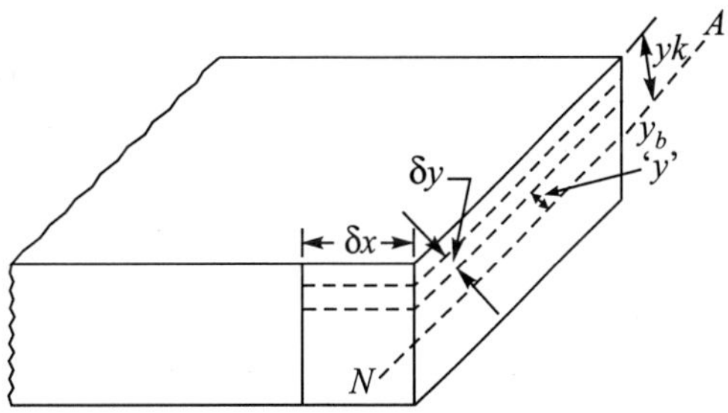

Fig. 8.3

The strain energy due to bending in a member of length "l" is given by.

$$\int U = \boxed{U = \int_0^l \frac{M^2 ds}{2\,EI},}$$

where M is the $B.M.$ at a section, having $M.I.$ "I" and "ds" is the elementary length of the section. Now if 'H' is the redundant horizontal force at any section, then we have,

$$\frac{\partial U}{\partial H} = \int_0^l \frac{M}{EI} \times \left(\frac{\partial M}{\partial H}\right) ds = 0$$

Also, the strain energy due to direct stresses is given by $\int U = \sum \frac{p^2 \delta l}{2\,AE}$, where '$P$' is the direct force acting on length 'δl' and 'A' is the area of X-section.

Strain energy due to shear stresses is given by,

$$U = \int_0^l \int_{=yb}^{yk} \frac{qy^2}{2N} dy \cdot dx$$

where 'qy' is the shear stress at distance 'y' from $N.A.$; 'b' is the breadth of the section and 'N' is the modulus of rigidity.

8.1.4. Problems Based on Strain Energy

Example 8.1: *To find R_A and M_B.*

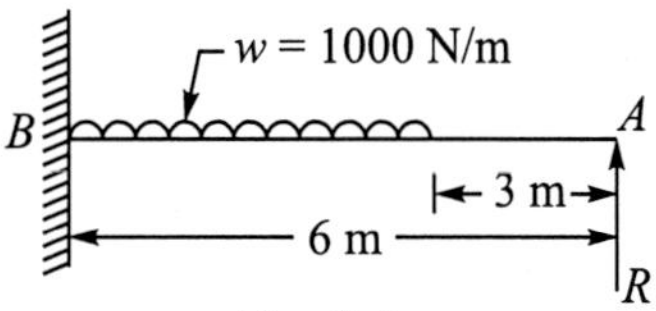

Fig. 8.4

Solution: As the support 'A' does not yield after loading, partial derivative of strain energy with respect to 'R' will be zero.

If 'U' is the strain energy of the beam due to bending,

then,
$$\frac{\partial U}{\partial R} = 0$$

$$U = \int_0^l \frac{M^2 dx}{2EI}$$

$$\therefore \quad \frac{\partial U}{\partial R} = \int_0^l \frac{2M}{2EI} \times \frac{\partial M}{\partial R} dx = \frac{1}{EI} \int_0^l M \cdot \frac{\partial M}{\partial R} dx$$

For the purpose of finding the value of $\frac{\partial U}{\partial R}$, the beam is divided in two portions AC and CB. The equation of $B.M.$, $\frac{\partial M}{\partial R}$ and limits of integration for the two portions are given in tables below:

Portion	M	$\partial M/\partial R$	Limit
AC	$R \times x$	x	0 – 3
BC	$R \times x - \frac{1000\,(x-3)^2}{2}$	x	3 – 6

$$\frac{\partial U}{\partial R} = \frac{1}{EI} \int_0^3 R \times x \times x \cdot dx + \frac{1}{EI} \int_3^6 [Rx - 500\,(x-3)^2] x \cdot dx = 0$$

$$\therefore \quad \frac{1}{EI} \int_3^6 Rx^2\, dx - \frac{1}{EI} \int_3^6 500\,(x^3 - 6x^2 + 9x)\, dx = 0$$

$$\therefore \quad R\left[\frac{x^3}{3}\right]_0^6 - 500\left[\frac{x^4}{4} - \frac{6x^3}{3} + \frac{9x^2}{2}\right]_3^6 = 0$$

$$\therefore \quad 72\,R - 500\left[\left(\frac{6^4 - 3^4}{4}\right) - 2(6^3 - 3^3) + \frac{9}{2}(6^2 - 3^2)\right] = 0$$

$$\therefore \qquad 72\,R - 500\left[\frac{1215}{4} - 2\times189 + 9\times27\right] = 0$$

$$72\,R - 500 \times \frac{184}{4} = 0$$

$$\therefore \qquad R = 328.1 \text{ N} \quad M_{BA} = 328.1 \times 6 - 1000 \times 3 \times \frac{3}{2} = (-)\ 2531.4 \text{ N-m.}$$

Example 8.2: *A continuous beam ABC of uniform section has two equal spans AB and BC each of length "l". During loading support 'B' sinks by δ_1 and support C sinks by δ_2. Find the reactions at supports interms of δ_1 and δ_2 and l and flexual rigidity EI of the beam.*

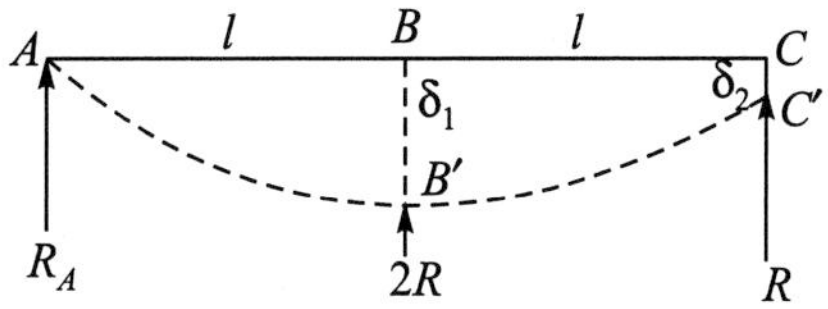

Fig. 8.5

Solution: Let M_B be the moment at B,

$$M^B,\ R_A \times l + M_B = 0$$

(for *AB portion*) $\qquad R_A = -\dfrac{M_B}{l}$

M^B (for *BC portion*) $\qquad R_C = -\dfrac{M_B}{l}$

and $\qquad R_B = -2R$

$\therefore$ For A_B, $\qquad M = R \times x$

Strain energy of beam $\qquad ABC = 2\int_0^l \dfrac{M^2\,dx}{2\,EI} = 2\int_0^l \dfrac{(Rx)^2\,dx}{2\,EI}$

$$= \frac{R^2}{EI}\left[\frac{x^3}{3}\right]_0^l = \frac{R^2 l^3}{3\,EI}$$

Total work done on the

$$\text{Structure} = \text{“}W\text{”} = \frac{R^2 l^3}{3\,EI} - 2R \times \delta_1 + R \times \delta_2$$

For work to be minimum

$$\frac{\partial W}{\partial R} = 0$$

$$\therefore \qquad \frac{2\,Rl^3}{3\,EI} - 2\,\delta_1 + \delta_2 = 0$$

$$\therefore \qquad R = \frac{3\,EI\,(2\,\delta_1 - \delta_2)}{2\,l^3}$$

$$\therefore \qquad R_A = R_C = \frac{3\,EI\,(2\,\delta_1 - \delta_2)}{2\,l^3}$$

$$\therefore \qquad R_B = -2\,R = -\frac{3\,EI\,(2\,\delta_1 - \delta_2)}{l^3}$$

8.1.5. Case for Portal Frames

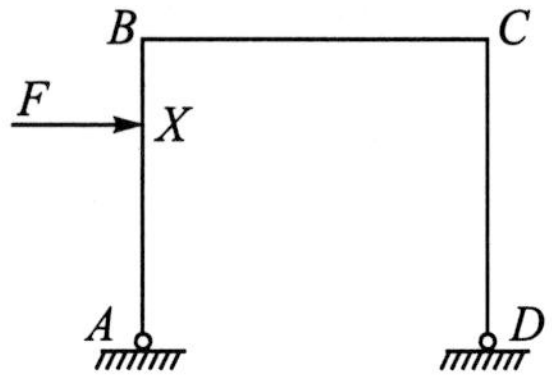

Fig. 8.6

Note: For the Portal frame, *ABCD*, consider the portions *AB*, *BC* and *CD* separately with the proper limits of integration.

For instance, if there is a force '*F*' acting somewhere in between the member *AB* as shown, then treat the member *AB* into two portions '*AX*' and '*XB*' and apply proper limits of integration (Refer Fig. 8.6).

For example,

Steps: 1. Find V_A and V_D by taking moments about one of the two points (two supports) *viz*, M^D,

$$M_D + V_A \times 4 + H \times 3 - 4000 \times 2 - 6 \times 1000 \times 3 = 0$$

$$\therefore \quad \text{"}V_A\text{"} = 6500 - \frac{3}{4}H - \frac{M_D}{4}$$

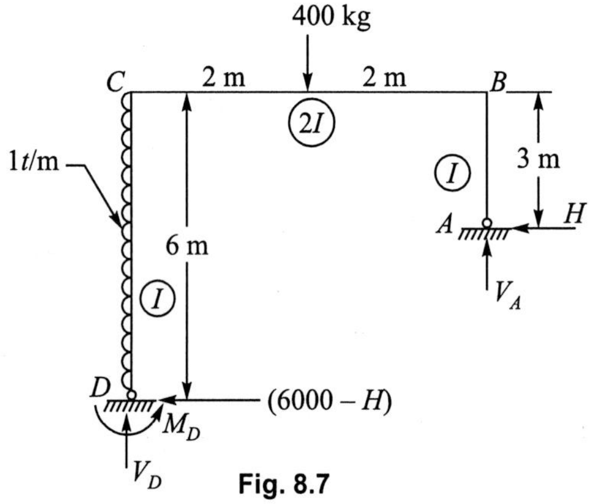

Fig. 8.7

From the table as shown below:

Table 8.1

Origin	Portion	M	$\frac{\partial M}{\partial H}$	$\frac{\partial M}{\partial M_D}$	Limits
A	AB	$H \times y$	y	–	0 – 3
B	BE	$\left(6500 - \frac{3}{4}H - \frac{M_D}{4}\right)x - 3H$	$-\left(\frac{3}{4}x + 3\right)$	$-\frac{x}{4}$	0 – 2
		$= 6500x - \left(\frac{3}{4}x + 3\right)H - \frac{M_D}{4}x$			
B	EC	$6500x - \left(\frac{3}{4}x + 3\right)H$	$-\left(\frac{3}{4}x + 3\right)$	$-\frac{x}{4}$	2 – 4
		$-\frac{M_D}{4}x - 4000 \times (x - 2)$			
D	DC	$(6000 - H)y$ $- M_D - 1000y^2/2$	$-y$	-1	0 – 6

Next find the values of $\frac{\partial U}{\partial H} = 0$

and $\frac{\partial U}{\partial M_D} = 0$, separately, and from these two equations evaluate for H and M_D.

Example 8.3: *Analyse the hinged base portal frame shown in Fig. 8.8 below by strain energy method.*

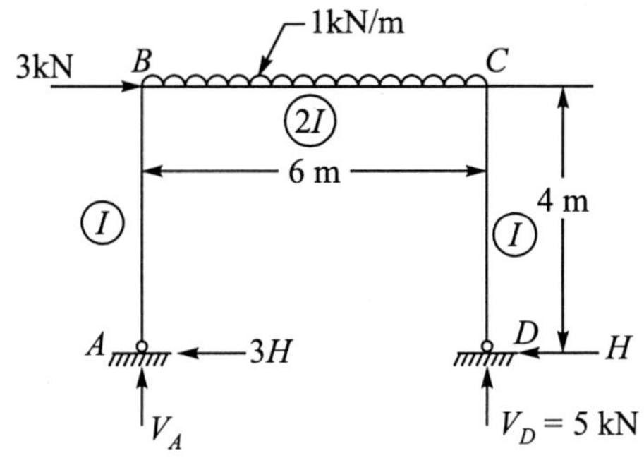

Fig. 8.8

Solution: Let 'H' be the redundant horizontal force at 'D'. Horizontal force at A,

$$H_A = 3 - H$$

Taking moments about D, M^D,

$$V_A \times 6 + 3 \times 4 - 6 \times 1 \times 3 = 0$$

$$\therefore \quad V_A = 3 - 2 = 1 \text{ kN}$$

$$\therefore \quad V_D = 6 \times 1 - 1 = 5 \text{ kN.}$$

As the ends 'A' and 'D' of the frame do not yield, the partial derivatives of strain energy of the whole frame with respect of 'H' will be zero.

For the purpose of finding the value of $\frac{\partial U}{\partial H}$ due to bending, the frame is divided in portions AB, BC and CD.

The equations for *B.M.*, values of $\frac{\partial M}{\partial H}$ and limits of integration are shown in the table below.

Table 8.2

Origin at	Portion	Moment	$\frac{\partial M}{\partial H}$	Limits
A	AB	$(3 - H) \times y$	$-y$	0 – 4
B	BC	$x - H \times 4 - 1 \times \frac{x^2}{2}$	-4	0 – 6
C	DC	$H \times y$	y	0 – 4

Strain energy
$$U = \int \frac{M^2 ds}{2\,EI}$$

$$\frac{\partial U}{\partial H} = \int \frac{M}{EI} \times \frac{\partial M}{\partial H}\, ds = 0$$

$$\frac{\partial U}{\partial H} = \int_0^4 \frac{(3-H)\, y\times(-y)\, dy}{EI} + \int_0^6 \frac{\left[5x-4H-\frac{x^2}{2}\right]\times(-4)\, dx}{E\times 2I} + \int_0^4 \frac{Hy}{EI}\times y\cdot dy = 0$$

$$\int_0^2 \frac{(-3y-Hy^2)}{EI} dy + \int_0^6 \frac{(-20x-16H+2x^2)}{2\,EI} dx + \int_0^4 \frac{Hy^2}{EI} dy = 0$$

$$\left[-\frac{3y^3}{3}+\frac{Hy^3}{3}\right]_0^4 + \frac{1}{2}\left[-10x^2+16\,Hx+\frac{2x^3}{2}\right]_0^6 + H\left[\frac{y^3}{3}\right]_0^4 = 0$$

$$\therefore \quad -64 + \frac{64}{3}H + \frac{1}{2}[-360 + 96\,H + 144] + \frac{64}{3}H = 0$$

$$\therefore \quad -64 + \frac{128}{3}H - 180 + 48\,H + 72 = 0$$

$$\therefore \quad \text{“}H\text{”} = \frac{516}{272} = \mathbf{1.9\ kN.}$$

$$M_{CD} = 1.9 \times 4 = \mathbf{7.6\ kN\text{-}m}$$

$$M_{BA} = (3-1.9)\times 4 = \mathbf{4.4\ kN\text{-}m}$$

8.1.6. Problems on Castigliano's Second Theorem

Example 8.4: *A beam of span 'l' is fixed at one end and simply supported at the other end. It carries a udl of w/per mtr. run over the whole span. Find the reaction at the simply supported end by the Principle of least work.*

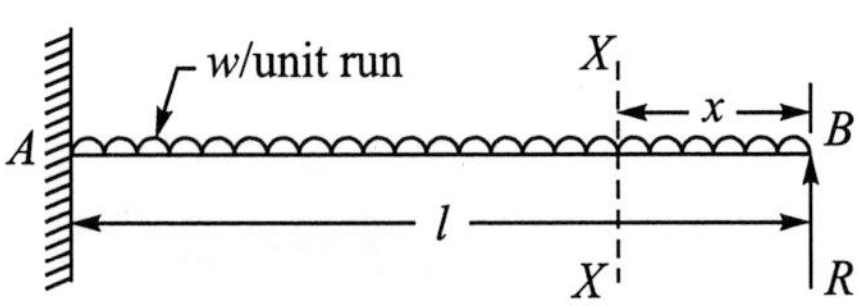

Fig. 8.9

Solution: Let 'R' be the reaction at 'B'.

The *B.M.* at any section *X-X*, distant 'x' from B is given by,

$$M_{XX} = R\times x - wx\times\frac{x}{2}$$

or

$$M_X = Rx - \frac{wx^2}{2}$$

$\therefore$ Strain Energy stored by the beam is

$$W_i = \int \frac{M^2 dx}{2\,EI} = \int_0^l \left[Rx - \frac{wx^2}{2}\right]^2 \frac{dx}{2\,EI}.$$

By second theorem of Castigliano, the reaction 'R' will have a value that strain energy stored is a minimum.

i.e., 'R' is given by the condition $\frac{\partial w_i}{\partial R} = 0$

$$\therefore \quad \int_0^l 2\left[Rx - \frac{wx^2}{2}\right] x \cdot \frac{dx}{2\,EI} = 0$$

$$= \int_0^l \left[2\,Rx \times x \times \frac{dx}{2\,EI} - \frac{2\,wx^2}{2} \times x \times \frac{dx}{2\,EI}\right]$$

or $$\frac{R}{EI}\int_0^l x^2 \cdot dx - \frac{w}{2\,EI}\int_0^l x^3 \cdot dx = 0$$

or $$\frac{Rl^3}{3\,EI} - \frac{wl^4}{8\,EI} = 0.$$

$$\therefore \quad \boxed{R = \frac{3}{8}wl.}$$

Example 8.5: *Find the reaction at the prop for the loaded propped cantilever shown.*

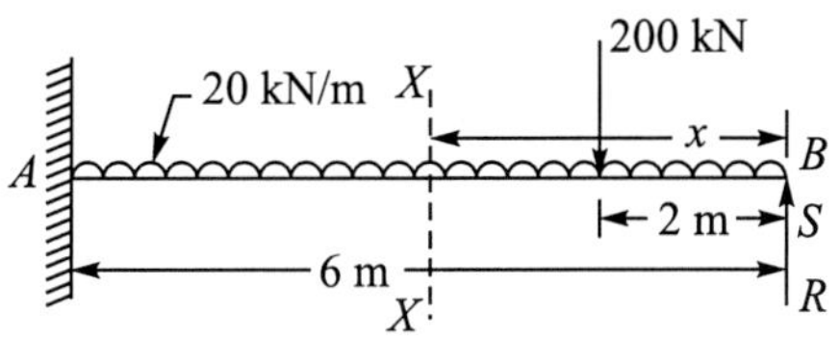

Fig. 8.10

Solution: Let 'R' be the reaction of the prop. at B, $B.M.$ at any section X-X, distant 'x' from B, is given by $Mx = R \times x - w \times \dfrac{x^2}{2} - 200\,(x - 2)$. strain energy stored.

$$W_i = \sum \frac{M^2 dx}{2\,EI}$$

$$\therefore \quad W_i = \int_0^2 (Rx - 10x^2)^2 \frac{dx}{2\,EI} + \int_0^6 [Rx - 10x^2 - 200\,(x-2)]^2 \frac{dx}{2\,EI}$$

The prop. reaction 'R' is given by the condition, $\dfrac{\delta W_i}{\delta R} = 0.$

$$\therefore \quad \frac{\delta W_i}{\delta R} = \int_0^2 2\,(Rx - 10x^2)\frac{x \cdot dx}{2\,EI} + \int_0^6 2\,[Rx - 10x^2 - 200\,(x-2)]\frac{x \cdot dx}{2\,EI} = 0$$

$$\therefore \quad \int_0^2 (Rx - 10x^2)\, x \cdot dx + \int_2^6 [Rx - 10x^2 - 200\,(x-2)]\, x \cdot dx = 0$$

$$\therefore \quad R \times \frac{2^3}{3} - 10 \times \frac{2^4}{16} + \frac{R}{3}(6^3 - 2^3) - \frac{10}{4}(6^4 - 2^4) - \frac{200}{3}(6^3 - 2^3) + \frac{400}{2}(6^2 - 2^2) = 0.$$

or $$\frac{216}{3}R = \frac{32120}{3}$$

$$\therefore \quad \text{"}R\text{"} = 148.7037 \text{ kN.}$$

Example 8.6: *Find the value of reaction "R" and end reactions in case of the loaded beam shown below:*

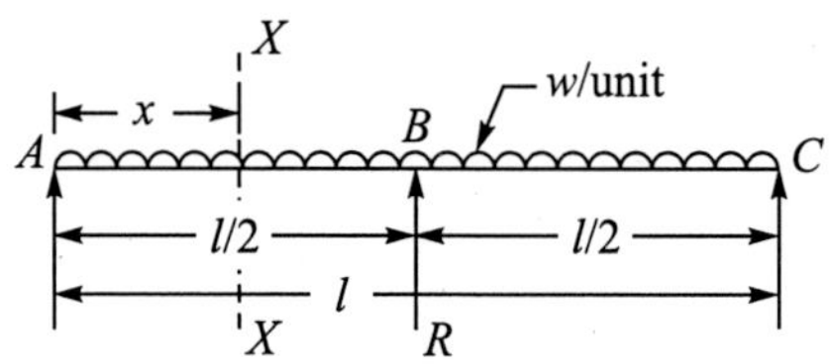

Fig. 8.11

Solution: Let the reaction at 'B' be R.

$\therefore$ Reaction at each end $= \dfrac{wl - R}{2}$

$\therefore$ Total strain energy stored,

$W_i =$ Twice the strain energy stored by AB.

$B.M.$ at any section in AB, distant 'x' from A,

$$= \frac{wl - R}{2}x - \frac{wx^2}{2}$$

$\therefore$ Total strain energy stored $= W_i = \sum \int \dfrac{M^2 dx}{2\,EI}$

$$= 2\int_0^{l/2}\left[\frac{wl-R}{2}x - \frac{wx^2}{2}\right]^2 \frac{dx}{2\,EI} = \frac{1}{EI}\int_0^{l/2}\left[\frac{wl-R}{2}x - \frac{wx^2}{2}\right]^2 dx$$

By principle of least work,
'R' is given by the condition

$$\frac{\partial w_i}{\partial R} = 0$$

$$\frac{\partial w_i}{\partial R} = \frac{1}{EI}\int_0^{l/2} 2\left[\frac{wl-R}{2}x - \frac{wx^2}{2}\right]\left(-\frac{x}{2}\right) dx = 0$$

$$= \frac{1}{EI}\int_0^{l/2}\left[\frac{wx^3}{2} - \frac{wl-R}{2}x^2\right] dx = 0$$

$$= \frac{1}{EI}\left[\frac{1}{2}\cdot\frac{1}{4}\,\frac{wl^4}{16} - \frac{wl+R}{2}\times\frac{1}{3}\times\frac{l^3}{8}\right] = 0$$

$$\therefore \quad \frac{wl^4}{128} - \frac{wl^4}{48} + \frac{Rl^3}{48} = 0$$

$$\therefore \quad \frac{Rl^3}{48} = \frac{5}{384}\,wl^4$$

$$\therefore \quad \boxed{R = \frac{5}{8} wl}$$

$$\therefore \quad \text{Each end reaction} = \frac{wl - \frac{5}{8} wl}{2} = \frac{3}{16} wl.$$

Example 8.7: *A uniform continuous bar ABCD is built in at A and laterally supported at B as shown in the Fig. 8.12. Find the reactive force R at B due to the action of a vertical load W at D. Neglect the effect of direct compression in the vertical portion of the bar. Joint C is stiff.*

Solution: Let the reactive force at '*B*' be '*R*', since the support '*B*' does not yield, we have

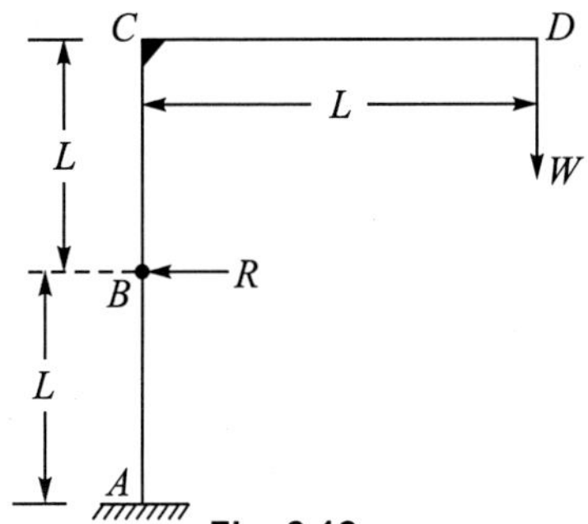

Fig. 8.12

$$\frac{\partial U_{DA}}{\partial R} = 0$$

and
$$\frac{\partial U_{DC}}{\partial R} + \frac{\partial U_{CB}}{\partial R} + \frac{\partial U_{BA}}{\partial R} \quad \ldots(8.9)$$

(*i*) For member *DC*

$M_X = - W \cdot x$, where '*x*' being measured from *D*.

$$\frac{\partial M_X}{\partial R} = 0$$

$$\therefore \quad \frac{\partial U_{DC}}{\partial R} = \frac{1}{EI} \int_0^L M_x \frac{\partial M_x}{\partial R} dx = 0 \quad \ldots(8.10)$$

(*ii*) For member *CB*, measuring '*x*' from *C* towards *B*,

$$M_X = - W \times L$$

$$\frac{\partial M_X}{\partial R} = 0$$

$$\therefore \quad \frac{\partial U_{CB}}{\partial R} = \frac{1}{EI} \int_0^L M_x \frac{\partial M_x}{\partial R} dx = 0 \quad \ldots(8.11)$$

(*iii*) For member *BA*, measuring '*x*' from *B* towards *A*,

$$M_X = - W \times L + R \times x$$

$$\frac{\partial M_X}{\partial R} = (+)\, x$$

$$\therefore \quad \frac{\partial U_{BA}}{\partial R} = \frac{1}{EI} \int_0^L (-WL + R \cdot x)(+x)\, dx$$

$$= \frac{1}{EI} \left[\frac{Rx^3}{3} - \frac{WLx^2}{2} \right]_0^L = \frac{1}{EI} \left(\frac{RL^3}{3} - \frac{WL^3}{2} \right) \quad \ldots(8.12)$$

Substituting the values in eq. (8.9), we get,

$$\frac{\partial U_{DA}}{\partial R} = \frac{1}{EI}\left(\frac{RL^3}{3} - \frac{WL^3}{2}\right) = 0.$$

from which,

$$\boxed{R = \frac{3W}{2}}$$

Example 8.8: *Using Castigliano's second theorem. Analyse the portal frame shown in Fig. 8.13.*

Solution: The frame is statically inderterminate to first degree.

Let the horizontal reaction at '*A*' be '*H*'.

Since there is no external horizontal force acting, the horizontal reaction at *D*, will also be '*H*'. Let us treat '*H*' as the redundant reaction. Since the support '*A*' does not move,

$$\frac{\partial U}{\partial H} = 0 = \frac{\partial U_{AB}}{\partial H} + \frac{\partial U_{BC}}{\partial H} + \frac{\partial U_{DC}}{\partial H} \quad \ldots(8.13)$$

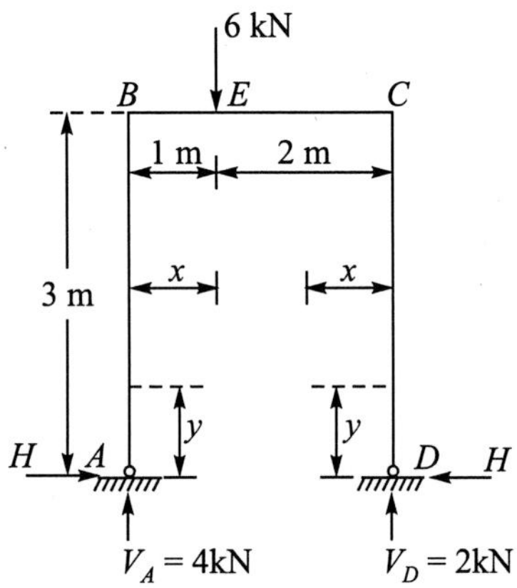

Fig. 8.13

Taking moments about *D*,

$$V_A = \frac{6 \times 2}{3} = 4 \text{ kN}$$

Hence, $$V_D = 6 - 4 = 2 \text{ kN}$$

Tabulating these values as under:

Member	M	$\delta M/\delta H$	Limits
AB	$-H \times y$	$-y$	0 – 3
DC	$-H \times y$	$-y$	0 – 3
BE	$+4x - 3H$	-3	0 – 1
CE	$+2x - 3H$	-3	0 – 2

For *AB*:

$$\frac{\partial U_{AB}}{\partial H} = \frac{1}{EI}\int_0^3 (H \cdot y)\,(y) \cdot dy$$

$$= \frac{H}{3\,EI}(3)^3 = \frac{9\,H}{EI}$$

For *DC*:

$$\frac{\partial U_{DC}}{\partial H} = \frac{1}{EI}\int_0^3 (H \cdot y)\,(y)\,dx$$

$$= \frac{H}{3\,EI}(3)^3 = \frac{9\,H}{EI}$$

For BC:

$$\frac{\partial U_{BC}}{\partial H} = \frac{1}{EI}\left[\int_0^1 (+4x-3H)\,(-3)\,dx + \int_0^2 (+2x-3H)\,(-3)\,dx\right]$$

$$= \frac{1}{EI}\left[(-6+9H)+(-12+18H)\right] = \frac{9}{EI}\,(-2+3H)$$

Substituting the values in eq. (8.13), we have

$$\frac{9H}{EI} + \frac{9H}{EI} + \frac{9}{EI}\,(-2+3+1) = 0$$

∴ "H" = **0.4 kN.**

Hence, $M_A = 0,\ M_B = -3H = (-)\ 0.4 \times 3 = -1.2$ kN-m

$M_C = -3H = -1.2$ kN-m.

∴ $M_D = 0$

Note: For solving strain energy problems, following integration formula are found common in use.

1. $\int_0^l (x)^n\,dx = \int_0^l n\,(x)^{n-1}\,.\,dx$

2. $\int (ax+b)^n \cdot dx = \dfrac{(ax+b)^{n+1}}{(n+1)\times a}$

3. $\int_0^2 x^2 \cdot dx = \left[\dfrac{x^3}{3}\right]_0^2 = \dfrac{(2)^3}{3} - \dfrac{(0)^3}{0} = \dfrac{8}{3}.$

SUMMARY

- When an elastic body is loaded it undergoes deformation *i.e.*, its dimensions change and when it is relieved of the load it regains its original shape. For the time loaded energy is stored in it, the same is given up or released by the loading when the load is removed. This energy is called strain energy.
- The strain energy stored by the body 'within' elastic limit, when loaded externally is called 'Resilience' and the maximum energy which a body stores upto elastic limit is called 'Proof Resilience'.
- In pinjointed structures, where the members are in tension or compression, the energy stored depends on direct forces only. However, in beams and frames having rigid joints, shear stress and bending stress may also occur at any section and the total strain energy stored depends upon the magnitudes of direct forces, shear and moment.
- The concept of elastic strain energy can be very useful in the study of deflections of the various points of structure under load. Instead of directly equating the external work to the internal strain energy, considerable simplification is obtianed by Castigliano's first theorem which states that the deflection caused by any external force is equal to the partial derivative of the strain energy with respect to that force.
- Castigliano extended the principle of least work to the self straining systems. For example, if λ is a small strain or displacement, within the elastic limit, in the direction of the redundant force T, we have

$$\frac{\partial U}{\partial T} = \lambda,$$

This equation is known as Castigliano's second theorem.

SHORT ANSWER QUESTIONS

1. Find the strain energy stored by an elastic prismatic rod of length l, sectional area A, modulous of elasticity E subjected to a tension S.
2. A simply supported beam of span 'l' carries a concentrated load P at distances a and b from the two ends. Find the strain energy stored by the beam.
3. For a uniformly loaded beam AB with built in ends, determine the end moments by using the theorem of castigliano.
4. A cantilever of length L is loaded with udl 'w' per unit length over the whole span. It is propped at the free end. Calculate the prop reaction if: (*a*) the prop is rigid, (*b*) the prop yields by an amount 'λ' under unit load.

EXERCISE FOR PRACTICE

1. Find the fixed end moments for the fixed beam loaded as shown in Fig. 8.14. Use Castigliano's theorem.

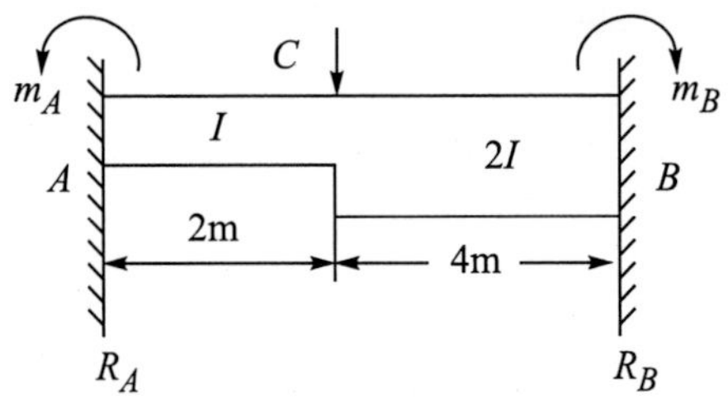

Fig. 8.14.

2. Determine the vertical and horizontal displacement at the free end D in the frame as shown in Fig. 8.15. Take $EI = 12 \times 10^{13}$ N-mm^2. Use Castigliano's theorem.

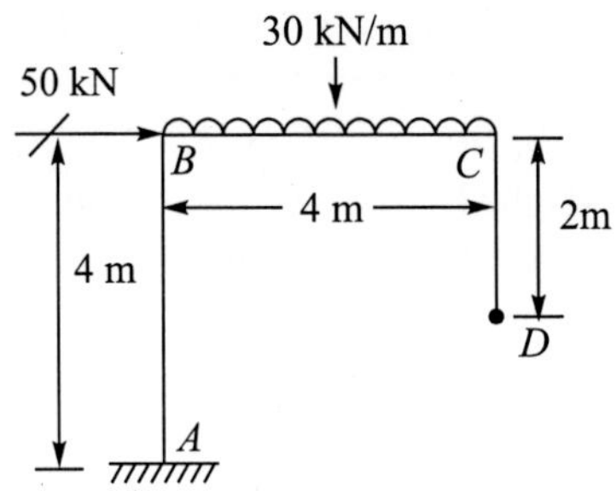

Fig. 8.15.

3. Using Castigliano's theorem, determine the deflection of the free end of the cantilever beam shown in Fig 8.16. Take $EI = 4.9$ mNm^2.

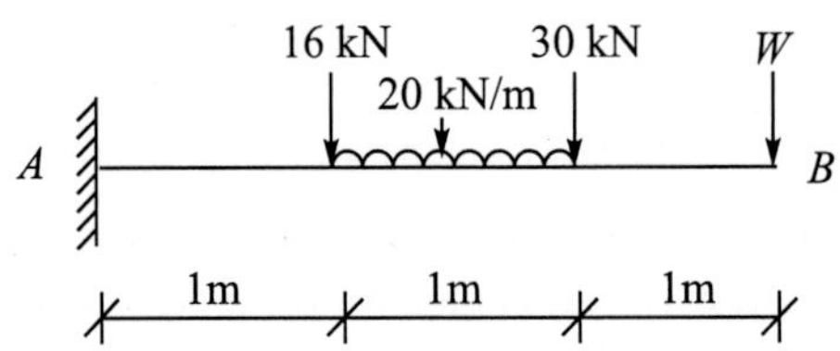

Fig. 8.16.

CHAPTER 9

Slope Deflection Method

9.1. INTRODUCTION

This method was first proposed by Prof. G.A. Maney in 1915, as a general method for the analysis of indeterminate beams and frames. This method has been widely used, till Prof. Hardy Cross introduced moment distribution method in 1930, which is very simple to understand and has scientific and systematic approach.

9.1.1. Assumptions in the Slope Deflection Method

1. All the joints of the frame are rigid, *i.e.*, the angles (q) between the members at the joints do not change, when the frame is loaded.

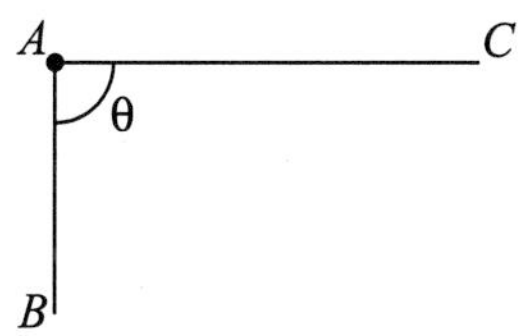

Fig. 9.1 (*a*)

2. Distortions due to axial deformations are neglected. Thus in the frame shown, $BB' = CC' = \Delta$

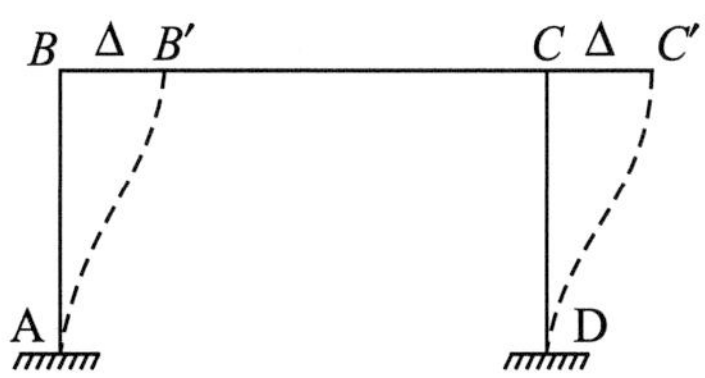

Fig. 9.1 (*b*)

3. Shear deformations are neglected.

9.1.2 Sign Conventions

(*a*) **Moments:** All the clockwise moments at the ends of a member will be taken as positive, where as all the anticlockwise moments at the ends will be taken as negative.

i.e., M_{BA} = (+) 20 kNm and M_{BC} = (–) 20 kNm.

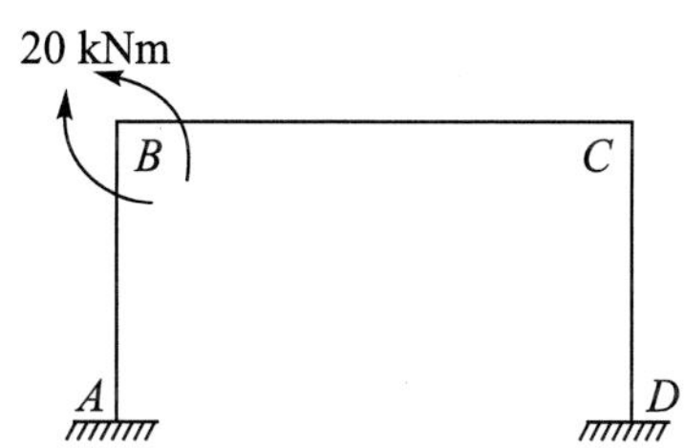

Fig. 9.1 (*c*)

(*b*) **Rotation:** Clockwise rotation of a tangent to elastic curve is positive rotation and anticlockwise rotation is a negative rotation. In the Fig. 9.2, θ_1 and θ_3 are positive rotations and θ_2 is negative rotation.

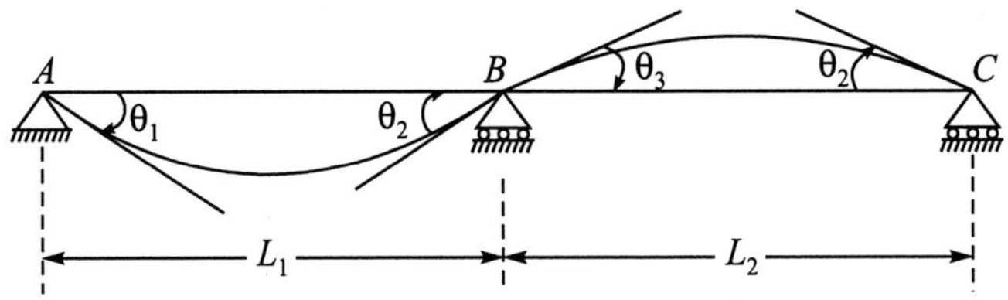

Fig. 9.2

(*c*) **Settlement:** Settlement Δ is positive if right side support is below left side support. Δ is negative if left side support is below the right side support.

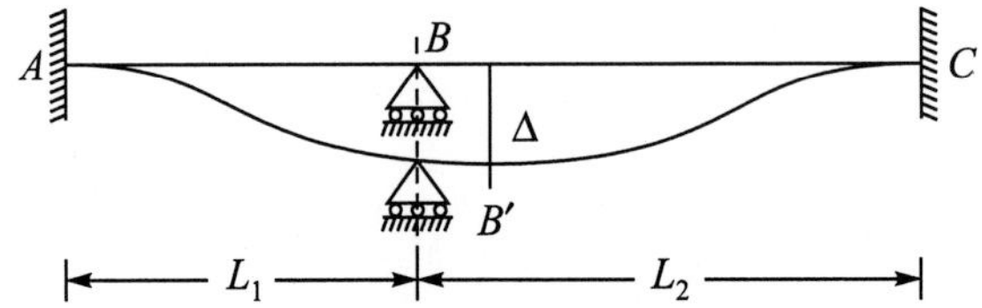

Fig. 9.3

9.1.3. Analysis of Continuous Beams and Frames

Procedure:

1. Treat each span as fixed beam and calculate the fixed end moments.
2. Write down slope deflection equations in terms of end moments, fixing moments, joint rotations and joint translation for each span.
3. Write down equilibrium equations for the individual joints.
4. Substitute the rotations back into the slope deflection equations and solve for the end-moments.

Contineous Beams: Consider the continuous beam *ABCD* as shown.

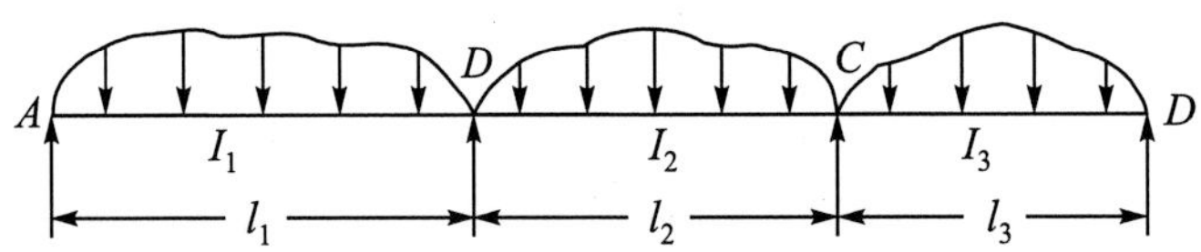

Fig. 9.4

The beam will have rotations θ_A, θ_B, θ_C and θ_D at respectively.

Let (+) δ_1, (+) δ_2 and (+) δ_3 be the relative displacements of the end '*B*' with respect to end '*A*', the end '*C*' with respect to end '*B*' and the end '*D*' with respect to the end '*C*', respectively. The equations for moments M_{AB}, M_{BA}, M_{BC}, M_{CB}, M_{CD} and M_{DC} can be written in terms of slope and displacements. If the ends '*A*' and '*D*' are simply supported, the moments at these ends will be zero. The sum of the moments at '*B*', *i.e.*, $M_{BA} + M_{BC}$, and the sum of the moments at '*C*' *i.e.*, $M_{CB} + M_{CD}$ will each be zero since the joints '*B*' and '*C*' are in equilibrium. The free body diagram of the beam *ABCD* is shown below:

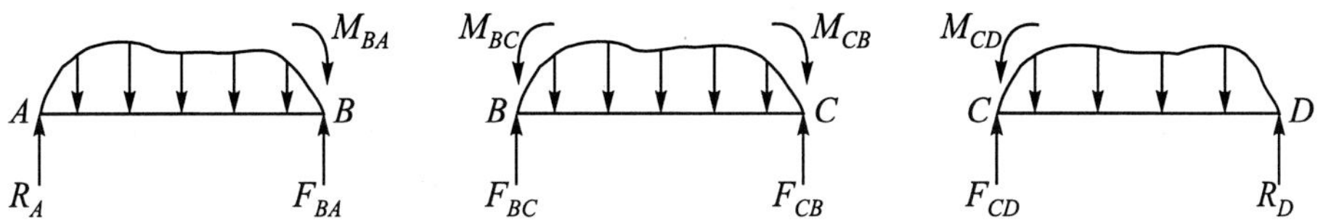

$M_{AB} = 0$
$M_{DC} = 0$
$M_{BA} + M_{BC} = 0$
$M_{CD} + M_{CB} = 0$

With the help of these four equations, θ_A, θ_B, θ_C and θ_D can be solved and moments determined.

In the above continuous beam, if the end '*A*' is fixed, the rotation $\theta_A = 0$ and there will be only three unknow θ_B, θ_C and θ_D which can be evaluated from the equations,

$$M_{DC} = 0$$
$$M_{BA} + M_{BC} = 0$$
$$M_{CB} + M_{CD} = 0$$

Also incase the ends '*A*' and '*D*' are fixed,

$\theta_A = 0$, $\theta_D = 0$ and there will be

only two unknowns θ_B and θ_C. These can be evaluated from the equations,

$$M_{BA} + M_{BC} = 0$$

and $$M_{CB} + M_{CD} = 0$$

9.1.4. Analysis with Different End Conditions

Case I: Beam with Fixed Ends

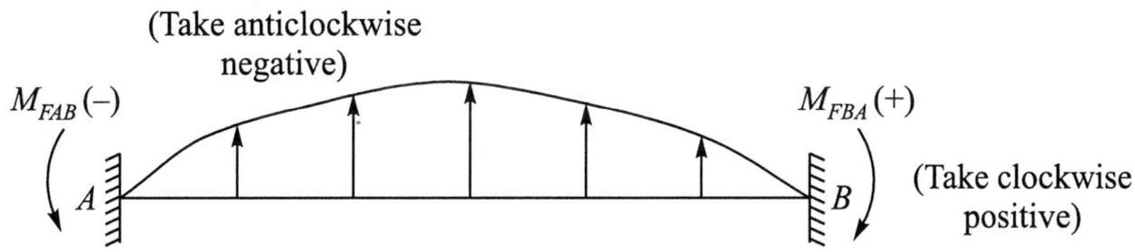

Fig. 9.5

Case II: Beam with end '*B*' fixed and End '*A*' is rotated clockwise

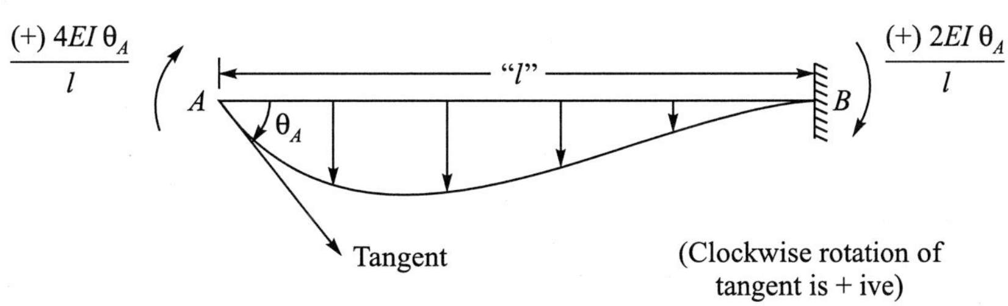

Fig. 9.6

Note: If the moment at the free end is 4 M, take moment at fixed end = $\frac{1}{2}$ of 4 M

Case III: End 'A' is fixed and the end 'B' is rotated clockwise.

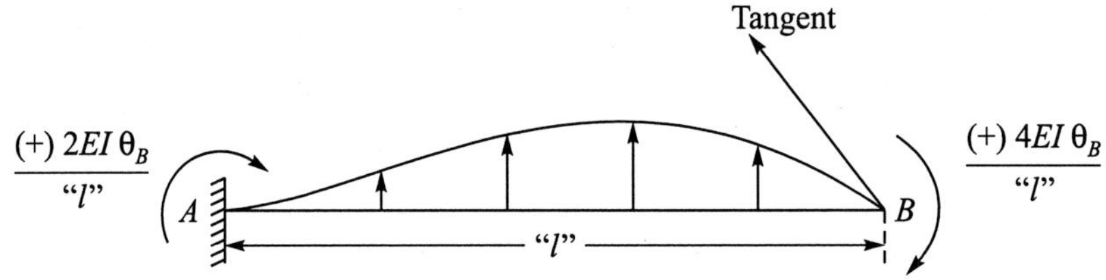

Fig. 9.7

Note: Both the moments applied must be in the clockwise direction.

Case IV: Case of sway (Sinking of support)

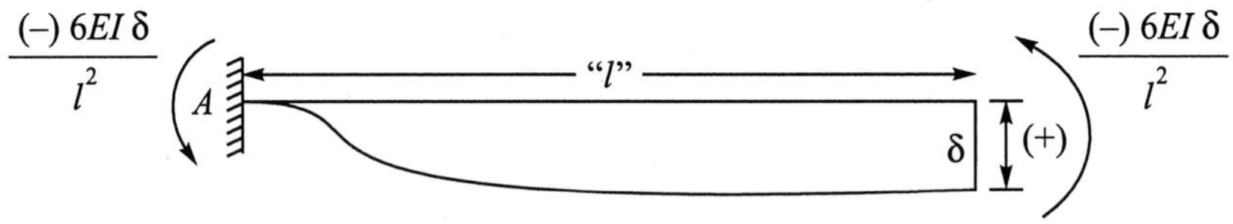

Fig. 9.8

"δ" is positive, it the right hand support sinks down with respect to left.

9.1.5. Formulation of Slope Deflection Equations

From all the four cases above, **algebraically add all values of moments separately for support 'A' and 'B'**

$$M_{AB} = \left\{\frac{4\,EI\,\theta_A}{l} + \frac{2\,EI\,\theta_B}{l} - \frac{6\,EI\,\delta}{l^2}\right\} + (-\,M_{FAB})$$

$$\therefore \quad M_{AB} = \frac{2\,EI}{l}\left\{2\,\theta_A + \theta_B - \frac{3\,\delta}{l}\right\} + (-\,M_{FAB}) \quad \ldots(9.1)$$

Similarly

$$M_{BA} = \left\{\frac{2\,EI\,\theta_A}{l} + \frac{4\,EI\,\theta_B}{l} - \frac{6\,EI\,\delta}{l^2}\right\} + M_{FBA}$$

$$\therefore \quad M_{BA} = \frac{2\,EI}{l}\left\{\theta_A + 2\,\theta_B - \frac{3\,\delta}{l}\right\} + M_{FBA}. \quad \ldots(9.2)$$

Note: $\frac{I}{l}$ = "k" (Stiffness Factor)

9.1.6. Some Important Values of Fixed End Moments for Continuous Beams

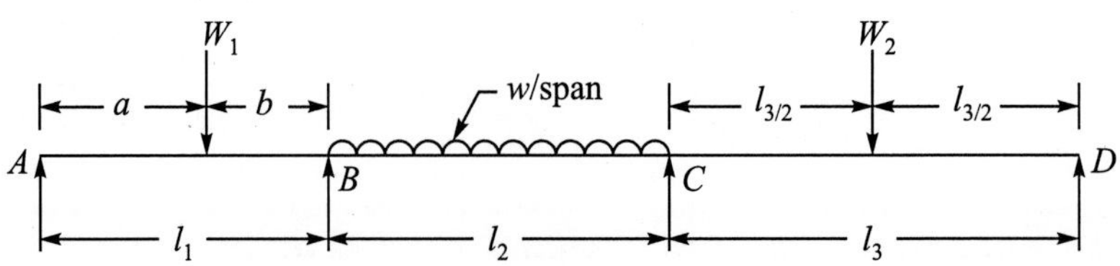

Fig. 9.9

$$M_{FAB} = (-)\,\frac{W_1\,ab^2}{l_1^2},\; M_{FBA} = (+)\,\frac{W_1\,ba^2}{l_1^2}$$

$$M_{FBC} = (-)\,\frac{wl_2^2}{12},\; M_{FCB} = (+)\,\frac{wl_2^2}{12}$$

$$M_{FCD} = (-)\,\frac{W_2\,l_3}{8},\; M_{FDC} = (+)\,\frac{W_2\,l_3}{8}$$

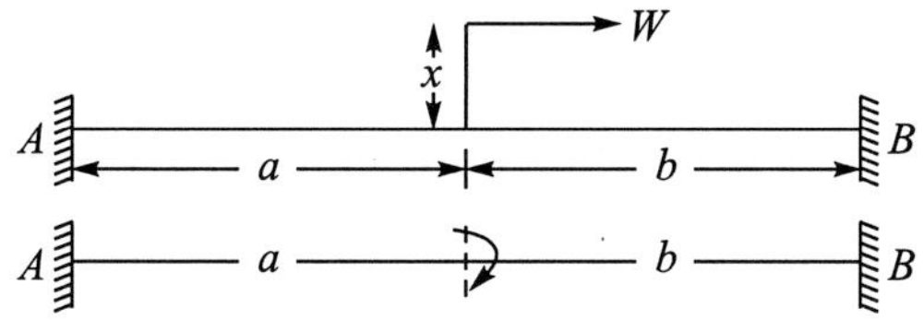

Fig. 9.10

$$M_{FAB} = (-)\,\frac{M_b}{l^2}\,(b - 2a) \text{ and } M_{FBA} = (-)\,\frac{M_a}{l^2}\,(2b - a)$$

9.2. PROBLEMS ON SLOPE DEFLECTION METHOD

Example 9.1: *Find the support moments and draw the B.M. diagram for the continuous beam shown below:*

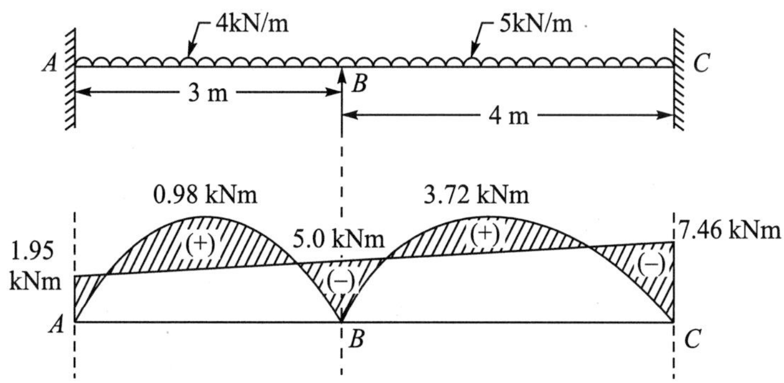

B.M. Diag.

Solutions: The fixed end moments for the spans are as follows:

$$M_{FAB} = (-)\,\frac{4 \times 3^2}{12} = (-)\ 3 \text{ kNm}$$

$$M_{FBA} = (+)\,\frac{4 \times 3^2}{12} = (+)\ 3 \text{ kNm}$$

$$M_{FBC} = (-)\,\frac{5 \times 4^2}{12} = (-)\ 6.67 \text{ kNm}$$

$$M_{FCB} = (+)\,\frac{5 \times 4^2}{12} = (+)\ 6.67 \text{ kNm}$$

We know that end rotations θ_A and θ_C are zero, since the beam is fixed at *A* and *C*. Also ends do not settle down, hence 'δ' for each span in zero. Using slope deflection equations for each span,

Span *AB*

$$M_{AB} = \frac{2\,EI}{l}\left\{2\,\theta_A + \theta_B - \frac{3\,\delta}{l}\right\} + (-M_{FAB})$$

$$= \frac{2\,EI}{l}\{0 + \theta_B - 0\} - M_{FAB}$$

$$= \frac{2\,EI}{l} \times \theta_B - (3) = \frac{2\,EI}{3} \times \theta_B - 3 = -3 + \frac{2\,EI}{3}\,\theta_B \qquad \ldots(i)$$

$$M_{BA} = \frac{2\,EI}{l}\left\{\theta_A + 2\,\theta_B - \frac{3\,\delta}{l}\right\} + M_{FBA}$$

$$= \frac{2\,EI}{l}\{0 + 2\theta_B - 0\} + 3$$

$$= \frac{4\,EI\,\theta_B}{3} + 3 = (+)\,3 + \frac{4\,EI\,\theta_B}{3} \qquad \ldots(ii)$$

Span *BC*

$$M_{BC} = \frac{2\,EI}{l}\{2\theta_B + \theta_C\} - M_{FBC}$$

$$= \frac{2\,EI}{4}\{2\theta_B + 0\} - 6.67 = -\,6.67 + EI_{\;B} \qquad \ldots(iii)$$

$$M_{CB} = \frac{2\,EI}{l}\{2\theta_C + \theta_B\} + M_{FCB}$$

$$= \frac{2\,EI}{l}\{0 + \theta_B\} + 6.67$$

$$= \frac{2\,EI}{4} \times \theta_B + 6.67 = \frac{EI\,\theta_B}{2} + 6.67 \qquad \ldots(iv)$$

For equilibrium condition at *B*, we have

$$M_{BA} + M_{BC} = 0$$

$$3 + \frac{4}{3}\,EI\,\theta_B - 6.67 + EI\,\theta_B = 0$$

or $$\frac{7}{3}\,EI\,\theta_B = 3.67$$

or $$EI\,\theta_B = 1.57286$$

Substituting for $EI\,\theta_B$ in eqs. (*i*), (*ii*), (*iii*) and (*iv*)

$$M_{AB} = (-)\,3 + \frac{2}{3}\,(1.57\,286) = (-)\,1.95 \text{ kNm}$$

$$M_{BA} = (+)\,5.10 \text{ kN}$$

$$M_{BC} = (-)\,5.10 \text{ kN}$$

and $$M_{CB} = (+)\,7.46 \text{ kNm}$$

with the above values of *B.Ms.*, the *B.M.* diagram can be easily drawn.

Example 9.2: *Find the support moments and draw the S.F. and B.M. diagram for the beam shown below:*

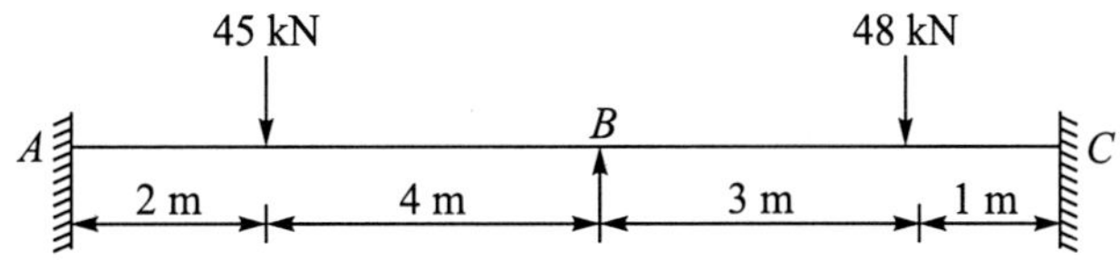

Fig. 9.11

Solution: Fixed end moments

$$M_{FAB} = (-)\ \frac{45 \times 2 \times 4^2}{6^2} = (-)\ 40 \text{ kN-m}$$

$$M_{FBA} = (+)\ \frac{45 \times 2^2 \times 4}{6^2} = (+)\ 20 \text{ kN-m}$$

$$M_{FBC} = (-)\ \frac{48 \times 3 \times 1^2}{4^2} = (-)\ 9 \text{ kN-m}$$

$$M_{FCB} = (+)\ \frac{48 \times 3^2 \times 1}{4^2} = (+)\ 27 \text{ kN-m}$$

Applying slope-deflection equations for each span.

Span *AB*

$$M_{AB} = \frac{2\ EI}{6}(0 + \theta_B) - 40 = (-)\ 40 + \frac{1}{3}\ EI\,\theta_B \qquad \ldots(i)$$

$$M_{BA} = \frac{2\ EI}{6}\ (2\theta_B + 0) = (+)\ 20 + \frac{2}{3}\ EI\,\theta_B \qquad \ldots(ii)$$

Span *BC*

$$M_{BC} = \frac{2\ EI}{4}\ (2\theta_B + 0) - 9 = (-)\ 9 + EI\,\theta_B \qquad \ldots(iii)$$

and

$$M_{CB} = \frac{2\ EI}{4}(0 + \theta_B) + 27 = (+)\ 27 + \frac{1}{2}\ EI\,\theta_B \qquad \ldots(iv)$$

For equilibrium condition at *B*,

$$M_{BA} + M_{BC} = 0$$

$$(+)\ 20 + \frac{2}{3}\ EI\,\theta_B - 9 + EI\,\theta_B = 0$$

$$\frac{5}{3}\ EI\,\theta_B = -11$$

$$\therefore \qquad EI\,\theta_B = -\ 6.60$$

Substituting the value of $EI\,\theta_B$ in all the four equations above,

$$M_{AB} = -\ 40 + \frac{1}{3}\ (-\ 6.60) = (-)\ 42.20 \text{ kN-m}$$

$$M_{BA} = (+)\ 20 + \frac{2}{3}\ (-\ 6.60) = (+)\ 15.60 \text{ kN-m}$$

$$M_{BC} = -9 + (-6.60) = (-)\ 15.60 \text{ kN-m}$$

$$M_{CB} = (+)\ 27 + \frac{1}{2}\ (-6.60) = (+)\ 23.70 \text{ kNm}$$

Calculations for reactions

$$V_a \times 6 - 45 \times 4 - 42.20 = (-)\ 15.60$$

$\therefore$ $V_a = 34.433$ kN ↑

Similarly, $V_c \times 4 - 48 \times 3 - 23.70 =$ (–) 15.60

$\therefore$ $V_c = 38.025$ kN ↑

and $V_B = (45 + 48) - \{34.433 + 38.025\}$

$= 20.542$ kN ↑

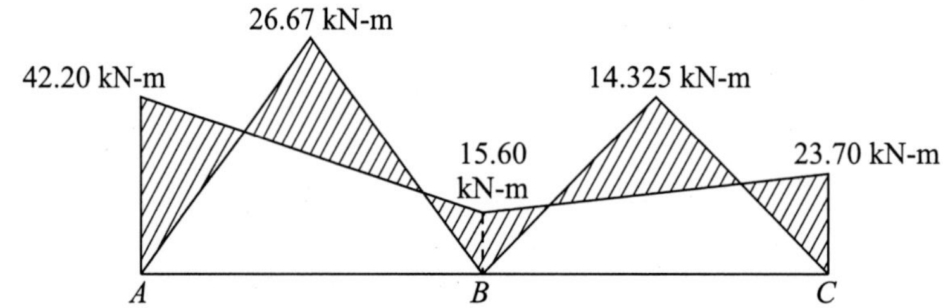

B.M. Diag

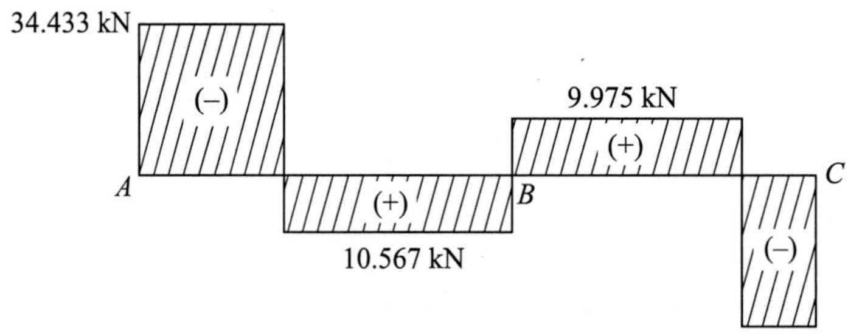

S.F. Diag

Fig. 9.12

Example 9.3: *A continuous beam ABCD 18 m long is loaded as shown in the Fig. 9.13. During loading support 'B' sinks by 1 cm. Find the support moments, $I = 800$ cm^4, $E = 2 \times 10^5$ N/mm^2.*

Solution:

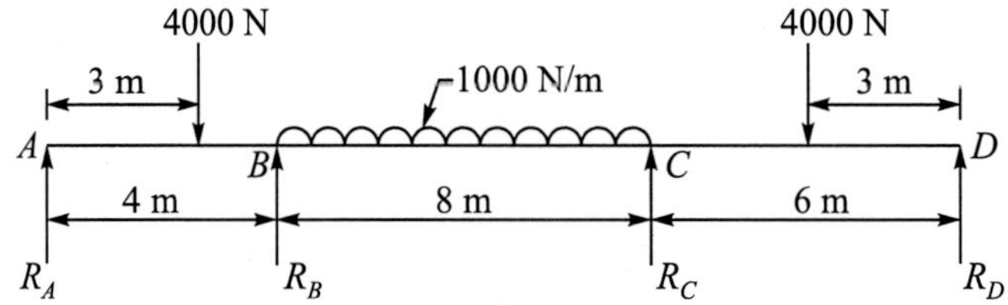

Fig. 9.13

Let θ_A, θ_B, θ_C and θ_D be the rotations at *A, B, C* and *D* respectively.

$$EI = 800 \times \frac{2 \times 10^5}{(100)^4} \times (1000)^2 = 16 \times 10^5 \text{ N-m}^2$$

The fixed end moments are

$$M_{FAB} = (-)\ \frac{W\,ab^2}{l^2} = (-)\ \frac{4000\times 3\times 1\times 1}{(4)^2} = -\ 750 \text{ N-m}$$

$$M_{FBA} = (+)\ \frac{W\,ab^2}{l^2} = (+)\ \frac{4000\times 1\times 3^2}{(4)^2} = (+)\ 2250 \text{ N-m}$$

$$M_{FBC} = (-)\ \frac{Wl^2}{12} = (-)\ \frac{1000\times (8)^2}{12} = -\ 5333 \text{ N-m}$$

$$M_{FCB} = (+)\ \frac{Wl^2}{12} = (+)\ 5333 \text{ N-m}$$

$$M_{FCD} = (-)\ \frac{Wl}{8} = -\frac{4000\times 6}{8} = (-)\ 3000 \text{ N-m}$$

$$M_{FDC} = +\frac{Wl}{8} = +\frac{4000\times 6}{8} = (+)\ 3000 \text{ N-m}$$

Support 'B' sinks by 1 cm = $\frac{1}{100}$ m

Applying slope deflection equations, we have,

$$M_{AB} = \frac{2\,EI}{4}\left(2\,\theta_A + \theta_B - \frac{3\,\delta}{4}\right) - 750$$

$$= \frac{EI}{2}\left(2\theta_A + \theta_B - \frac{3}{400}\right) - 750$$

$$M_{BA} = \frac{2\,EI}{4}\left(2\theta_B + \theta_A - \frac{3\,\delta}{4}\right) + 2250$$

$$= \frac{EI}{2}\left(\theta_A + 2\theta_B - \frac{3}{400}\right) + 2250$$

$$M_{BC} = \frac{2E\times 2I}{8}\left(2\theta_B + \theta_C + \frac{3\,\delta}{8}\right) - 5333$$

$$= \frac{4\,EI}{8}\left(2\theta_B + \theta_C + \frac{3\,\delta}{8}\right) - 5333$$

$$M_{CB} = \frac{2E\times 2I}{8}\left(\theta_B + 2\theta_C + \frac{3\,\delta}{8}\right) + 5333$$

$$= \frac{EI}{2}\left(2\theta_B + 2\theta_C + \frac{3}{800}\right) + 5333$$

$$M_{CD} = \frac{2E\times 1.5I}{6}\left(2\theta_C + \theta_D - 0\right) - 3000$$

$$= \frac{EI}{2}(2\theta_C + \theta_D) - 3000$$

$$M_{DC} = \frac{2E \times 1.5I}{6}(\theta_C + 2\theta_D - 0) + 3000$$

$$= \frac{EI}{2}(\theta_C + 2\theta_D) + 3000$$

The end 'A' is simply supported. Hence, $M_{AB} = 0$

$$\therefore \quad \frac{EI}{2}\left(2\theta_A + \theta_B - \frac{3}{400}\right) - 750 = 0$$

$$\therefore \quad 2\theta_A + \theta_B - \frac{3}{400} = \frac{1500}{EI}$$

$$2\theta_A = \frac{1500}{16 \times 10^5} + \frac{3}{400} - \theta_B$$

$$= \frac{27}{3200} - \theta_B$$

$$\theta_A = \frac{27}{6400} - \frac{\theta_B}{2} \qquad \ldots(i)$$

The end 'D' is also simply supported. Hence, $M_{DC} = 0$

$$\therefore \quad \frac{EI}{2}(\theta_C + 2\theta_D) + 3000 = 0$$

$$\theta_C + 2\theta_D = (-)\frac{6000}{EI}$$

$$\therefore \quad 2\theta_D = (-)\frac{6000}{16 \times 10^5} - \theta_C$$

$$\therefore \quad \theta_D = (-)\frac{3}{1600} - \frac{\theta_C}{2} \qquad \ldots(ii)$$

For equilibrium, the sum of the moments at support 'B' is zero,

$$\therefore \quad M_{BA} + M_{BC} = 0$$

$$\therefore \quad \frac{EI}{2}\left(\theta_A + 2\theta_B - \frac{3}{400}\right) + 2250 + \frac{EI}{2}\left(2\theta_B + \theta_C + \frac{3}{800}\right) - 5333 = 0$$

$$\theta_A + 4\theta_B + \theta_C - \frac{3}{800} = \frac{3083 \times 2}{EI}$$

or $$\theta_A + 4\theta_B + \theta_C = \frac{3}{800} + \frac{3083 \times 2}{16 \times 10^5}$$

or $$\theta_A + 4\theta_B + \theta_C = \frac{6.083}{800}$$

Substituting the value of θ_A from eq. (*i*), above

$$\frac{27}{6400} - \frac{\theta_B}{2} + 4\theta_B + \theta_C = \frac{6.083}{800}$$

or $$\frac{7}{2}\theta_B + \theta_C = \frac{6.083}{800} - \frac{27}{6400}$$

or $$\frac{7}{2}\theta_B + \theta_C = \frac{2.708}{800}$$

or $$7\theta_B + 2\theta_C = \frac{2.708}{400} \quad ...(iii)$$

For equilibrium the sum of the moments at joint '*C*' is zero.

∴ $$M_{CB} + M_{CD} = 0$$

$$\frac{EI}{2}\left(\theta_B + 2\theta_C + \frac{3}{800}\right) + 5333 + \frac{EI}{2}(2\theta_C + \theta_D) - 3000 = 0$$

or $$\theta_B + 4\theta_C + \theta_D = -\frac{3}{800} - \frac{2333\times 2}{EI}$$

or $$\theta_B + 4\theta_C + \theta_D = -\frac{3}{800} - \frac{2333\times 2}{16\times 10^5}$$

or $$\theta_B + 4\theta_C + \theta_D = (-)\frac{5.333}{800}$$

Substituting the value of θ_D eq. (*iii*), we have

$$\theta_B + 4\theta_C - \frac{3}{1600} - \frac{\theta_C}{2} = (-)\frac{5.333}{800}$$

or $$\theta_B + \frac{7}{2}\theta_C = (-)\frac{3.833}{800}$$

or $$2\theta_B + 7\theta_C = (-)\frac{3.833}{400}$$

...(*iv*)

Multiplying eq. (*iv*) by 7, $14\theta_B + 49\theta_C = (-)\frac{26.831}{400}$ ...(*v*)

Multiplying eq. (*iii*) by 2, $14\theta_B + 4\theta_C = (+)\frac{5.461}{400}$ ...(*vi*)

eq. (*v*) – eq. (*vi*), gives,

$$45\theta_C = (-)\frac{32.247}{400}$$

∴ $$\theta_C = (-)\frac{0.7166}{400}$$

$$2\theta_B + 7\theta_C = -\frac{3.833}{400}$$

∴ $$2\theta_B = (-)\frac{3.833}{400} - 7\theta_C = (-)\frac{3.833}{400} + 7\times\frac{0.7166}{400} = \frac{1.1832}{400}$$

$$\therefore \quad \theta_B = \frac{0.5916}{400}$$

$$\therefore \quad \theta_D = (-)\frac{3}{1600} - \frac{\theta_C}{2} = (-)\frac{3}{1600} + \frac{0.7166}{2\times 400}$$

$$\therefore \quad \theta_D = (-)\frac{0.3917}{400}$$

$$\theta_A = \frac{27}{6400} - \frac{\theta_B}{2} = \frac{1.6875}{400} - \frac{0.2958}{400}$$

$$\therefore \quad \theta_A = \frac{1.3917}{400}$$

$$\therefore \quad M_{BA} = \frac{EI}{2}\left[\theta_A + 2\theta_B - \frac{3}{400}\right] + 2250$$

$$= \frac{16\times 10^5}{2}\left[\frac{1.3917}{400} + \frac{0.5916\times 2}{400} - \frac{3}{400}\right] + 225$$

$$= (-)\ 850.2 + 2250 = 1399.8 \text{ N-m}$$

$$M_{CD} = \frac{EI}{2}(2\theta_C + \theta_D) - 3000$$

$$= \frac{16\times 10^5}{2}\left((-)\ 2\times\frac{0.7166}{400} - \frac{0.3917}{400}\right) - 3000$$

$$= (-)\ 3649.8 - 3000 = -\ 6649.8 \text{ N-m.}$$

9.2.1. Analysis of Portal Frames by Slope Deflection Equations

A simple portal frame consists of a beam resting over two columns. The junction of the beams and columns behaves like a rigid joint. In general, the portal frames are of the following two types:

1. Symmetrical portal frames
2. Unsymmetrical frames

A symmetrical portal frame is that, in which both the columns are of the same length, having similar end conditions (Either hinged or Fixed), moment of intertie, modulus of electricity as well as subjected to symmetrical loading.

A little consideration will show, that the joints of such a portal frame will not be subjected to any translation on side sway.

All the single bay and single storey symmetrical portal frames are analysed by opening them out, and treating them exactly like a three span contineous beam.

Example 9.4: *A portal frame ABCD as shown in Fig. 9.14 is loaded with a uniformally distributed load (udl) of 20 kN/m on the horizontal member. The M.I. of member AB = that of CD = I and that of BC = 3I. Find the support reactions and B.M.S in the frame and draw the B.M. diagram.*

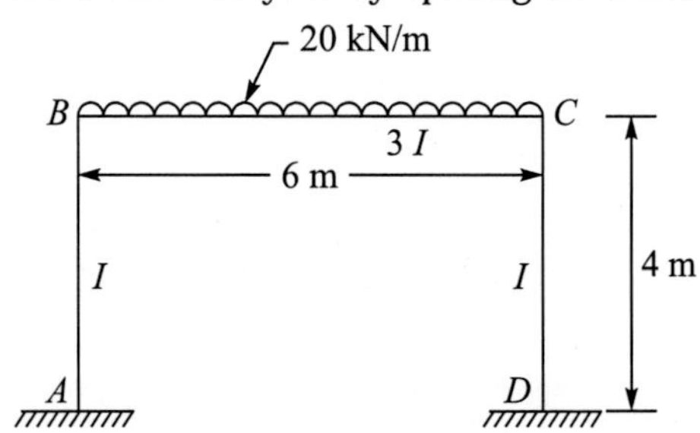

Fig. 9.14

Solution: Support reactions,

Let M_{AB} = Moment at *A*.

M_{BA} = Moment at *B* in span *BA*.

M_{BC} = Moment at *B* in span *BC*.

M_{CB} = Moment at *C* in span *CB*.

M_{CD} = Moment at *C* in span *CD*.

and M_{DC} = Moment at *D*.

Calculation of Fixed End Moments: First of all, let us assume the frame to be made up of fixed beams *AB*, *BC* and *CD*. From the geometry of the figure. We find that the *F.E.M* at *A*,

$$M_{FAB} = M_{BA} = 0 \qquad (\because AB \text{ is not carrying any load})$$

$$M_{FBC} = -\frac{wl^2}{12} = (-)\frac{20 \times 6^2}{12} = (-)\ 60 \text{ kN-m}$$

$$M_{FCB} = (+)\frac{wl^2}{12} = (+)\frac{20 \times 6^2}{12} = (+)\ 60 \text{ kN-m}$$

$$M_{FCD} = M_{FDC} = 0 \qquad (\because CD \text{ is not carrying any load})$$

Slope Deflection Equations: Since the frame is fixed at *A* and *D*, therefore the slopes i_A and i_B will be equal to zero.

We know that the moment at *A* in the span *AB*,

$$M_{AB} = \frac{2\ EI_{AB}}{l}(2i_A + i_B) + M_{FAB} = \frac{2\ EI}{4}(0 + i_B) + 0 \qquad (\because i_A \text{ and } M_{FAB} = 0)$$

$$= \frac{EI \times i_B}{2} \qquad \ldots(i)$$

and

$$M_{BA} = \frac{2\ EI_{BA}}{2}(2i_B + i_A) + M_{FBA}$$

$$= \frac{2\ EI}{l}(2i_B + 0) + 0 \qquad (\because i_A \text{ and } M_{FBA} = 0)$$

$$= EI \times i_B \qquad \ldots(ii)$$

Similarly moment at *B* in span *BC*,

$$M_{BC} = \frac{2E \times I_{BC}}{l}(2i_B + i_C) + M_{FBC}$$

$$= \frac{2E \times 3I}{6}(2i_B + i_C) - 60$$

$$= EI\,(2i_B + i_C) - 60 \qquad \ldots(iii)$$

and

$$M_{CB} = \frac{2E \times I_{CB}}{l}(2i_C + i_B) + M_{FCB}$$

$$= \frac{2E \times 3I}{6}(2i_C + i_B) + 60$$

$$= EI\,(2i_C + i_B) + 60 \qquad \ldots(iv)$$

and moment at C in span CD,

$$M_{CD} = \frac{2E \times I_{CD}}{l}(2i_C + i_D) + M_{FCD}$$

$$= \frac{2\,EI}{4}(2i_C + 0) + 0 \qquad (\because i_D \text{ and } M_{FCD} = 0)$$

$$= EI \times i_C \qquad \ldots(v)$$

and

$$M_{DC} = \frac{2\,EI_{DC}}{l}(2i_D + i_C) + M_{FDC}$$

$$= \frac{2\,EI}{4}(0 + i_C) + 0 \qquad (\because i_D \text{ and } M_{DC} = 0)$$

$$= \frac{EI \times i_C}{2} \qquad \ldots(vi)$$

Equilibrium Conditions: Since the joints B and C are in equilibrium, therefore first equating $M_{BA} + M_{BC}$ equal to zero.

$$EI \times i_B + EI\,(2i_B + i_C) - 60 = 0$$

$$(EI \times i_B) + (2EI \times i_B) + (EI \times i_C) - 60 = 0$$

$$3\,(EI \times i_B) + (EI \times i_C) = 60 \qquad \ldots(vii)$$

Now equating M_{CB} and M_{CD} equal to zero,

$$EI\,(2i_C + i_B) + 60 + (EI \times i_C) = 0$$

$$2\,(EI \times i_C) + (EI \times i_B) + 60 + (EI \times i_C) = 0$$

$$3\,(EI \times i_C) + (EI \times i_B) + 60 = 0 \qquad \ldots(viii)$$

we know that by symmetry, $i_B = i_C$, therefore, substituting these values in eqs. (*vii*) and (*viii*)

$$2\,EI \times i_B = 60$$

$\therefore$ $$2\,EI \times i_B = 30$$

and $$2\,EI \times i_C = -60$$

$\therefore$ $$EI \times i_C = -30$$

Final Moments: Substituting the above values in eq. (*i*) to eq. (*vi*),

$$M_{AB} = \frac{EI \times i_B}{2} = \frac{30}{2} = 15 \text{ kN-m}$$

$$M_{BA} = EI \times i_B = 30 \text{ kN-m}$$

$$M_{BC} = EI\,(2i_B + i_C) - 60$$

$$= 2\,EI \times i_B + EI \times i_C - 60$$

$$= 60 - 30 - 60 = -\,30 \text{ kN-m.}$$

$$M_{CB} = EI\,(2i_C + i_B) + 60$$

$$= (2EI \times i_C) + (EI \times i_B) + 60$$

$$= -\,60 + 30 + 60 = (+)\,30 \text{ kN-m}$$

$$M_{CD} = EI \times i_C = (-)\,30 \text{ kN-m}$$

$$M_{DC} = \frac{EI \times i_C}{2} = -\frac{30}{2} = (-)\,15 \text{ kN-m}$$

Vertical reaction at A, $V_A = V_D = \dfrac{20 \times 6}{2} = 60$ kN.

Horizontal reaction at A, $H_A = \dfrac{15 + 30}{4} = 11.25$ kN ($\rightarrow$)

$11^{ly}\ H_D = 11.25$ kN ($\leftarrow$)

The *B.M.* at the mid of span *BC*, by consider it as a *S.S.*Beam $= \dfrac{wl^2}{8} = \dfrac{20 \times 6^2}{8} = 90$ kN-m

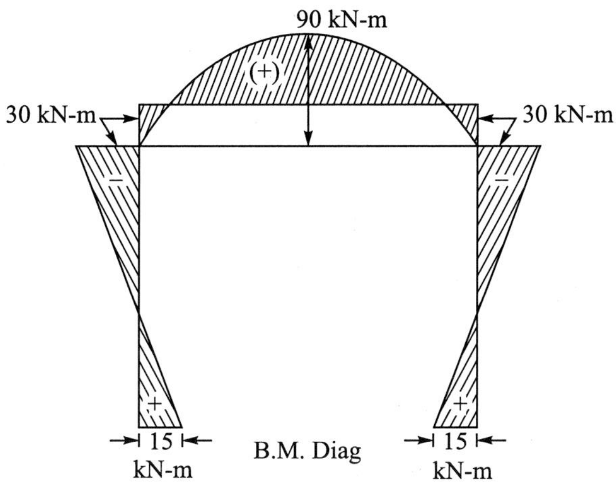

Fig. 9.15

Example 9.5: *Analyse the portal frame shown in the Fig. 9.16 by slope deflection equations.*

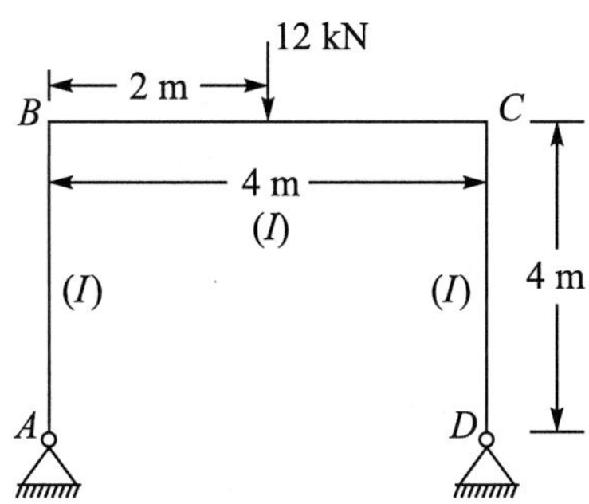

Fig. 9.16

Solution: Calculation of fixed end moments

$$M_{FAB} = M_{FBA} = M_{FCD} = M_{FDC} = 0$$

$$M_{FBC} = (-)\ \frac{12 \times 4}{8} = (-)\ 6 \text{ kN-m}$$

$$M_{FCB} = (+)\ \frac{12 \times 4}{8} = (+)\ 6 \text{ kN-m}$$

Due to symmetry, $\theta_C = -\theta_B$

Span *AB*

Note that '*A*' is hinged, $M_{AB} = 0$

$$\therefore \quad M_{BA} = M_{FBA} - \frac{M_{FBA}}{2} + \frac{3\ EI\ \theta_B}{l}$$

$$= 0 - 0 + \frac{3\ EI\ \theta_B}{4}$$

$$\therefore \quad M_{BA} = \frac{3}{4}\ EI\theta_3 \qquad \ldots(i)$$

Span *BC*

$$M_{BC} = M_{FBC} + \frac{2\ EI}{l}\ (2\theta_B + \theta_C)$$

$$= -6 + \frac{2\ EI}{4}\ (2\theta_B - \theta_B)$$

$$\therefore \quad M_{BC} = -6 + \frac{1}{2}\ EI\ \theta_B$$

For equilibrium condition at *B*,

$$M_{BA} + M_{BC} = 0$$

$$\frac{3}{4}\ EI\ \theta_B - 6 + \frac{1}{2}\ EI\ \theta_B = 0$$

or $\quad \frac{5}{4}\ EI\ \theta_B = 6, \therefore \quad EI\ \theta_B = 4.8$

Substituting for $EI\ \theta_B$, $\quad M_{AB} = 0$

$$\therefore \quad M_{BA} = \frac{3}{4}\ (4.8) = (+)\ 3.60 \text{ kN-m}$$

$$M_{BC} = -6 + \frac{1}{2}\ (4.8) = (-)\ 3.60 \text{ kN-m}$$

Reactions

$$V_A = V_D = \frac{12}{2} = 6 \text{ kN.}$$

$$H_A = 0. + \frac{3.60}{4} = 0.90 \text{ kN } (\rightarrow)$$

By symmetry, $\quad H_D = 0.90 \text{ kN } (\leftarrow)$

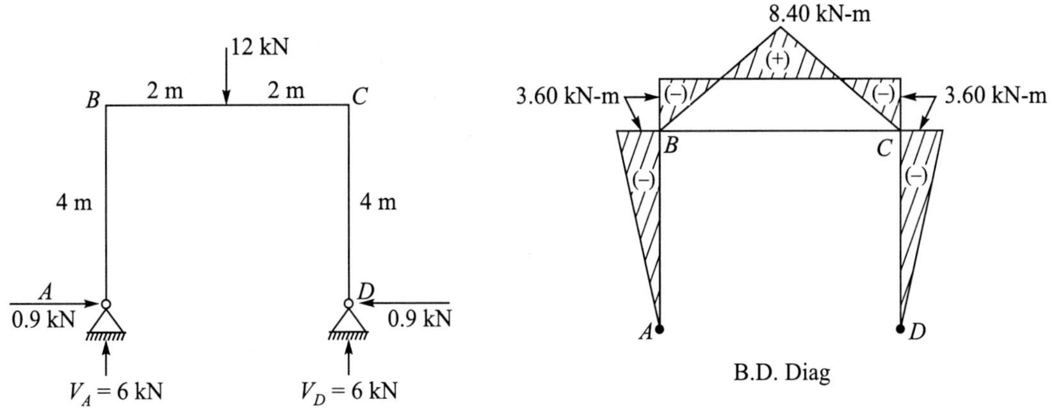

Fig. 9.17

9.2.2 Analysis of Portal Frame with Inclined Legs

The joints '*B*' and '*C*' of the portal frame *ABCD* will deflect at right angles to *AB* and *DC* respectively. Let δ_1 and δ_2 be the displacements of *AB* and *DC* respectively.

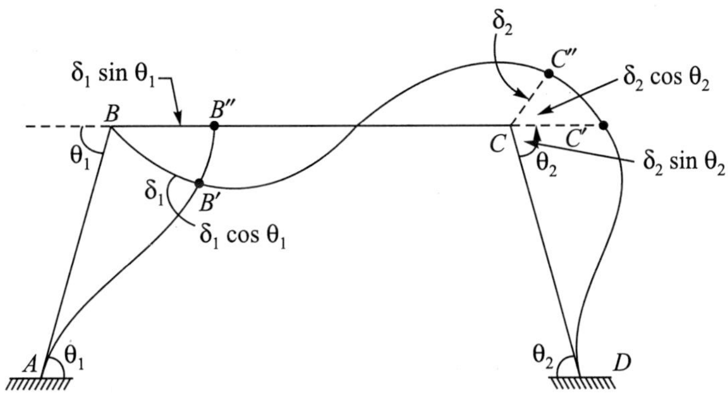

Fig. 9.18

The horizontal component of δ_1 will be equal to horizontal component of δ_2, since the length of member *BC* does not change.

$$\therefore \quad BB'' = CC''$$

or $$\delta_1 \cos (90 - \theta_1) = \delta_2 \cos (90 - \theta_2)$$

or $$\delta_1 \sin \theta_1 = \delta_2 \sin \theta_2$$

Relative displacement of '*C*' with respect to '*B*' will be

$$B'B'' + C'C'' = \delta_1 \sin (90 - \theta_1) + \delta_2 \sin (90 - \theta_2)$$

$$= \delta_1 \cos \theta_1 + \delta_2 \cos \theta_2$$

The additional equation is obtained from *S.F.* considerations.

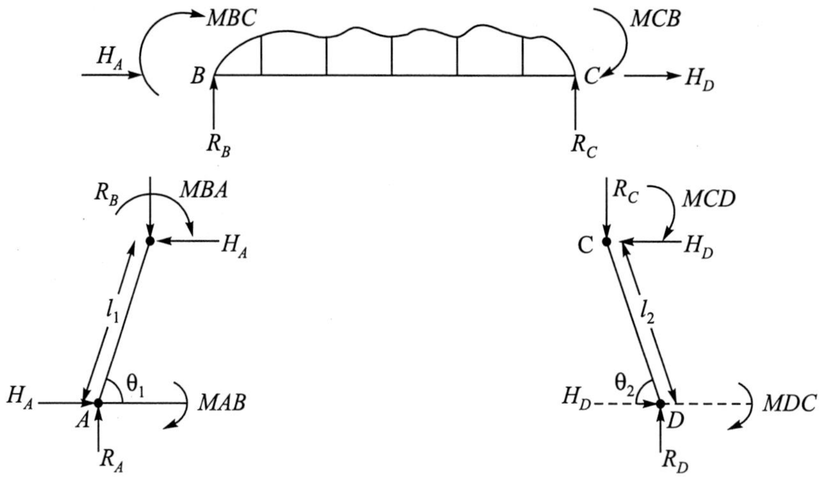

Fig. 9.19

If H_A and H_D are horizontal reactions at '*A*' and '*D*' respectively, $H_A + H_D + P = 0$, where *P* is the horizontal force acting on the frame.

Forces acting towards right will be taken as positive, consider the horizontal member BC_1. The vertical reactions at the joints '*B*' and '*C*' are calculated. From the free body diagrams of the members *AB* and *DC*, the horizontal reactions are calculated.

$$H_A\, l_1 \sin\theta_1 = M_{AB} + M_{BA} - R_A\, l_1 \cos\theta_1$$

$$\therefore \qquad \text{“}H_A\text{”} = \frac{M_{AB} + M_{BA} - R_A\, l_1 \cos\theta_1}{l_1 \sin\theta_1}$$

$$H_D\, l_2 \sin\theta_2 = M_{DC} + M_{CD} - R_D\, l_2 \cos\theta_2$$

$$\therefore \qquad \text{“}H_D\text{”} = \frac{M_{DC} + M_{CD} - R_D\, l_2 \cos\theta_2}{l_2 \sin\theta_2}$$

Example 9.6: *Analyse the portal frame shown in Fig. 9.20 using slope deflection equations.*

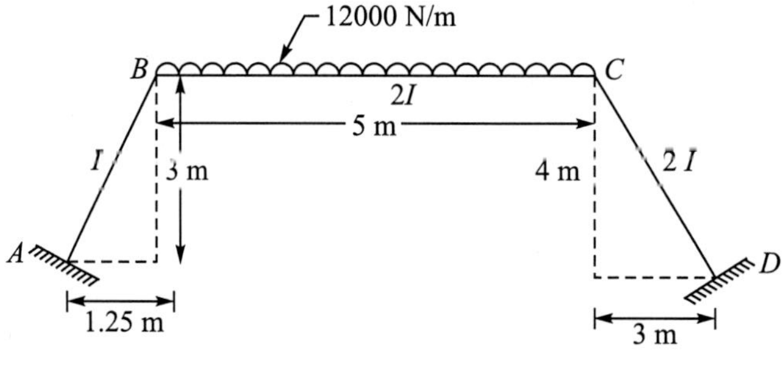

Fig. 9.20

Solution: Let θ_B and θ_C be the slopes at *B* and *C* respectively. Slopes at the fixed supports *A* and *D* will be zero.

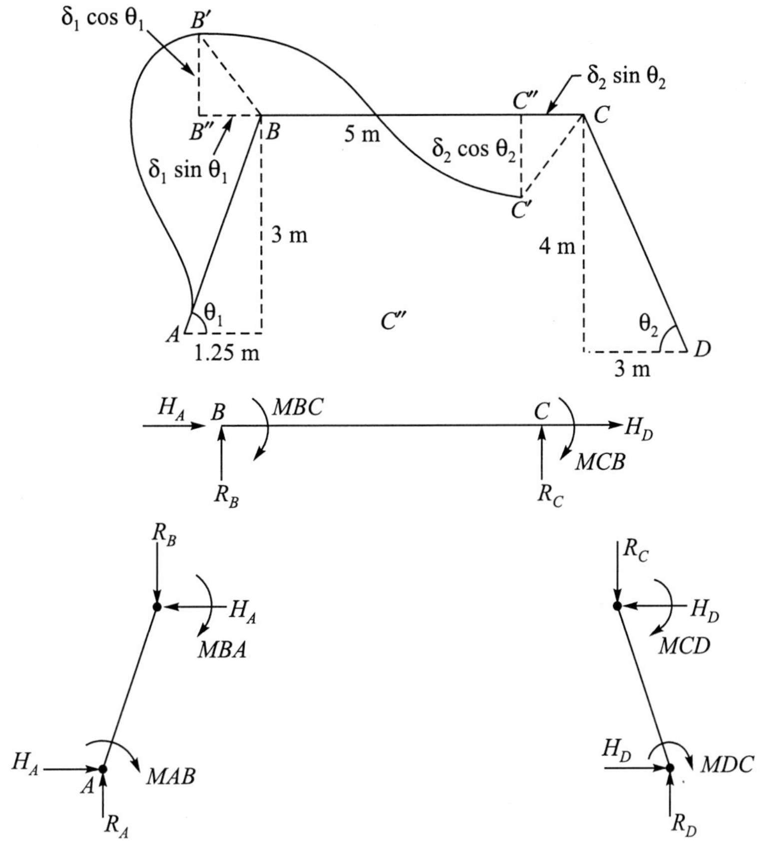

Fig. 9.21

The deflected form of the frame is shown in Fig. 9.21. If is assumed that frame sways towards left so that value of 'δ' in the slope deflection equation for members *AB* and *CD* will be negative and that for member *BC* will be positive from Fig. 9.21.

$$BB' = CC''$$

$$\therefore \quad \delta_1 \cos (90 - \theta_1) = \delta_2 \cos (90 - \theta_2)$$

$$\delta_1 \sin \theta_1 = \delta_2 \sin \theta_2$$

$$\delta_1 \times \frac{3}{3.25} = \delta_2 \times \frac{4}{5}$$

$$\therefore \quad \delta_2 = \frac{3}{3.25} \times \frac{4}{5} \times \delta_1 = \frac{15}{13} \delta_1$$

Relative displacements of the joints '*B*' and '*C*' will be,

$$\delta_1 \cos \theta_1 + \delta_2 \cos \theta_2 = \frac{1.25}{3.25} \delta_1 + \frac{3}{5} \delta_2$$

$$= \frac{5}{13} \delta_1 + \frac{3}{5} \times \frac{15}{13} \delta_1 = \frac{5}{13} \delta_1 + \frac{9}{13} \delta_1 = \frac{14}{13} \delta_1$$

The fixed end moments are:

$$M_{FBC} = (-)\frac{wl^2}{12} = (-)\frac{12000 \times 5 \times 5}{12}$$

$$= (-)\ 25{,}000 \text{ N-m.}$$

$$M_{FCB} = (+)\ 25{,}000 \text{ N-m.}$$

The slope deflection equations interms of unknown are,

$$M_{AB} = \frac{2\,EI}{3.25}\left[\theta_B - \frac{3\,(-\delta_1)}{3.25}\right] = \frac{2\,EI}{3.25}\left(\theta_B + \frac{3\delta_1}{3.25}\right)$$

$$M_{BA} = \frac{2\,EI}{3.25}\left(2\theta_B + \frac{3\delta_1}{3.25}\right)$$

$$M_{BC} = \frac{2E \times 2I}{5}\left(2\theta_B + \theta_C - 3\times\frac{14}{13}\times\frac{\delta_1}{5}\right) - 25{,}000$$

$$= \frac{4\,EI}{5}\left(2\theta_B + \theta_C - \frac{42}{65}\delta_1\right) - 25{,}000$$

$$M_{CB} = \frac{4\,EI}{5}\left(\theta_B + 2\theta_C - \frac{42}{65}\delta_1\right) + 25{,}000$$

$$M_{DC} = \frac{2E \times 2I}{5}\left(\theta_C + 3\times\frac{15}{13}\times\frac{\delta_1}{5}\right)$$

$$= \frac{4\,EI}{5}\left(\theta_C + \frac{9}{13}\delta_1\right)$$

$$M_{CD} = \frac{4\,EI}{5}\left(2\theta_C + \frac{9}{13}\delta_1\right).$$

For equilibrium, sum of moments at joint '*B*' = 0

$$M_{BC} + M_{BA} = 0$$

$$\frac{4\,EI}{5}\left(2\theta_B + \theta_C - \frac{42}{65}\delta_1\right) - 25000 + \frac{2\,EI}{3.25}\left(2\theta_B - \frac{3\delta_1}{3.25}\right) = 0$$

$$\therefore \quad 3.538\,\theta_B + \theta_C + 0.0638\,\delta_1 = \frac{31250}{EI} \qquad \ldots(i)$$

Similarly for equilibrium, sum of moments at joint '*C*' = 0

$$M_{CB} + M_{CD} = 0$$

$$\frac{4\,EI}{5}\left(1\theta_B + 2\theta_C - \frac{42}{65}\delta_1\right) + 25000 + \frac{4\,EI}{5}\left(2\theta_C + \frac{9}{13}\delta_1\right) = 0$$

$$0.25\,\theta_B + \theta_C + 0.0115\,\delta_1 = -\frac{7812.5}{EI} \qquad \ldots(ii)$$

Equation (*i*) – eq. (*ii*), gives,

$$3.28\ \theta_B + 0.0523\ \delta_1 = \frac{3.096205}{EI} \qquad \ldots(iii)$$

Next consider free body diagram of member *BC*,

$$R_C = \frac{12{,}000 \times 5 \times \frac{5}{2}}{5} + \frac{M_{BC} + M_{CB}}{5}$$

$$\therefore \qquad R_C = 30{,}000 + \frac{M_{BC} + M_{CB}}{5} = R_D$$

$$\therefore \qquad R_B = 30{,}000 + \frac{M_{BC} + M_{CB}}{5} = R_A$$

Next consider member *AB*,

Taking moments about *B*,

$$H_A \times 3 = R_A \times 1.25 + M_{AB} + M_{BA}$$

$$\therefore \qquad H_A = \frac{1.25}{3} R_A + \frac{M_{AB} + M_{BA}}{3}$$

Consider member *CD*, taking moments about *C*,

$$4\ H_D = -\ 3\ R_C + M_{CD} + M_{DC}$$

$$\therefore \qquad H_D = -\ \frac{3}{4}\ R_C + \frac{M_{CD} + M_{DC}}{4}$$

But $\qquad H_A + H_D = 0.$

$$\therefore \qquad \frac{1.25}{3}\ R_A + \frac{M_{AB} + M_{BA}}{3} - \frac{3}{4} R_C + \frac{M_{CD} + M_{DC}}{4} = 0$$

$$\frac{1.25}{3}\left(30{,}000 - \frac{M_{BC} + M_{CB}}{5} + \frac{M_{AB} + M_{BA}}{3}\right) - \frac{3}{4}\left(30{,}000 + \frac{M_{BC} + M_{CB}}{5}\right) + \frac{M_{CD} + M_{DC}}{4} = 0$$

$$20\ (M_{AB} + M_{BA}) + 15\ (M_{CD} + M_{DC}) - 14\ (M_{BC} + M_{CB}) = 60{,}0000$$

$$20\left\{\frac{2\ EI}{3.25}\left(3\theta_B + \frac{6\delta_1}{3.25}\right)\right\} + 15\left\{\frac{4\ EI}{5}\left(3\theta_C + \frac{18}{13}\delta_1\right)\right\} - 14\left\{\frac{4\ EI}{5}\left(3\theta_B + 3\theta_C - \frac{84}{65}\delta_1\right)\right\} = 60{,}0000$$

$$1.384\ \theta_B + \theta_C + 22.438\ \delta_1 = \frac{250000}{EI} \qquad \ldots(iv)$$

Equation (*iv*) (–) (*ii*) gives,

$$1.134\ \theta_B + 22.418\ \delta_1 = \frac{25{,}781{,}205}{EI} \qquad \ldots(v)$$

Dividing eq. (*v*) by 22.418,

$$0.0506\ \theta_B + \delta_1 = \frac{11{,}500}{EI} \qquad \ldots(vi)$$

Also dividing eq. (*vi*) by 0.0523.

$$62.75\ \theta_B + \delta_1 = \frac{746893}{EI} \quad \ldots(vii)$$

Equation (*vii*) – eq. (*vi*), gives,

$$62.691\ 4\theta_B = \frac{735393}{EI}$$

$$\therefore \quad \theta_B = \frac{11,730}{EI}$$

$$\delta_1 = \frac{11500}{EI} - 0.0506\ \theta_B = \frac{10,977}{EI}$$

∴ From eq. (*i*),

$$\theta_C = \frac{31250}{EI} - 3.538 \times \frac{11730}{EI} - 0.0638 - \frac{10907}{EI}$$

$$\therefore \quad \theta_C = (-)\ \frac{10,870}{EI}$$

$$\therefore \quad M_{BA} = \frac{2\ EI}{3.25}\left(2\theta_B + \frac{3\delta_1}{3.25}\right)$$

$$= \frac{2}{3.25}\left(23460 + 3 \times \frac{10907}{3.25}\right) = 20.630 \text{ N-m}$$

$$M_{AB} = \frac{2}{3.25}\left(\frac{11730 + 3 \times 10907}{3.25}\right) = 1341.2 \text{ Nm}$$

$$M_{CD} = \frac{4\ EI}{5}\left(-\frac{21740}{EI} + \frac{9}{13} \times \frac{10907}{EI}\right) = (-)\ 11351 \text{ N-m}$$

$$M_{DC} = \frac{4\ EI}{5}\left(-\frac{10876}{EI} + \frac{9}{13} \times \frac{10907}{EI}\right) = (-)\ 2655 \text{ N-m}$$

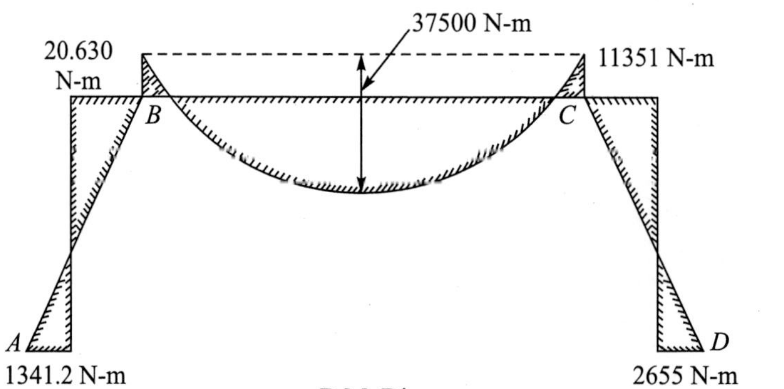

B.M. Diag.

Fig. 9.22

EXERCISE FOR PRACTICE

1. Analyse the continuous beam shown in Fig. 9.23 by slope deflection method and draw the B.M. diagram.

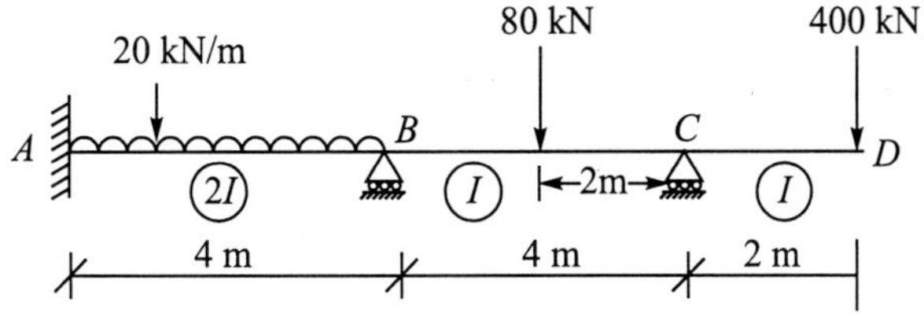

Fig. 9.23.

2. Analyse the continuous beam shown in Fig. 9.24 by slope deflection method, if joint *B* sinks by 10 mm. Given $EI = 4000$ kNm^2. Draw S.F. and B.M. Diagram. **(MDU 2009)**

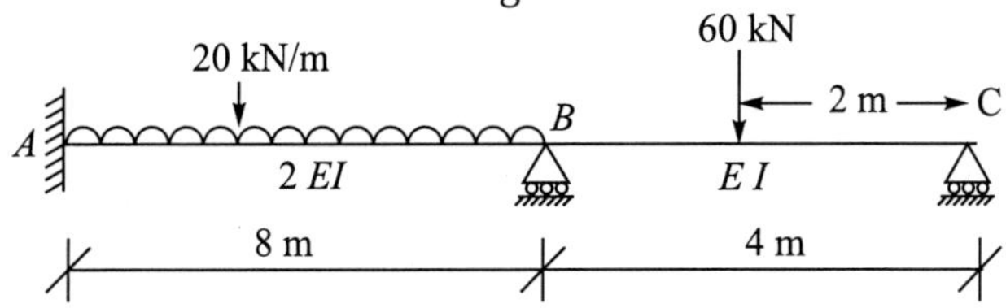

Fig. 9.24.

3. Analyse the frame shown in Fig. 9.25 by slope deflection method and draw B.M. Diagram *EI* is same for all the members.

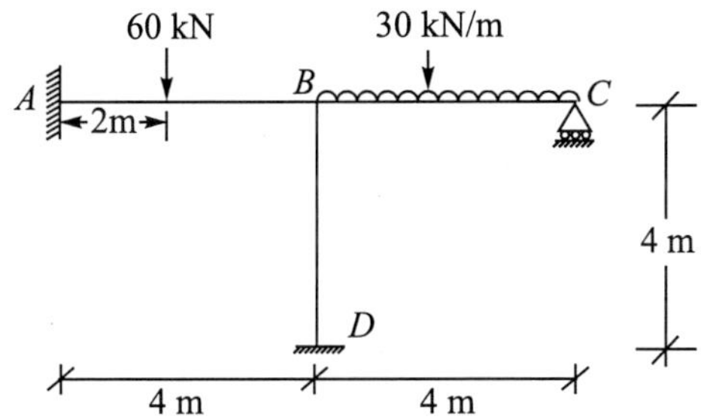

Fig. 9.25.

4. Analyse the portal frame *ABCD* as shown in Fig. 9.26 by slope deflection method. Plot the B.M. diagram.
5. Analyse completely the portal frame shown in Fig. 9.27 by slope deflection method. Draw the B.M. diagram.

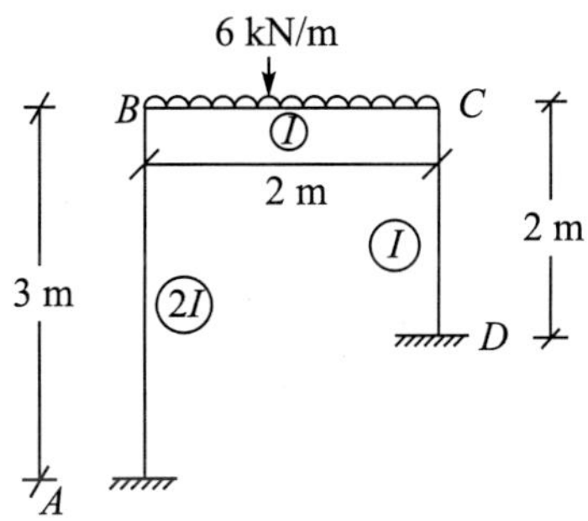

Fig. 9.26.

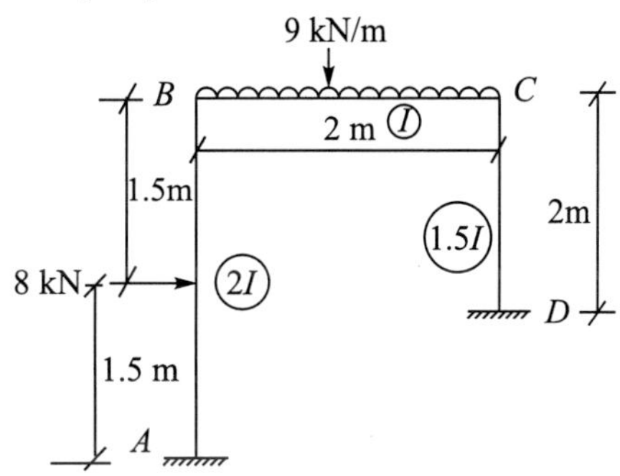

Fig. 9.27.

❑❑❑

CHAPTER

10 Moment Distribution Method

This method used for analysis of all types of indeterminate structures was first introduced by Prof. Hardy Cross in 1930.

In this method, all the members of a structure are first assumed to be fixed in position and direction and Fixed End Moments (*FEMs*) due to external loads are obtained. Next all the hinged joints are released by applying an equal and opposite moment and their effects are evaluated on the opposite joints. The unbalanced moment at a joint is distributed in the two spans in the ratio of their distribution factors (*D.F.*).

Sign Conventions:

1. All the clockwise moments at the ends are taken as "POSITIVE".
2. All the anticlockwise moments at the ends are taken as "NEGATIVE".

Carryover Factor: The ratio of moments produced at a joint to the moment applied at the other joint, without displacing it, is called carry over factor.

[For End Fixed, $C.O.F. = (+)\ \frac{1}{2}$]

Stiffness Factor: It is the moment required to rotate the end while acting on it through a unit angle without translation of the far end.

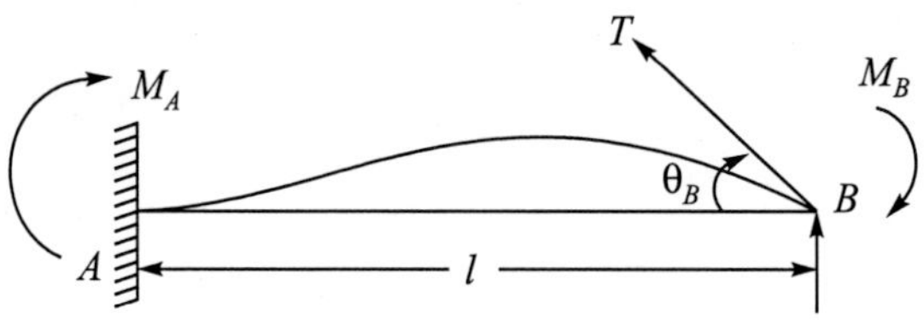

Fig. 10.1

(*i*) In the case of a beam having one end fixed and the other end simply supported, the clockwise moment "μ" = $\frac{4\,EI\,\theta_B}{l}$ (Refer Fig. 10.1)

Put $\theta_B = 1$ (unit angle rotation)

∴ Stiffness factor "*k*" in this case

$$\text{“}k\text{”} = \frac{4\,EI}{l}$$

(*ii*) In the case of a beam having simply supported ends (Hinged), the clockwise moment

$$\text{“}\mu\text{”} = \frac{3\,EI\,\theta_B}{l} \qquad \text{(Refer Fig. 10.2)}$$

Put $\theta_B = 1$ (∵ unit angle rotation)

$\therefore$ Stiffness factor "k" in this case

$$\text{“}k\text{”} = \frac{3\,EI}{l}$$

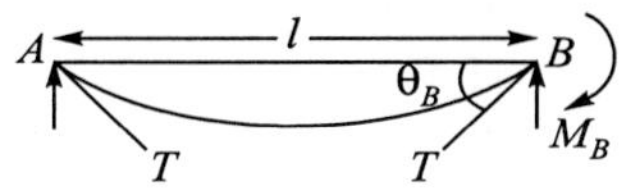

Fig. 10.2

Distribution Factor:

$$D.F. = \frac{\text{Stiffness of member}}{\text{Total stiffness}} = \frac{k_1}{\Sigma k}$$

Proof: Consider four members *OA, OB, OC* and *OD* meeting at '*A*'. Let the members *OA* and *OC* be fixed at '*A*' and '*C*', whereas the members *OB* and *OD* be hinged at '*B*' and '*D*'. Considering length and *EI* being constant for all the members. '*O*' is the rigid joint.

Let 'θ' be the angle through which each member is rotated.

$$\therefore \quad \left.\begin{aligned} M_1 &= k_1\theta \\ M_2 &= k_2\theta \\ M_3 &= k_3\theta \\ M_4 &= k_4\theta \end{aligned}\right\}$$

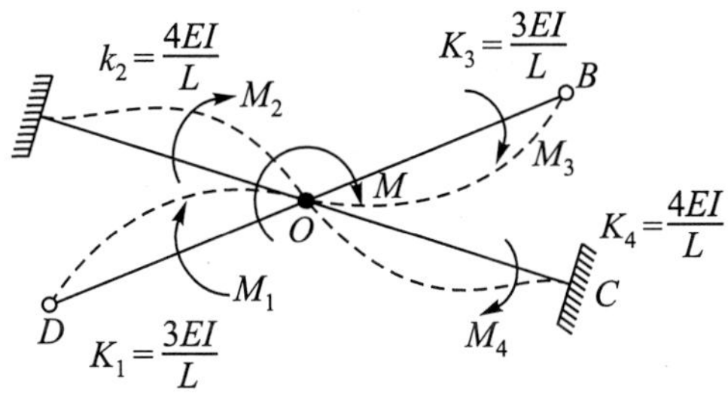

Fig. 10.3

$$\therefore \quad \text{“}M\text{”} = \theta\,(k_1 + k_2 + k_3 + k_4)$$

$$(M_1 + M_2 + M_3 + M_4) = \theta\,\Sigma\,k$$

$$\therefore \quad \theta = \frac{M}{\Sigma k}$$

$$\therefore \quad M_1 = k_1 \times \frac{M}{\Sigma k}$$

$$M_2 = k_2 \times \frac{M}{\Sigma k}$$

$$M_3 = k_3 \times \frac{M}{\Sigma k}$$

and

$$M_4 = k_4 \times \frac{M}{\Sigma k}$$

This $\frac{k\text{ member}}{\Sigma k}$ is called the *D.F.*

Certain General Results to be Remembered in M.D. Analysis

F.E.Ms

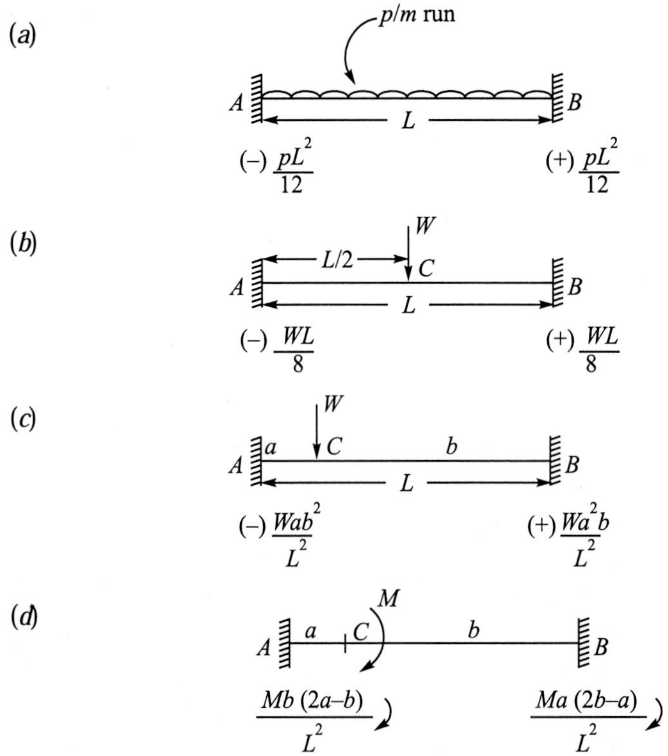

Fig. 10.4

10.1. PROCEDURE FOR SOLVING MOMENT DISTRIBUTION PROBLEMS

1. Find out the *F.E.Ms.*
2. Calculate the distribution factors *D.Fs.*
3. Look into the given Fig. 10.5, simply supported or fixed end type.
4. Tabulate the values and commence moment distribution.

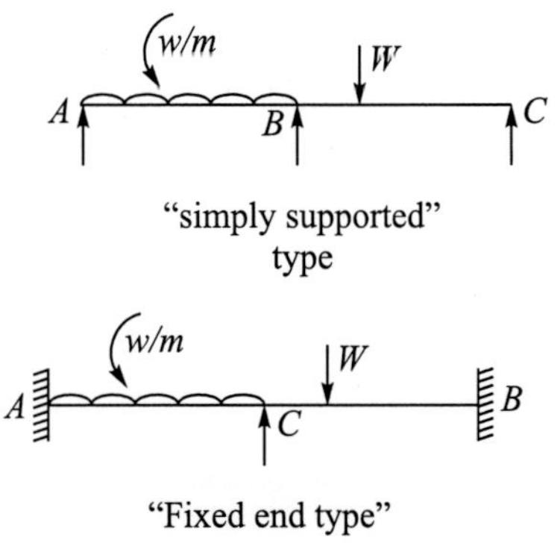

Fig. 10.5

"Table for S.S. type"

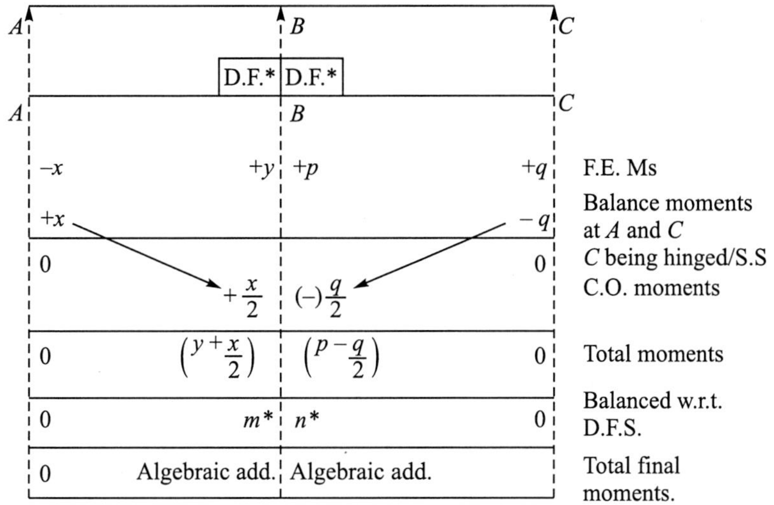

Fig. 10.6

Note: If initial moments to be balanced is having (+) sign, then put (–) sign for '*m*' and '*n*' values.

"Table for fixed end types"

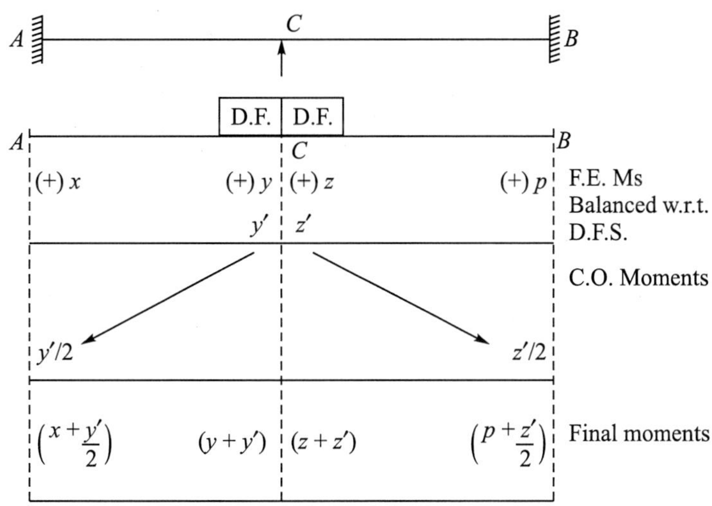

Example 10.1: *Analyse the beam shown by M.D. method and draw the S.F. and B.M. diagrams.*

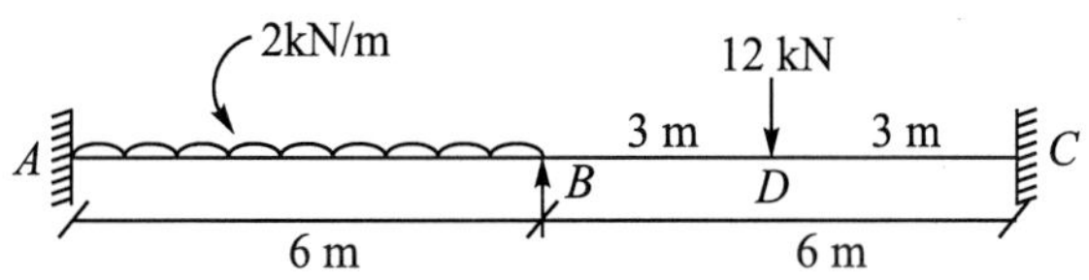

Fig. 10.7

Solution: Let us consider the continuous beam *ABC* to be split into two fixed beams *AB* and *BC*.

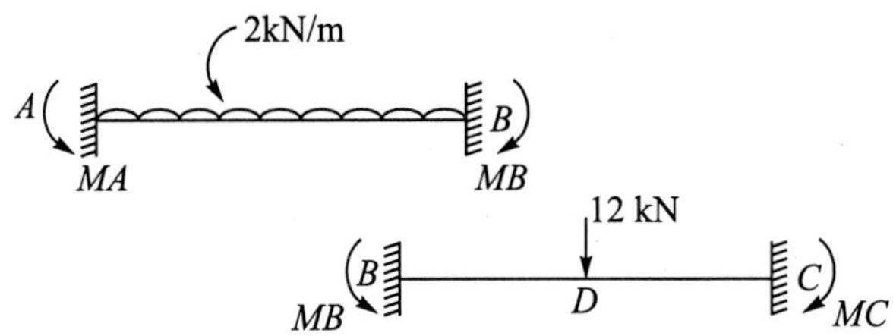

Fig. 10.8

Span *AB*: *F.E.M.* at $A = (-)\ \dfrac{wl^2}{12}$ (–ive sign due to anticlockwise moments)

$$= (-)\ \frac{2\times 6^2}{12} = (-)\ 6 \text{ kN-m}$$

F.E.M. at $B = (+)\ \dfrac{wl^2}{12} = (+)\ 6$ kN-m (+ sign due to clockwise moments)

Span *BC*: *F.E.M.* at $B = (-)\ \dfrac{Wl}{8}$ (–ive sign due to anticlockwise moments)

$$= -\ \frac{12\times 6}{8} = (-)\ 9.0 \text{ kN-m}$$

F.E.M. at $C = (+)\ \dfrac{Wl}{8} = (+)\ 9.0$ kN-m (+ive sign due to anticlockwise moments)

Next find *D.F.* at '*B*', from the geometry of the figure, we find that,

Stiffness factor for *BA*

$$k_B = \frac{4\ EI}{L} = \frac{4\ EI}{6} = \frac{2}{3}\ EI \qquad (\because \text{ Beam is fixed at } A)$$

Similarly,

$$k_B = \frac{4\ EI}{L} = \frac{4\ EI}{6} = \frac{2}{3}\ EI \qquad (\because \text{ Beam is fixed at } C)$$

$\therefore$ *D.F.* for the members *BA* and *BC*,

$$= \frac{\frac{2\ EI}{3}}{\frac{2\ EI}{3} + \frac{2\ EI}{3}} \text{ and } \frac{\frac{2\ EI}{3}}{\frac{2\ EI}{3} + \frac{2\ EI}{3}}$$

$$= \frac{1}{2} \text{ and } \frac{1}{2}$$

We find that the fixing moment at *B* in span *AB* is (+) 6.0 kN-m and that in span *BC* is (–) 9.0 kN-m.

Thus there is an unbalanced moment = – 9 + 6 = (–) 3 kN-m

Now distribute this unbalanced moment of –3 kN-m into the spans *AB* and *BC* in the ratio of their distribution factors *i.e.*, (+) 1.5 kN and (+) 1.5 kN respectively. Next carryover the effects of these distributed moments at '*A*' and '*D*' = $\frac{1}{2} \times 1.5 = (+)\ 0.75$ kN. Then distribute the unbalanced moment at *B* (In this case there is no carryover moment from '*A*' or '*D*' at '*B*', so the distribution of moments at *B* is zero.)

Now find the final moments at *A*, *B*, and *C* in the spans *AB* and *BC* by algebraic adding the respective values in the table.

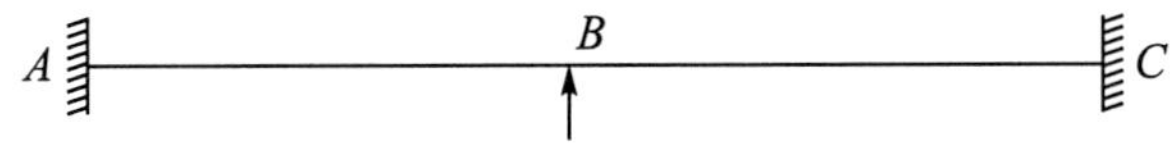

A	B (AB)	B (BC)	C	
	0.5	0.5		D.F.S.
– 6.0	+ 6.0	– 9.0	+ 9.0	F.E.Ms.
	+ 1.5	+ 1.5		
+ 0.75			+ 0.75	C.O.
0	0	0	0	distribute
– 5.25	+ 7.5	–7.5	+ 9.75	Final moments

Fig. 10.9

We know that *B.M.* at the centre of the span *AB*, treating it as *S.S.* beam

$$= \frac{wl^2}{8} = \frac{2 \times 6^2}{8} = 9.0 \text{ kN-m}$$

Similarly, *B.M.* under the load in the span *BC*,

$$= \frac{WL}{4} = \frac{12 \times 6}{4} = 18 \text{ kN-m}$$

B.M. diagram can thus be drawn.

Taking moments about *B*,

$$M^B,\ R_A \times 6 + M_B = M_A + 2 \times 6 \times 3$$

$$R_A \times 6 - 7.5 = -5.25 + 36 = 30.75$$

$\therefore$ $$R_A = 6.375 \text{ kN.}$$

Again, M^B, $$R_C \times 6 + M_B = M_C + 12 \times 3$$

$$R_C \times 6 - 7.5 = (-)\ 9.75 + 36 = 26.25$$

$\therefore$ $$R_C = 5.625 \text{ kN.}$$

and $$R_B = (2 \times 6 + 12) - (6.375 + 5.625)$$

$$= 12.0 \text{ kN.}$$

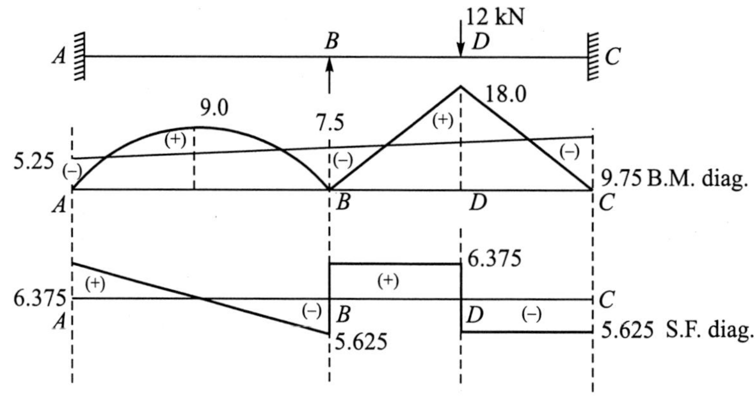

Fig. 10.10

Example 10.2: *Beam with simply supported ends.*

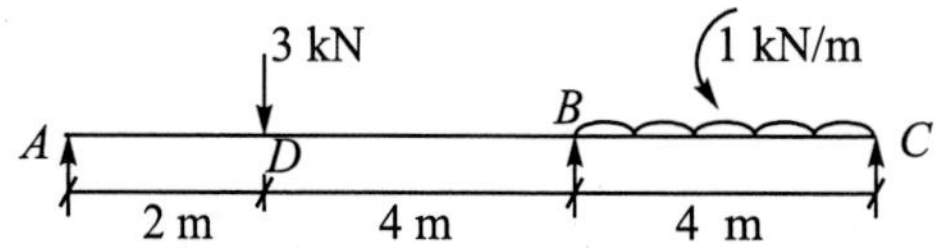

Fig. 10.11

Solution: Span *AB*:

$$F.E.Ms \text{ at } A, = (-)\ \frac{Wab^2}{l^2} \qquad \text{(–ive sign due to anticlockwise moments)}$$

$$= (-)\ \frac{3\times 2\times 4^2}{6^2} = (-)\ 2.67 \text{ kN-m}$$

$$F.E.Ms \text{ at } B, = (+)\ \frac{Wa^2b}{l^2} \qquad \text{(+ive sign due to anticlockwise moments)}$$

$$= (+)\ \frac{3\times 2^2\times 4^2}{6^2} = (+)\ 1.33 \text{ kN-m}$$

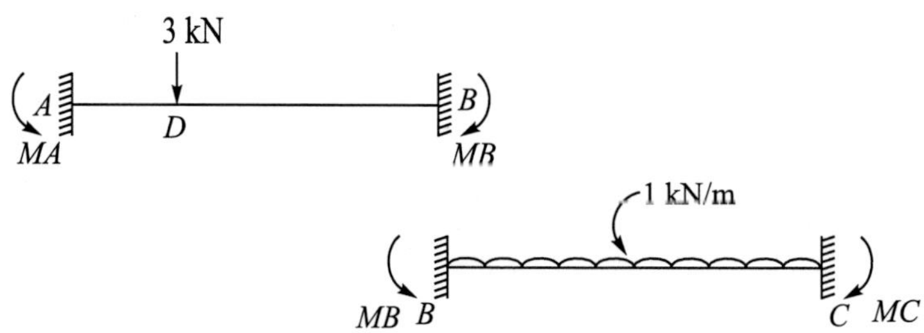

Fig. 10.12

Span *BC*:

$$F.E.Ms \text{ at } B = (-)\ \frac{wl^2}{12} \qquad \text{(–ive sign due to anticlockwise moments)}$$

$$= (-)\ \frac{1\times 4^2}{12} = (-)\ 1.33 \text{ kN-m}$$

$$F.E.M. \text{ at } C, = (+)\ \frac{wl^2}{12} \qquad \text{(+ive sign due to anticlockwise moments)}$$

$$= (+)\ 1.33 \text{ kN-m}$$

Next calculate the distribution factors at *B*. From the geometry of the figure, we find that,

Stiffness factor for *BA*,

i.e., $$K_{BA} = \frac{3\,EI}{l} = \frac{3\,EI}{6} = \frac{EI}{2} \qquad (\because \text{ Beam is hinged at } A)$$

Similarly, stiffness factor for *BC*,

i.e., $$K_{BC} = \frac{3\,EI}{l} = \frac{3\,EI}{4} = \frac{3\,EI}{4} \qquad (\because \text{ Beam is hinged at } C)$$

$\therefore$ *D.F.S.* for *BA* and *BC*,

$$= \frac{\frac{EI}{2}}{\frac{EI}{2} + \frac{3\,EI}{4}} \quad \text{and} \quad \frac{\frac{3\,EI}{4}}{\frac{EI}{2} + \frac{3\,EI}{4}}$$

$$= \frac{2}{5} \quad \text{and} \quad \frac{3}{5}$$

Note that the fixing moments on a simply supported end is zero. Here release the ends *A* and *C* by applying equal and opposite moments. Next carryover moments from *A* to *B* and *C* to *B*. Prepare the table distribute the unbalanced moments in the ratio of their *D.F.S.* and find the final moments.

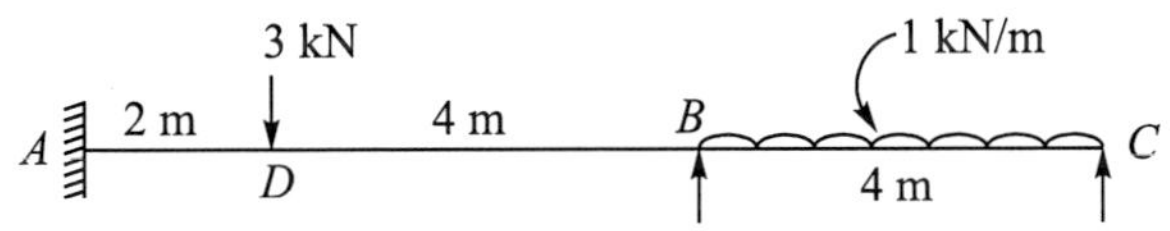

A	B (BA)	B (BC)	C	
	2/5	3/5		D.F.S
– 2.67	+ 1.33	– 1.33	+ 1.33	F.E.Ms
+ 2.67	0	0	– 1.33	Balance
	(+)1.33	(–)0.67		C.O
0	+ 2.66	– 2.00	0	Total moments
	– 0.26	– 0.40		Distribute w.r.t D.F.S
0	+ 2.40	– 2.40	0	Final moments

Fig. 10.13

Calculation of reactions:

M^B, $$R_A \times 6 - 3 \times 4 = (-)\ 2.4 \qquad (\because M_B = 2.4 \text{ kN-m})$$

$\therefore$ $$R_A = 1.6 \text{ kN.}$$

Similarly, $R_C \times 4 - 4 \times 2 = -2.4$ $(\because M_B = -2.4 \text{ kN-m})$

$$\therefore \quad R_C = 1.4 \text{ kN}.$$

$$R_B = (3 + 4) - (1.6 + 1.4) = 4.0 \text{ kN}$$

Next *S.F.* and *B.M.* diagram can be drawn easily.

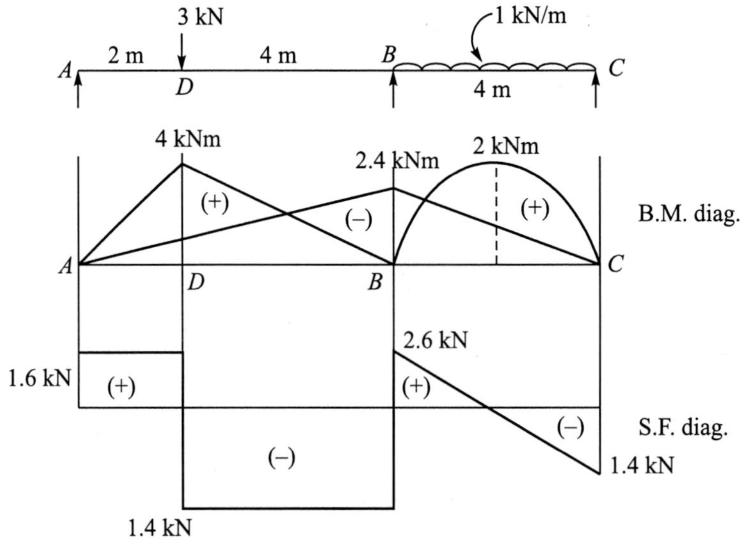

Fig. 10.14

Example 10.3: *Beam with end span overhanging*

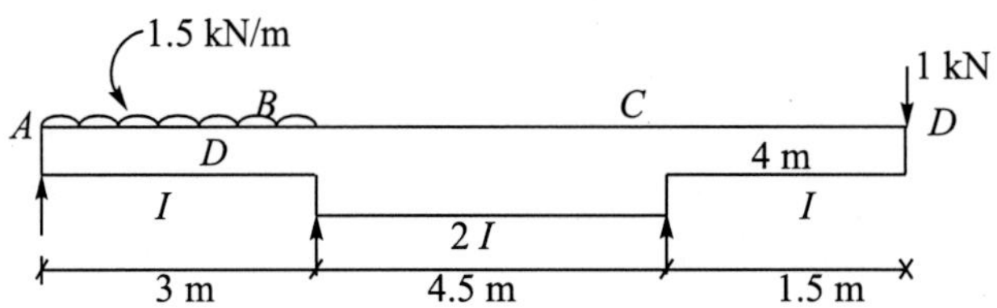

Fig. 10.15

Solution: Split entire beam into 3 parts, *AB*, *BC*, as fixed beams and *CD* as a cantilever.

Span *AB*: *F.E.M.* at *A*, $= (-)\dfrac{wl^2}{12} = (-)\dfrac{1.5 \times 3^2}{12} = (-)\ 1.125$ kN-m

F.E.M. at *B*, $= (+)\dfrac{wl^2}{12}$

$$= (+)\ 1.125 \text{ kN-m}$$

Since the span *BC* is not carrying any load, *F.E.Ms* at *B* and *C* will be zero. The moment at '*C*' for the cantilever *CD* will be 1 × 1.5 = (–) 1.5 kN-m. From the geometry of the figure, stiffness factor for *BA*,

i.e., $k_{BA} = \dfrac{3\ EI}{l} = \dfrac{3\ EI}{3} = EI$ $(\because$ end '*A*' is *S.S.* at *A*)

Similarly, $$k_{BC} = \frac{3\,EI}{l} = \frac{3E \times 2I}{l} = \frac{6EI}{4.5} = \frac{4}{3}\,EI$$

(∵ Beam is overhanging beyond *C*)

Distribution factors for *BA* and *BC*,

$$= \frac{2\,EI}{1\,EI + \frac{4}{3}\,EI} \quad \text{and} \quad \frac{\frac{4\,EI}{3}}{1\,EI + \frac{4}{3}\,EI}$$

$$= \frac{3}{7} \quad \text{and} \quad \frac{4}{7}$$

We know that *D.F.S.* for *CA* and *CD* will be 1 and 0, because the beam is overhanging at *C*.

A	B (BA)	B (BC)	C (CB)	C (CD) — D	
	$\frac{3}{7}$	$\frac{4}{7}$	1	0	D.F.S
− 1.125	+ 1.125	0	0	− 1.5	F.E.Ms.
+ 1.125	0	0	+ 1.5	0	Release 'A' and balance 'C'
	(+) 0.562	+0.75	0	0	C.O.
0	+ 1.687	+ 0.75	+ 1.5	− 1.5	Total moments
	− 1.037	− 1.40	−	−	Distribute w.r.t. D.F.S
0	+ 0.65	− 0.65	+ 1.5	− 1.5	Final moments

Fig. 10.16

B.M. at the mid of span *AB* due to *udl*

$$= \frac{wl^2}{8} = \frac{1.5 \times 3^2}{8} = 1.69 \text{ kN-m}$$

Reactions:

M^B, $R_A \times 3 - (1.5 \times 3 \times 1.5) = (-)\ 0.65$ (∵ $M_B = (-)\ 0.65$ kNm)

∴ $R_A = 2.03$ kN.

M^B, $R_C \times 4.5 - (1 \times 6) = -\ 0.65$ (∵ $M_B = -\ 0.65$ kNm)

∴ $R_C = 1.19$ kN.

and $R_B = (3 \times 1.5 + 1) - (2.03 + 1.19)$

$= 2.28$ kN.

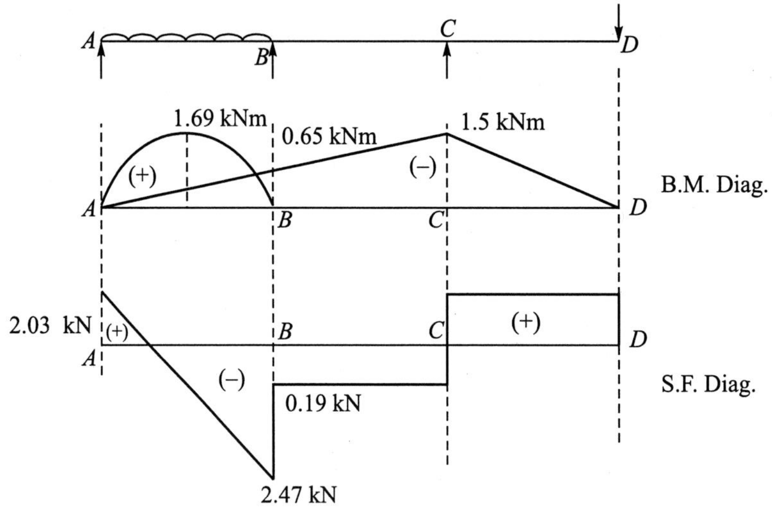

Fig. 10.17

10.2. BEAMS WITH SINKING SUPPORT

Consider a beam *ABC*, simply supported at *A, B* and *C* subjected to any loading. As a result, let the support '*B*' sinks down by an amount '*d*'. Now assume the beam *ABC* to be split up into two beams *AB* and *BC* as shown.

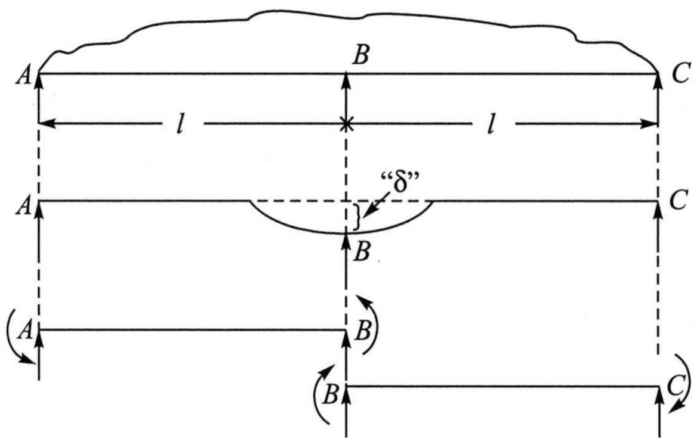

Fig. 10.18

Now from the geometry of the beam, we find the moments caused at *A* and *B* due to sinking of support '*B*'

$$= (-)\ \frac{6\ EI\ \delta}{l^2} \qquad \text{(–ive sign due to right support sinking down)}$$

Similarly, moment caused at '*B*' and '*C*' of the beam *BC* due to sinking of support *B*

$$= (+)\ \frac{6\ EI\ \delta}{l^2} \qquad \text{(+ive sign due to left support sinking down)}$$

Example 10.4: *Support 'B' sinks by 5 mm below A and C, Take E = 200 GPA*

$$I = 332 \times 10^6\ mm^4$$

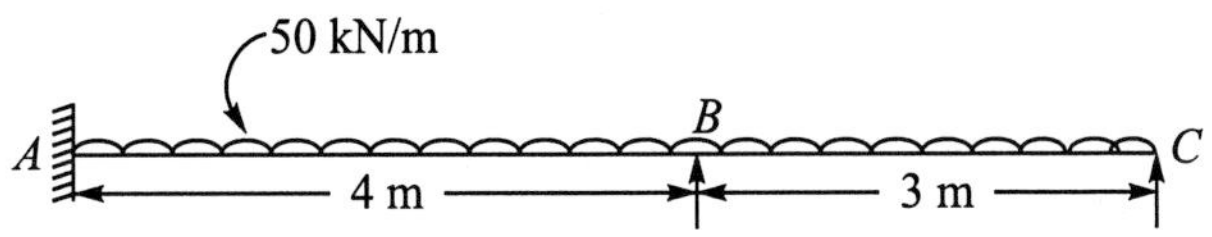

Fig. 10.19

Solution: Assume beam *ABC*, to be split up into fixed beams *AB* and *BC*.

Span *AB*: $\quad F.E.M.$ at 'A' $= (-)\ \dfrac{wl^2}{12}$

$$= (-)\ \frac{50 \times 4^2}{12} = (-)\ 66.7 \text{ kN-m}$$

$$F.E.M. \text{ at } `B' = (+)\ \frac{wl^2}{12}\ (+)\ 66.7 \text{ kN-m}$$

Now in span *AB*, moment at *A* due to sinking of support *B*

$$= (-)\ \frac{6\ EI\ \delta}{l^2} \qquad \text{(–ive sign due to rt. support sinking)}$$

$$= (-)\ \frac{6 \times (200 \times 10^6) \times (332 \times 10^{-6}) \times 0.005}{4^2}$$

$$= (-)\ 124.5 \text{ kN-m}$$

Similarly, moment at *B*, due to sinking of support "*B*"

$$= (-)\ 124.5 \text{ kN-m}$$

$\therefore$ Total moment at '*A*' in span *AB*

$$= -\ 66.7 - 124.5 = (-)\ 191.2 \text{ kN-m}$$

and Total moment at '*B*' in span

$$AB = (+)\ 66.7 - 124.5 = (-)\ 57.8 \text{ kN-m}$$

In span *BC*, fixing moment at "*B*" due to loading

$$= (-)\ \frac{wl^2}{12} = (-)\ \frac{50 \times 3^2}{12} = (-)\ 37.5 \text{ kN-m}$$

and $\quad F.E.M.$ at $\quad$ "C" $= (+)\ \dfrac{wl^2}{12} = (+)\ 37.5$ kN-m

Now, in span *BC*,

Moment at '*B*' due to sinking of support *B*

$$= (+)\ \frac{6\ EI\ \delta}{l^2} \qquad \text{(+ive sign due to left support sinking)}$$

$$= (+)\ \frac{6 \times (200 \times 10^6) \times (332 \times 10^{-6}) \times 0.005}{3^2}$$

$$= 221.3 \text{ kN-m}$$

Similarly, moment at *C*, due to sinking of support '*B*'

$$= (+)\ 221.3 \text{ kN-m.}$$

∴ Total moment at B, in span BC

$$= -37.5 + 221.3$$
$$= 183.8 \text{ kN-m}$$

and total moment at C, in span BC

$$= (+)\ 37.5 + 221.3 = 258.8 \text{ kN-m}$$

***D.F.S.* at *B*:** $k_{BA} = \dfrac{4\ EI}{l} = \dfrac{4\ EI}{4} = EI$ (∵ Beam is fixed at A)

and $k_{BC} = \dfrac{3\ EI}{l} = \dfrac{3}{3}\ EI = EI$ (∵ Beam is *S.S.* at C)

∴ *D.F.S.* for BA and BC will be $\frac{1}{2}$ and $\frac{1}{2}$.

A		B		C	
	$\frac{1}{2}$	$\frac{1}{2}$			D.F.S.
− 191.2	−57.8	+ 183.8	+ 258.8		F.E.Ms
			− 258.8		Release 'C'
		(−) 129.4			C.O.
− 191.2	−57.8	+ 54.4	0		Initial moments
	+ 1.7	+ 1.7			distribute w.r.t. D.F.S.
+ 0.9					C.O.
0	0	0	0		
(−) 190.3	(−) 56.1	(+) 56.1	0		Final moments

Fig. 10.20

We know that *B.M.* at mid of span AB, treating it as *S.S.* beam $= \dfrac{wl^2}{8} = \dfrac{50 \times 4^2}{8} = 100$ kN-m

B.M. at mid of span BC $= \dfrac{wl^2}{8} = \dfrac{50 \times 3^2}{8} = 56.25$ kN-m

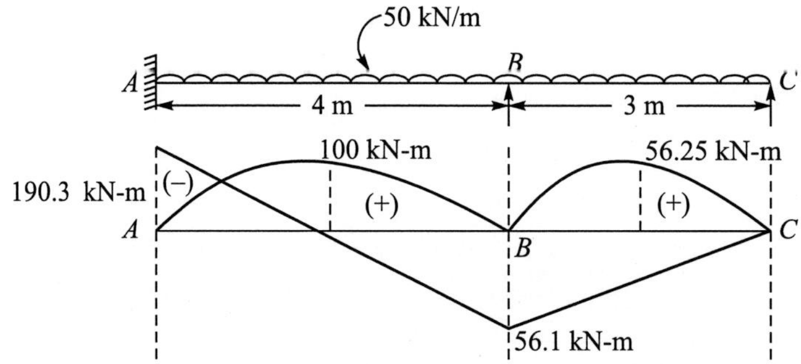

B.M. Diag.

Fig. 10.21

10.3. PROBLEMS ON PORTAL FRAMES

Example 10.5:

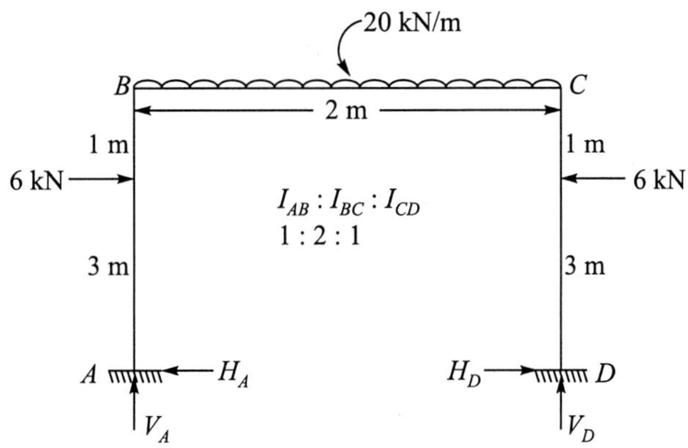

Fig. 10.22

Solution: Let $I_{AB} = I$, $I_{BC} = 2I$ and $I_{CD} = I$

The distribution factors are calculated as below:

Joint	Member	Relative stiffness	Total relative stiffness	*D.F.*
B	*BA*	$\frac{I}{4}$	$\frac{5I}{4}$	$\frac{1}{5}$
	BC	$\frac{2I}{2} = \frac{4I}{4}$		$\frac{4}{5}$
C	*CB*	$\frac{2I}{2} = \frac{4I}{4}$	$\frac{5I}{4}$	$\frac{4}{5}$
	CD	$\frac{I}{4}$		$\frac{1}{5}$

Calculation of *F.E.Ms.*

$$M_{FAB} = (-)\ \frac{6 \times 3 \times 1^2}{4^2} = (-)\ 1.12 \text{ kN-m}$$

$$M_{FBA} = (+)\ \frac{6 \times 3^2 \times 1}{4^2} = (+)\ 3.37 \text{ kN}$$

$$M_{FBC} = (-)\ \frac{20 \times 2^2}{12} = (-)\ 6.67 \text{ kN-m}$$

$$M_{FCB} = (+)\ 6.67 \text{ kN-m}$$

$$M_{FCD} = (-)\ 3.37 \text{ kN-m}$$

$$M_{FDC} = (+)\ 1.12 \text{ kN-m}$$

A	B		C		D	
	$\frac{1}{5}$	$\frac{4}{5}$	$\frac{4}{5}$	$\frac{1}{5}$		D.F.S.
− 1.12	+ 3.37	− 6.67	+ 6.67	− 3.37	+ 1.12	F.E.Ms
	+ 0.66	+ 2.64	− 2.64	− 0.66		Balance
− 0.33		− 1.32	+ 1.32		− 0.33	C.O.
	+ 0.26	+ 1.06	− 1.06	− 0.26		Balance
− 0.13		− 0.53	+ 0.53		− 0.13	C.O.
	+ 0.11	+ 0.42	− 0.42	− 0.11		Balance
+ 0.06		(−) 0.21	+ 0.21		− 0.06	C.O.
	− 0.04	(+) 0.17	(−) 0.17	− 0.04		Balance
− 0.60	+ 4.44	− 4.44	+ 4.44	− 4.44	+ 0.60	Final moments

Vertical reactions at A and D are equal,

$$\therefore \qquad V_A = V_D = \frac{20 \times 2}{2} = 20 \text{ kN}$$

Horizontal reaction at A, H_A

$$= \frac{M_{AB} + M_{BA} + \text{Moment of forces on } AB \text{ about } `B'}{h}$$

$$= (-)\ \frac{0.60 + 4.44 - 6 \times 1}{4} = (-)\ 0.54 \text{ kN } (\leftarrow)$$

Obviously, $H_D = (+)\ 0.54 \text{ kN } (\rightarrow)$

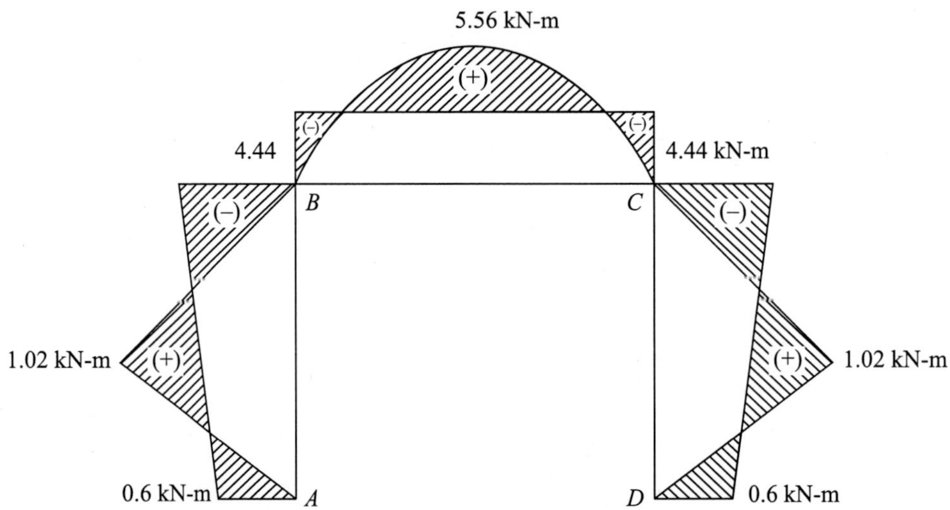

B.M. digram

Fig 10.23

Example 10.6: *"Portal Frame with Inclined legs"*

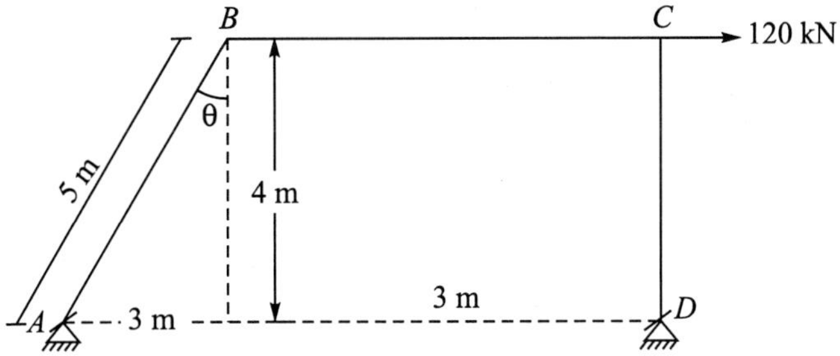

Fig. 10.24

Solution: Calculate the distribution factors as below:

Joint	Member	Relative stiffness	Total relative stiffness	D.F.
B	*BA*	$\frac{3}{4}\times\frac{I}{5}=\frac{9I}{60}$	$\frac{29I}{60}$	$\frac{9}{29}$
	BC	$\frac{I}{3}=\frac{20\,I}{60}$		$\frac{20}{29}$
C	*CB*	$\frac{I}{3}=\frac{16\,I}{48}$	$\frac{25\,I}{48}$	$\frac{16}{25}$
	CD	$\frac{3}{4}\times\frac{I}{4}=\frac{9I}{48}$		$\frac{9}{25}$

Let the frame *ABCD* deform to the position $A_1\ B_1\ C_1\ D$ as shown. Assuming fixed end conditions. Let us calculate the *F.E.Ms* corresponding to the deformed position of the frame. From the geometry of frame,

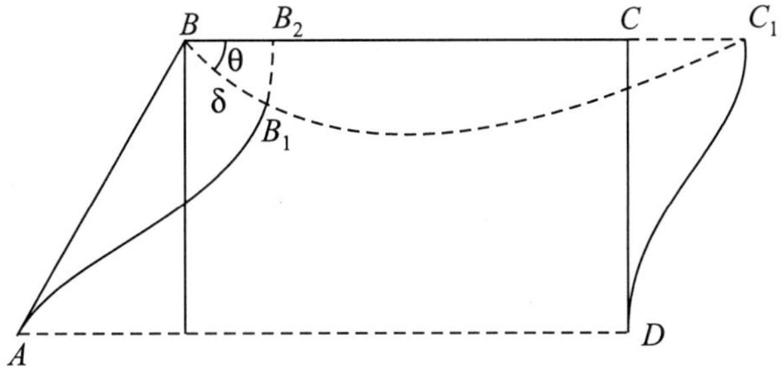

$$\tan\theta = \frac{3}{4}$$

$$\cos\theta = \frac{4}{5}$$

and $$\sin\theta = \frac{3}{5}$$

Let $$BB_1 = \delta$$

$$BB_2 = CC_1 = \delta\cos\theta = \frac{4}{5}\delta$$

$$B_1B_2 = \delta\sin\theta = \frac{3}{5}\delta$$

Ratio of the *F.E.Ms* for the members *AB, BC* and *CD*

$$= \frac{6\,EI\delta}{5^2} : \frac{6\,EI}{3^2}\cdot\frac{3}{5}\delta : \frac{6\,EI}{4^2}\cdot\frac{4}{5}\delta = \frac{1}{25} : \frac{1}{15} : \frac{1}{20}$$

$$= 1 : \frac{5}{3} : \frac{5}{3} = 12 : 20 : 15$$

A		B			C	D	
		9/29	20/29	16/25	9/25		D.F.S.
	−12.00	−12.00	+20.00	+20.00	−15.00	−15.00	F.E.Ms.
	+12.00	+6.00			(+) 7.50	+15.00	Release and C.O.
	0	−6.00	+20.00	+20.00	−7.50	0	Initially
		−4.34	−9.66	−8.00	−4.50		Balance
			4.00	−4.83			C.O.
		+1.24	+2.76	+3.09	+1.74		Balance
			+1.55	+1.38			C.O.
		−0.48	−1.07	−0.88	−0.50		Balance
			−0.44	−0.54			C.O.
		+0.14	+0.30	+.35	+0.19		Balance
			+0.18	+0.15			C.O.
		−0.66	−0.12	−0.10	−0.05		Balance
			−0.05	−0.06			C.O.
		+0.02	+0.03	+0.04	+0.02		Balance
Col (*a*)	0	(−) 9.48	(+) 9.48	(+) 10.60	(−)10.60 0		Final moment

Let the moments shown in col (*a*) be due to sway force '*S*'.

Horizontal reaction at $'D' = H_d = -\dfrac{10.60 + 0}{4} = (-)\ 2.65 \text{ kN } (\leftarrow)$

Vertical reaction at $'D',\ V_D = \dfrac{9.48 + 10.60}{3} = 6.693 \text{ kN } (\uparrow)$

Vertical reaction at $'A',\ V_A = 6.693 \text{ kN } (\downarrow)$

Horizontal reaction at $A,\ H_A = \dfrac{0 - 9.48 - 6.693 \times 3}{4} = (-)\ 7.39 \text{ kN } (\leftarrow)$

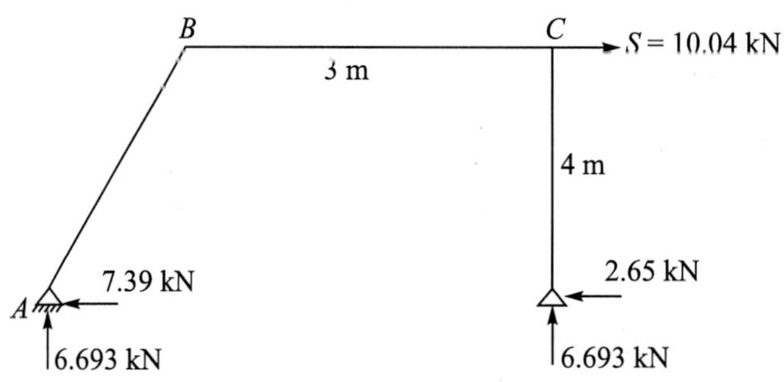

Fig. 10.25

Resolving the forces on the frame horizontally

$$S = 7.39 + 2.65$$
$$= 10.04 \text{ kN}$$

For a sway of 10.04 kN, the sway moments are as per col (*a*).

∴ For the actual sway force of 120 kN, the actual sway moments will be,

$$= \frac{120}{10.04} \times \text{col } (a) \text{ moments}$$

	A	*B*	*B*	*C*	*C*	*D*
Col (*a*)	0	(–) 9.48	(+) 9.48	(+) 10.60	(–) 10.60	0
Actual sway moments $= \frac{120}{10.04} \times$ col (*a*)	0	–113.31	+113.31	+126.69	+126.69	0

Actual Reactions at Supports:

Horizontal reaction at $D, = H_D = -\frac{126.69 + 0}{4} = 31.70$ kN

Vertical reaction at $D, V_D = \frac{113.31 + 126.69}{3} = 80$ kN (↑)

Vertical reaction at $A, V_A = 80$ kN (↓)

Horizontal reaction at $A, H_A = \frac{0 - 113.31 - 80 \times 3}{4} = 88.3$ kN

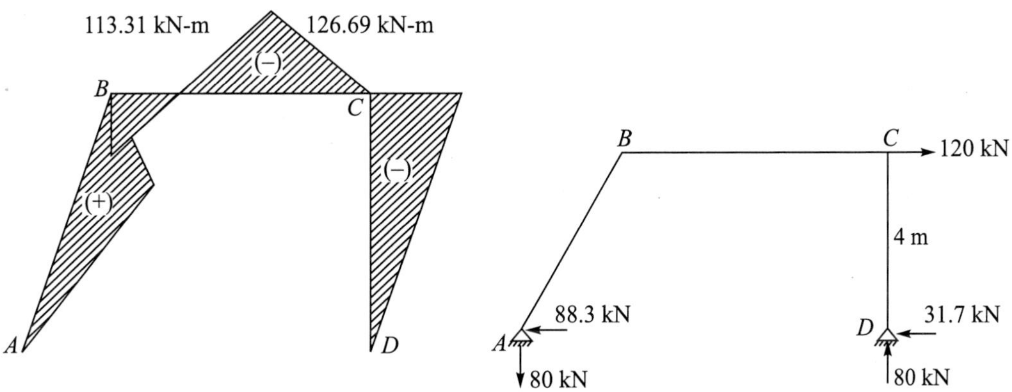

B.M. diagram

Fig. 10.26

EXERCISE FOR PRACTICE

1. Analyse the continuous beam shown in Fig.10.27 below and draw B.M. diagram. **(KU 2011)**

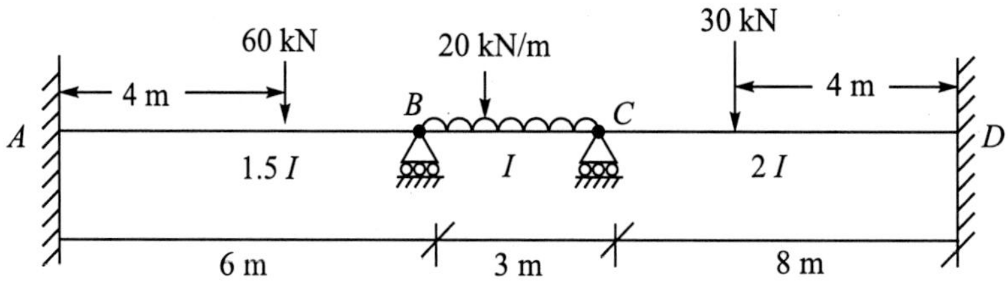

Fig. 10.27.

2. Analyse the continuous beam *ABCD* by moment distribution method.

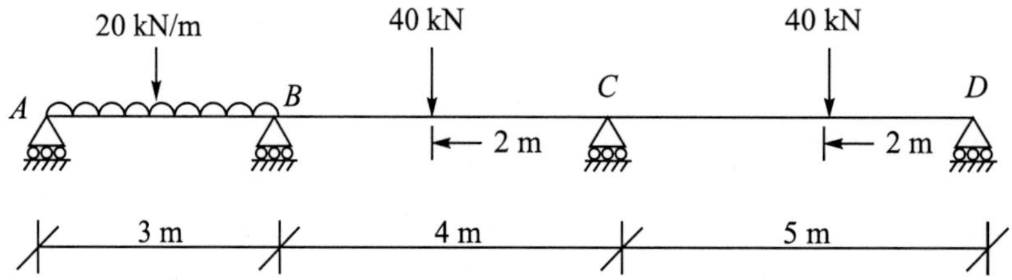

Fig. 10.28.

3. A continuous beam *ABC* is supported on an elastic column *BD*, and is loaded as shown in Fig. 10.29. Treating joint *B* as rigid, analyse the frame and plot the B.M.D. **(AMIE 2009)**

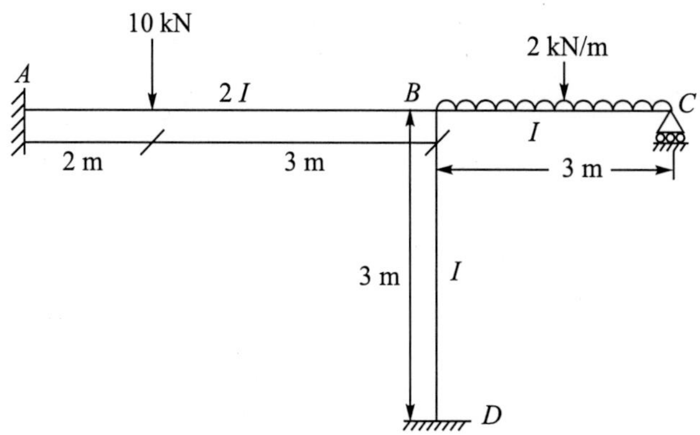

Fig. 10.29.

4. Analyse the portal frame shown in Fig. 10.30 by moment distribution method. The frame is fixed at *A* and *D* and has rigid joints at *B* and *C*. Draw the B.M. diagram.

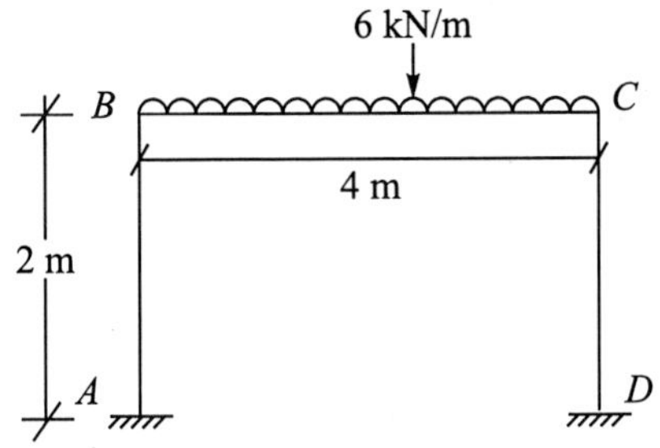

Fig. 10.30.

5. Analyse the rigid jointed frame shown in Fig. 10.31 by moment distribution method and draw the B.M. diagram.

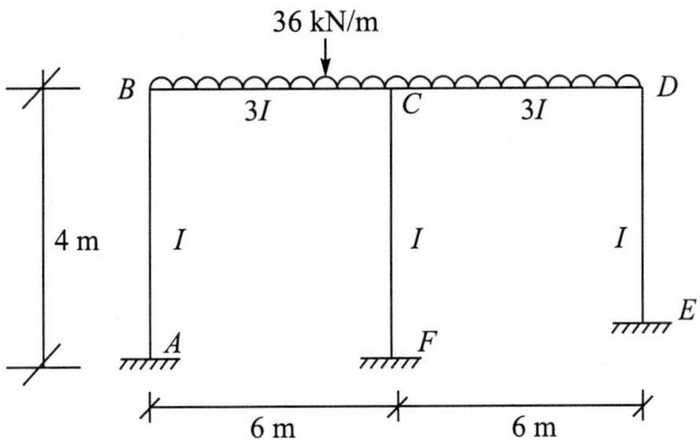

Fig. 10.31.

CHAPTER 11

Three Hinged Arch

11.1. DEFINITION

An arch is a curved girder or curved beam supported at its ends and carries transverse loads which are normally vertical. This forms part of a determinate structure.

Types: (*i*) Three hinged arch.

(*ii*) Two hinged arch.

(*iii*) Single hinged arch.

(*iv*) Fixed arch.

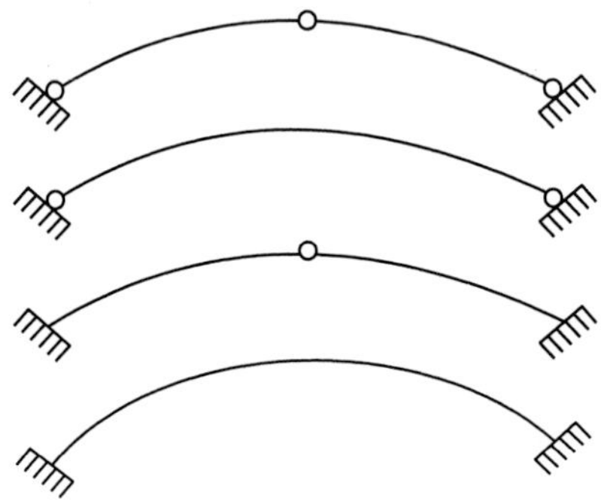

Fig. 11.1

In three hinged arch we will study about its horizontal thrust and drawing of S.F. and B.M. diagrams.

11.2. THREE HINGED ARCH

(*a*) **Case of a Parabolic Arch:** (Bottom hinges A and B are at the same-level)

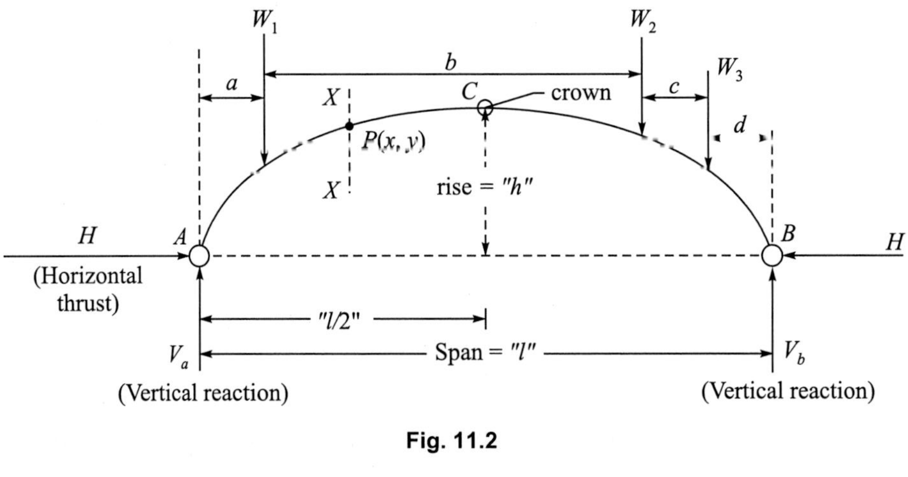

Fig. 11.2

Note: The tendency of the external loads will be to increase the span of the arch. Thus to maintain equilibrium of system, the hinge "*A*" and "*B*" will have two components of reaction viz. V_a and H or V_b and H.

Also when the external loads applied are vertical and hinges *A* and *B* are at same level, the horizontal thrust at the supports *A* and *B* will be the same. Also since the parts *AC* and *BC* are hinged at '*C*' *B.M.* at '*C*' must be zero.

The vertical reactions V_a and V_b can be determined by equating the the clockwise moments to the anticlockwise moments of the forces on the structure about either lower end. The horizontal thrust '*H*' can however be compulated by equating the *B.M.* at the crown '*C*' to zero. Referring to Fig. 11.2, taking moments about '*C*' of the forces on the left side of *C*, we have,

$$V_a \times \frac{l}{2} = Hh + W_1\left(\frac{l}{2} - a\right)$$

or taking moments about '*C*' of the forces on the right side of '*C*', we have,

$$V_b \times \frac{l}{2} = Hh + W_3\left(\frac{l}{2} - d\right) + W_2\left(\frac{l}{2} - d - c\right)$$

The value of '*H*' can be determined from any of the above equations. This is illustrated in the following example:

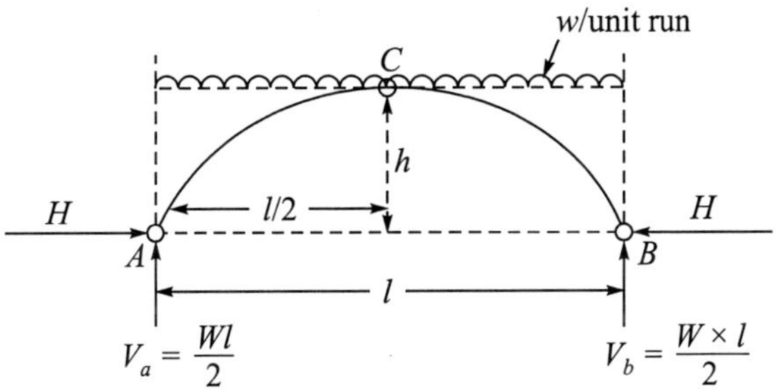

Fig. 11.3

Taking moments about '*C*', of the forces on the L.H.S. of '*C*', we have

$$\frac{Wl}{2} \times \frac{l}{2} = H \times h + \frac{Wl}{2} \times \frac{l}{4}$$

$$\therefore \qquad \text{"}H\text{"} = \frac{Wl^2}{8h}.$$

11.3. CASE OF A THREE HINGED ARCH, WHICH IS NOT PARABOLIC BUT SEMICIRCULAR IN SHAPE

$$V_a = V_b = (w \times 2R) \times \frac{1}{2} = w \times R$$

Taking moments about "*C*" of forces on left side of '*C*', we have,

$$wR \times R = H \times R + wR \times \frac{R}{2}$$

or

$$H = \frac{wR}{2}.$$

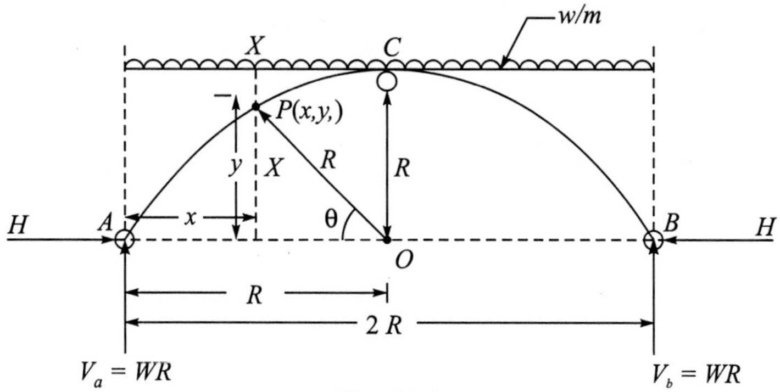

Fig. 11.4

Alternatively, $$H = \frac{wl^2}{8h} = \frac{w(2R)^2}{8R} = \frac{wR}{2}.$$

The *B.M.* at any section '*X*', making an angle 'θ' with the radius vector and horizontal is given by,

$$M_X = wRx - \frac{wx^2}{2} - Hy$$

From the figure above,

$$x = R - R\cos\theta = R(1 - \cos\theta)$$

and $$y = R\sin\theta.$$

Substituting these values of '*x*' and '*y*' in the above equation, we have,

$$M_X = wR \cdot R(1 - \cos\theta) - w\left[\frac{R(1-\cos\theta)}{2}\right]^2 - \frac{wR}{2} \times R\sin\theta.$$

or $$M_X = wR^2(1 - \cos\theta) - \frac{wR^2}{2}(1 - \cos\theta)^2 - \frac{wR^2}{2} \cdot \sin\theta$$

$$= \frac{wR^2}{2}(1 - \cos\theta)[2 - (1 - \cos\theta)] - \frac{wR^2}{2}\sin\theta$$

$$= \frac{wR^2}{2}(1 - \cos\theta)(1 + \cos\theta) - \frac{wR^2}{2} \cdot \sin\theta$$

$$= \frac{wR^2}{2}(1 - \cos^2\theta) - \frac{wR^2}{2} \cdot \sin\theta$$

$$= \frac{wR^2}{2}\sin^2\theta - \frac{wR^2}{2} \cdot \sin\theta$$

$$= \frac{wR^2}{2}(\sin^2\theta - \sin\theta)$$

∴ $$M_X = -\frac{wR^2}{2}[\sin\theta - \sin^2\theta]$$

Since sin θ being greater than $\sin^2 \theta$. The *B.M.* at any section of the semicircular arch is always negative *i.e.,* Hogging.

$$\text{Now for } M_X \text{ to be max, } \frac{\theta M_x}{d\theta} = 0$$

$$\therefore \quad \frac{d}{d\theta}\left[-\frac{wR^2}{2}(\sin\theta - \sin^2\theta)\right] = -\frac{wR^2}{2}[\cos\theta - 2\sin\theta \cdot \cos\theta] = 0$$

or $(\cos\theta - 2\sin\theta \cdot \cos\theta) = 0$

or $\cos\theta\,(1 - 2\sin\theta) = 0$

∴ $\cos\theta = 0$ or $\theta = 90°$

and $(1 - 2\sin\theta) = 0, \quad \therefore \quad \sin\theta = \frac{1}{2}$

or $\theta = 30°$

But at θ = 90°, *i.e.,* at crown, the *B.M.* = 0

∴ *B.M.* is max. at θ = 30°

$$\therefore \quad M_{max} = -\frac{wR^2}{2}(\sin 30° - \sin^2 30)$$

$$= -\frac{wR^2}{2}\left(\frac{1}{2} - \frac{1}{4}\right) = -\frac{wR^2}{2} \times \frac{1}{4}.$$

$$\therefore \quad M_{max} = (-)\frac{wR^2}{8}$$

and the distance of the point of max *B.M.* from crown = $R\cos 30° = R \times \frac{\sqrt{3}}{2}$

11.4. CASE OF A PARABOLIC THREE HINGED ARCH WHOSE ABUTMENTS VIZ., SUPPORTS *A* AND *B* ARE AT DIFFERENT DEPTH FROM THE CROWN

Type I: With Uniformaly distributed load (*udl*)

With crown "*C*" as origin, if (*x*, *y*) be the co-ordinates of a point on the centre live of the arch, we have by the property of parabola

$$\frac{x^2}{y} = \text{constant}$$

or

$$\frac{x}{\sqrt{y}} = \text{constant.}$$

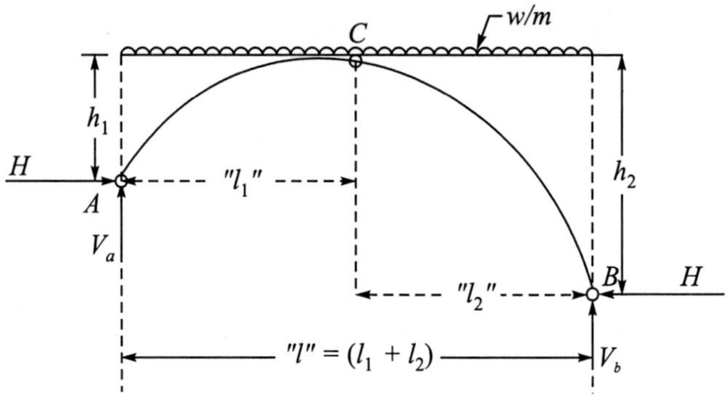

Fig. 11.5

Now applying this principle to the points '*A*' and '*B*' we have,

$$\frac{l_1}{\sqrt{h_1}} = \frac{l_2}{\sqrt{h_2}} = \frac{l_1 + l_2}{\sqrt{h_1} + \sqrt{h_2}} = \frac{l}{\sqrt{h_1} + \sqrt{h_2}}$$

or $$l_1 = \frac{l\sqrt{h_1}}{\sqrt{h_1} + \sqrt{h_2}} \text{ and } l_2 = \frac{l\sqrt{h_2}}{\sqrt{h_1} + \sqrt{h_2}}$$

Taking moments about '*C*' of the forces on the *L.H.S.* of '*C*', we have

$$V_a \times l_1 = \frac{wl_1^2}{2} + H \times h_1$$

$$\therefore \quad V_a = \frac{wl_1}{2} + H \times \frac{h_1}{l_1} \qquad \text{...(11.1)}$$

Similarly, taking moments about '*C*' of all the forces on the *R.H.S* of "*C*", we have

$$V_b \times l_2 = \frac{wl_2^2}{2} + Hh_1$$

$$\therefore \quad V_b = \frac{wl_2}{2} + H \cdot \frac{h_2}{l_2} \qquad \text{...(11.2)}$$

Adding equations (11.1) and (11.2), we have

$$V_a + V_b = \frac{w}{2}(l_1 + l_2) + H\left(\frac{h_1}{l_1} + \frac{h_2}{l_2}\right) = \frac{wl}{2} + H\left(\frac{h_1}{l_1} + \frac{h_2}{l_2}\right)$$

But $V_a + V_b$ = Total load on span = $w \times l$

$$\therefore \quad wl = \frac{wl}{2} + H\left(\frac{h_1}{l_1} + \frac{h_2}{l_2}\right)$$

$$\therefore \quad \text{'}H\text{'} = \frac{\frac{wl}{2}}{\left(\frac{h_1}{l_1} + \frac{h_2}{l_2}\right)}$$

But $$l_1 = \frac{l\sqrt{h_1}}{\sqrt{h_1} + \sqrt{h_2}} \quad \text{and} \quad l_2 = \frac{l\sqrt{h_2}}{\sqrt{h_1} + \sqrt{h_2}}$$

$$\therefore \qquad \text{"}H\text{"} = \frac{\dfrac{wl}{2}}{\dfrac{h_1(\sqrt{h_1}+\sqrt{h_2})}{l\sqrt{h_1}} + \dfrac{h_2(\sqrt{h_1}+\sqrt{h_2})}{l\sqrt{h_2}}}$$

$$= \frac{\dfrac{wl}{2}\times l}{\sqrt{h_1}(\sqrt{h_1}+\sqrt{h_2}) + \sqrt{h_2}(\sqrt{h_1}+\sqrt{h_2})}$$

or

$$\boxed{\text{"}H\text{"} = \frac{wl^2}{2(\sqrt{h_1}+\sqrt{h_2})^2}}$$

11.5. THE CASE OF A THREE HINGED PARABOLIC ARCH WITH DIFFERENT ABUTMENT HEIGHTS FROM THE CROWN AND CARRYING A CONCENTRATED LOAD AT CROWN

(Type-II: Case of concentrated load 'W')

We have,
$$\text{"}l_1\text{"} = \frac{l\sqrt{h_1}}{\sqrt{h_1}+\sqrt{h_2}} \quad \text{and} \quad \text{"}l_2\text{"} = \frac{l\sqrt{h_2}}{\sqrt{h_1}+\sqrt{h_2}}$$

Taking moments about 'C' of all the forces on *L.H.S.* of "C", we have,

$$V_a \times l_1 = H \times h_1$$

$$\therefore \qquad V_a = H \times \frac{h_1}{l_1} \qquad \ldots(11.3)$$

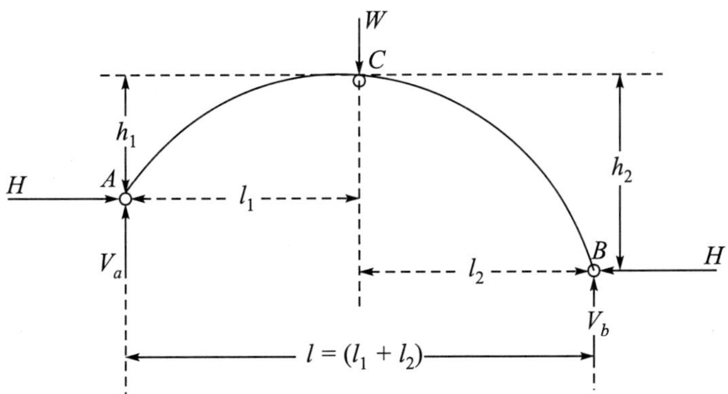

Fig. 11.6

Similarly, taking moments about 'C' of all the forces on the *R.H.S.* of 'C', we have

$$V_b \times l_2 = H \times h_2$$

or

$$V_b = H \times \frac{h_2}{l_2} \qquad \ldots(11.4)$$

Adding eqs. (11.3) and (11.4), we have,

$$V_a + V_b = H\left(\frac{h_1}{l_1} + \frac{h_2}{l_2}\right)$$

But $$V_a + V_b = W$$

$\therefore$ $$W = H\left(\frac{h_1}{l_1} + \frac{h_2}{l_2}\right)$$

$\therefore$ $$\text{"}H\text{"} = \frac{W}{\left(\frac{h_1}{l_1} + \frac{h_2}{l_2}\right)}, \text{ But } l_1 = \frac{l\sqrt{h_1}}{\sqrt{h_1} + \sqrt{h_2}} \quad \text{and} \quad l_2 = \frac{l\sqrt{h_2}}{\sqrt{h_1} + \sqrt{h_2}}$$

Substituting these values in above, we have

$\therefore$ $$\text{"}H\text{"} = \frac{W}{\frac{h_1(\sqrt{h_1} + \sqrt{h_2})}{l\sqrt{h_1}} + \frac{h_2(\sqrt{h_1} + \sqrt{h_2})}{l\sqrt{h_2}}}$$

$$= \frac{Wl}{\sqrt{h_1}(\sqrt{h_1} + \sqrt{h_2}) + \sqrt{h_2}(\sqrt{h_1} + \sqrt{h_2})}$$

or $$\boxed{\text{"}H\text{"} = \frac{Wl}{(\sqrt{h_1} + \sqrt{h_2})^2}}$$

Example 11.1: *A three hinged arch has a span of 30 meters and rise of 10 m. The arch carries a udl of 60 kN/m on the left half of its span. It also carries two concentrated loads of 160 kN and 100 kN at 5 m and 10 m from the right end. Determine the horizontal thrust at each support.*

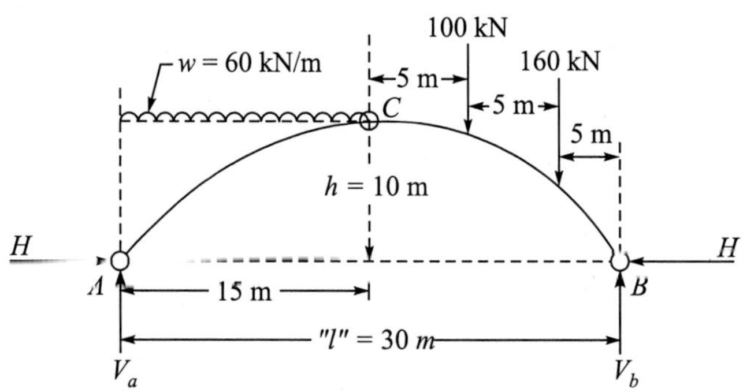

Fig. 11.7

Solution: For finding the support reactions V_a and V_b, take moments of all forces about the support 'A', we have

$$V_b \times 30 = 60 \times 15 \times \frac{15}{2} + 100 \times 20 + 160 \times 25$$

$$\text{"}V_b\text{"} = 425 \text{ kN}$$

$$\therefore \quad V_a = \{(60 \times 15) + 100 + 160\} - 425$$

$$\therefore \quad \text{“}V_a\text{”} = 735 \text{ kN.}$$

Next taking moments about '*C*' of the forces on the left side of '*C*', we have

$$735 \times 15 = 60 \times 15 \times \frac{15}{2} + H \times 10$$

$$\therefore \quad \text{“}H\text{”} = 427.5 \text{ kN}$$

Example 11.2: *A three hinged parabolic arch of span 20 mtr and rise 4 m, carries a udl of 20 kN/m run on the left half of the span. Find the max. B.M. for the arch.*

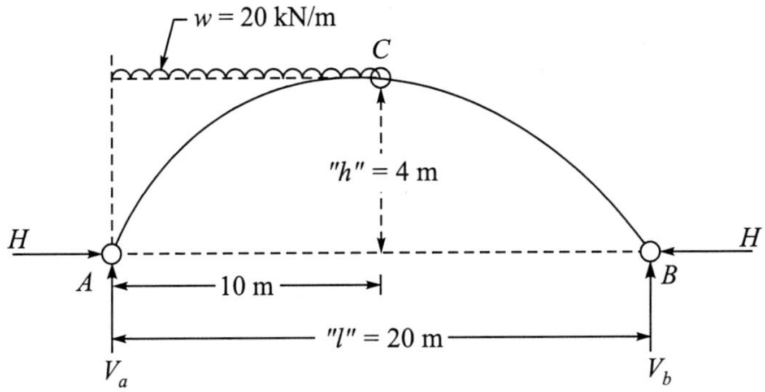

Fig. 11.8

Solution: For finding support reactions V_a and V_b take moments about A, we have

$$V_b \times 20 = 20 \times 10 \times \frac{10}{2}$$

$$\therefore \quad \text{“}V_b\text{”} = 50 \text{ kN}$$

and $\quad \text{“}V_a\text{”} = (20 \times 10) - 50 = 150 \text{ kN.}$

For finding the horizontal thrust "*H*", take moments of all the forces about '*C*' towards *R.H.S.* of "*C*",

$$V_b \times 10 = H \times 4$$

or $$H = \frac{10 \times V_b}{4} = \frac{10 \times 50}{4} = 125 \text{ kN.}$$

Now, from the property of Arched Parabola,

$$\text{“}y\text{”} = \frac{4h}{l^2} \times x \times (l - x)$$

or $$\text{“}y\text{”} = \frac{4 \times 4}{(20)^2} \times x \times (20 - x)$$

or $$\boxed{\text{“}y\text{”} = \frac{1}{25}\, x\,(20 - x)}$$

At any section "X" distant 'x' from A,

B.M. is given by,

$$M_X = V_a x - w_1 (x - a) - H \times y \qquad \text{(General equation)}$$

$$\therefore \quad M_X = 150 \times x - 20x \times \frac{x}{2} - 125 \times \frac{1}{25} x (20 - x)$$

$$= 150x - 10x^2 - 100x + 5x^2$$

$$= 50x - 5x^2$$

$$\therefore \quad \boxed{M_X = 5x(10 - x)}$$

For the condition of max. B.M.

$$\frac{dM_X}{dx} = 0.$$

$$\therefore \quad \frac{d}{dx}\{(5x)(10 - x)\} = 0$$

$$50 - 10x = 0$$

or $$x = 5 \text{ m}.$$

$$\therefore \quad M_{max} = (5 \times 5)(10 - 5) = 125 \text{ kN-m}$$

Example 11.3: *Figure 11.8 shows a three hinged arch consisting of two quadrantal parts AC and CB of radii 2 m and 4 m respectively. For the load system acting on the arch, calculate the reactions at the supports and the B.M. under loads.*

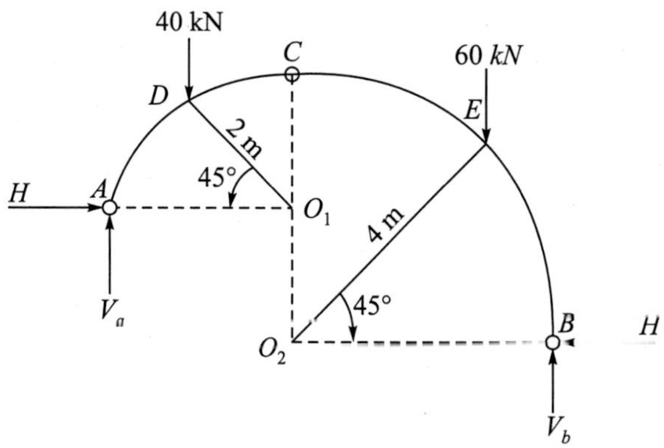

Fig. 11.9

Solution: Let V_a and V_b be the vertical reactions at the supports A and B respectively. Let 'H' be the horizontal thrust at each support.

Resolving the forces on the arch vertically,

$$V_a + V_b = 100 \text{ kN} \qquad \ldots(i)$$

Taking moments about 'C' of the forces on the left side of 'C',

$$V_a \times 2 = 2H + 40\sqrt{2}$$

$$\therefore \quad V_a = H + 20\sqrt{2} \qquad \ldots(ii)$$

Taking moments about 'C' of the forces on the right side of 'C',

$$V_b \times 4 = 4H + 60 \times 2\sqrt{2}$$

$$V_b = H + 30\sqrt{2} \qquad \ldots(iii)$$

Adding eqs. (*ii*) and (*iii*),

$$V_a + V_b = 2H + 50\sqrt{2} = 100$$

$$H = 50 - 25\sqrt{2} = 25(2 - \sqrt{2})$$

$$V_a = H + 20\sqrt{2} = 50 - 25\sqrt{2} + 20\sqrt{2}$$

$$= 50 - 5\sqrt{2} \text{ kN.}$$

$$V_b = H + 30\sqrt{2} = 50 - 25\sqrt{2} + 30\sqrt{2}$$

$$= 50 + 5\sqrt{2} \text{ kN}$$

B.M. at "D" $= M_d = V_a \times 2(1 - \cos 45°) - H \times 2 \sin 45°$

$$= (50 - 5\sqrt{2})\, 2\left(1 - \frac{1}{\sqrt{2}}\right) - 25(2\sqrt{2})\, 2 \times \frac{1}{\sqrt{2}}$$

$$= [110 - 60\sqrt{2}] - [50\sqrt{2} - 50]$$

$$= 160 - 110\sqrt{2} = 4.4365 \text{ kN-m}$$

B.M. at "E" $= M_e = V_b \times 4(1 - \cos 45°) - H \times 4 \sin 45°$

$$= (50 + 50\sqrt{2})\, 4\left[1 - \frac{1}{\sqrt{2}}\right] - 25(2 - \sqrt{2}) \times 4 \times \frac{1}{\sqrt{2}}$$

$$= [180 - 80\sqrt{2}] - [100\sqrt{2} - 100]$$

$$= 280 - 180\sqrt{2} = 25.4416 \text{ kN-m}$$

Example 11.4: *A three hinged arch ABC consisting of quadrantal parts AC and BC of radii 4 m and 6 m respectively carries two point loads of 100 kN each as shown in Fig. 11.9. Find the reactions at support A and B and B.M. at E and F.*

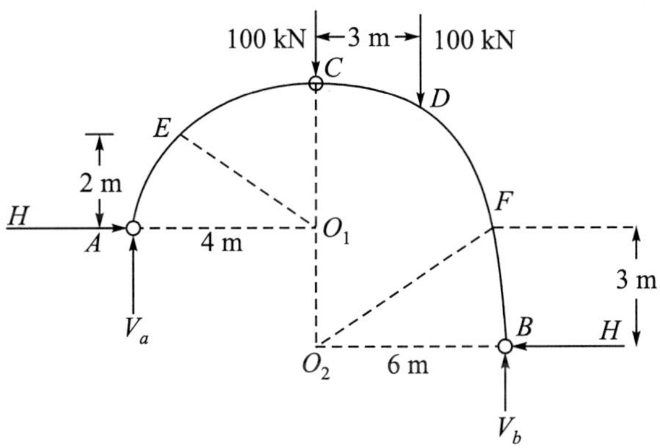

Fig. 11.10

Solution: Let V_a and V_b be the vertical reactions at *A* and *B*. Let '*H*' be the horizontal thrust at each support.

Taking moments about '*C*' of the forces on the *L.H.S.* of '*C*',

$$V_a \times 4 = H \times 4$$

$$\therefore \quad V_a = H \qquad \ldots(i)$$

Taking moments about "C" of the forces on the *R.H.S.* of "*C*",

$$V_b \times 6 = H \times 6 + 100 \times 3$$

$$\therefore \quad V_b = H + 50 \qquad \ldots(ii)$$

Resolving vertically,

$$V_a + V_b = 200 \qquad \ldots(iii)$$

Substituting for V_a and V_b,

$$H + H + 50 = 200$$

$$\therefore \quad \text{"}H\text{"} = 75 \text{ kN}$$

$$\therefore \quad V_a = 75 \text{ kN} \quad \text{and} \quad V_b = 75 + 50 = 125 \text{ kN.}$$

It can be observed that radii O_1E and O_2F are at 30° with the horizontal.

$$\therefore \quad B.M. \text{ at } E = 75 \times 4\,(1 - \cos 30°) - 75 \times 2$$

$$= (-)\ 109.81 \text{ kN-m}$$

Similarly,

$$B.M. \text{ at } F = 125 \times 6\,(1 - \cos 30°) - 75 \times 3$$

$$= (-)\ 124.52 \text{ kN-m}$$

Example 11.5: *Analyse the loaded frame and draw the B.M. diagram*

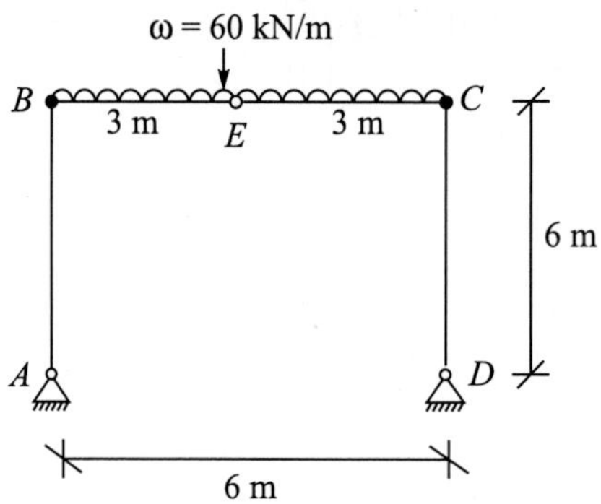

Fig. 11.11

Solution: Since the loading is symmetrical, the vertical reactions at supports are equal. Let '*H*' be the horizontal thrust at each support.

$$V_a = V_d = \frac{60 \times 6}{2} = 180 \text{ kN}$$

Taking moments about the hinge *E*, from *R.H.S.* of "*E*"

$$H \times 6 + 60 \times 3 \times 1.5 = 180 \times 3$$

$$\therefore \quad \text{"}H\text{"} = 45 \text{ kN.}$$

***B.M.* Calculations:**

B.M. at 'A' = *B.M.* at 'D' = 0

B.M. at 'B' = *B.M.* at 'C' = (–) 45 × 6 = (–) 270 kN-m

B.M. at 'E' = 0

At any section in *BE*, distant '*x*' from '*B*', the *B.M.* is given by,

$$M_X = 180x - 60 \times \frac{x^2}{2} - 45 \times 6$$

$$= 180x - 60\,\frac{x^2}{2} - 270$$

$$= 180x - 30x^2 - 270$$

At $x = 0$, *B.M.* at $B = (-)\ 270$ kN-m

At $x = 3$, *B.M.* at $E = 180 \times 3 - 30 \times 3^2 - 270 = 0$

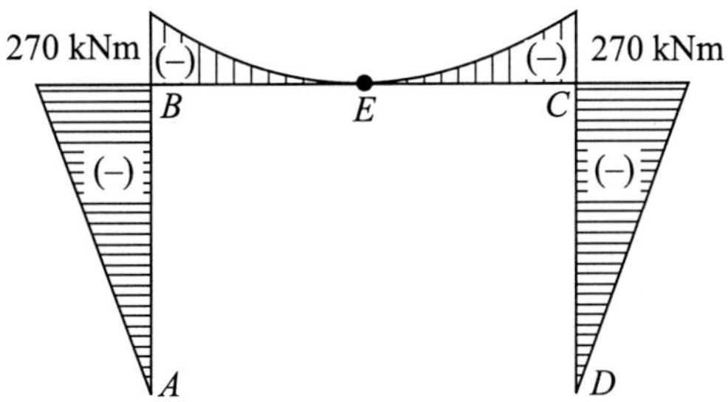

B.M. diag.

Fig. 11.12

Example 11.6: *Analyse the frame shown in Fig. 11.11 and draw B.M. diagram.*

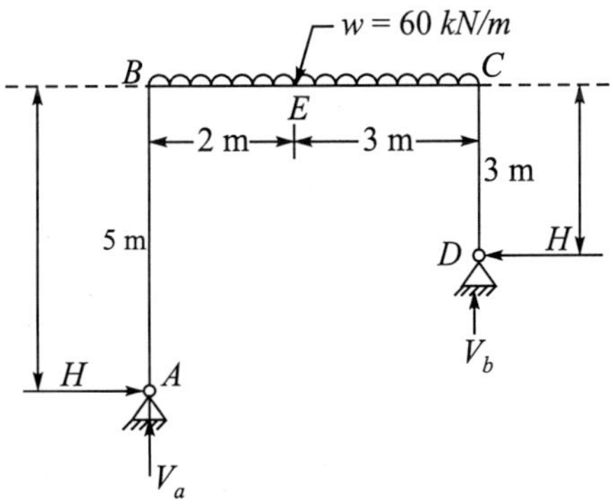

Fig. 11.13

Solution: Resolving vertically,

$$V_a + V_d = 60 \times 5 = 300 \text{ kN} \qquad \ldots(i)$$

Taking moments about '*E*' from left,

$$V_a \times 2 = H \times 5 + 60 \times 2 \times 1$$

$$\therefore \quad V_a = 2.5\,H + 60 \qquad \ldots(ii)$$

Taking moments about '*E*' from right,

$$V_d \times 3 = H \times 3 + 60 \times 3 \times 1.5$$

$$\therefore \quad V_d = H + 90 \qquad \ldots(iii)$$

Adding eqs. (*ii*) and (*iii*),

$$V_a + V_d = 3.5\,H + 150$$

But $$V_a + V_d = 300$$

$\therefore$ $$300 = 3.5\,H + 150$$

$\therefore$ $$H = 42.857 \text{ kN.}$$

Substituting for '*H*' $$V_d = 42.857 + 90 = 132.857 \text{ kN}$$

$\therefore$ $$V_a = 300 - 132.857 = 167.143 \text{ kN.}$$

***B.M.* Calculations:**

Member *AB*,

B.M. at 'A' = 0

B.M. at 'B' = (–) 42.857 × 5 = (–) 214.286 kNm

B.M. at 'E' = 0

Member *CD*,

B.M. at $D = 0$

B.M. at $C = -42.857 \times 3 = (-)\ 128.571$ kNm

Member *BE*,

B.M. at $B = -24.286$ kNm

B.M. at a distance '*x*' from '*B*',

$$M_X = 167.143x - 30x^2 - 42.857 \times 5$$
$$= 167.143x - 30x^2 - 214.286$$

B.M. at $E = 0$

Member *CE*,

B.M. at $C = -128.57$ kNm

B.M. at a distance '*x*' from *C*,

$$M_X = 132.857x - 30x^2 - 42.857 \times 3$$
$$= 132.857x - 30x^2 - 128.571$$

B.M. at $E = 0$

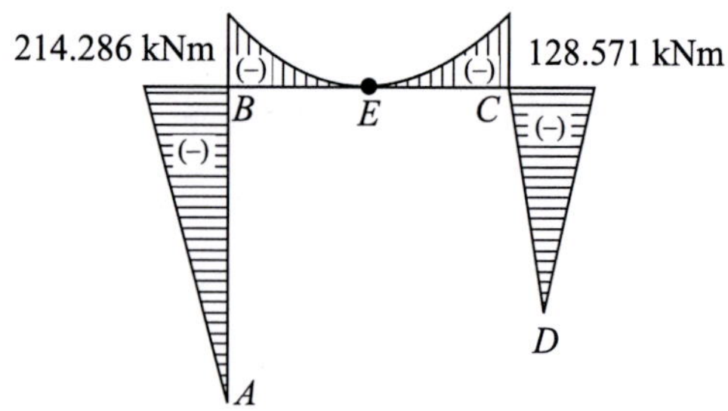

Fig. 11.14 B.M.

Example 11.7: *A symmetrical three hinged parabolic arch of span 'l' and rise 'h' carries a point load W which may be placed anywhere on the span. Find the section where the max. B.M. occurs.*

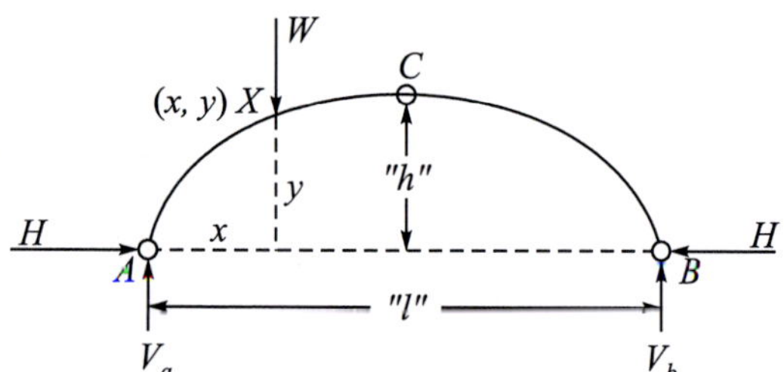

Fig. 11.15

Solution: Let the load '*W*' be at a distance '*x*' from *A*

$$\therefore \quad V_a = \frac{W(l-x)}{l}, \quad V_b = \frac{Wx}{l}$$

Taking moments about *C* of the forces on the *R.H.S.* of *C*

$$H \times h = \frac{Wx}{l} \times \frac{l}{2}$$

$$\therefore \quad H = \frac{Wx}{2h}$$

The max. *B.M.* occurs under the load,

B.M. under the load,

$$M_X = \frac{W(l-x)}{l}x - \frac{Wx}{2h}\cdot\frac{4h}{l^2}x(l-x)$$

$$\therefore \quad M_X = \frac{Wx(l-x)}{l} - \frac{2W}{l^2}x^2(l-x)$$

For the condition of absolute max. *B.M.*

$$\frac{dM_X}{dx} = 0$$

$$\therefore \quad \frac{dM_X}{dx} = \frac{W}{l}(l-2x) - \frac{2W}{l^2}(2lx - 3x^2) = 0$$

$$\therefore \quad l^2 - 2lx - 4lx + 6x^2 = 0$$

$$\therefore \quad 6x^2 - 6lx + l^2 = 0.$$

$$\therefore \quad x = \frac{6l \pm \sqrt{36l^2 - 24l^2}}{12} = \frac{6-2\sqrt{3}}{12}l, \text{ (Since } x < l)$$

$$\therefore \quad x = \frac{l}{2} - \frac{l}{2\sqrt{3}} \text{ from } A,$$

Hence, the absolute max. *B.M.* occurs at a distance of $\frac{l}{2\sqrt{3}}$ on either side of the crown.

Example 11.8: *A three hinged arch of span 16 m has its abutments A and B at depth of 4 m and 8 m below the crown C. The arch carries the load system as shown in Fig. 11.15. Determine the horizontal thrust and vertical reactions at supports.*

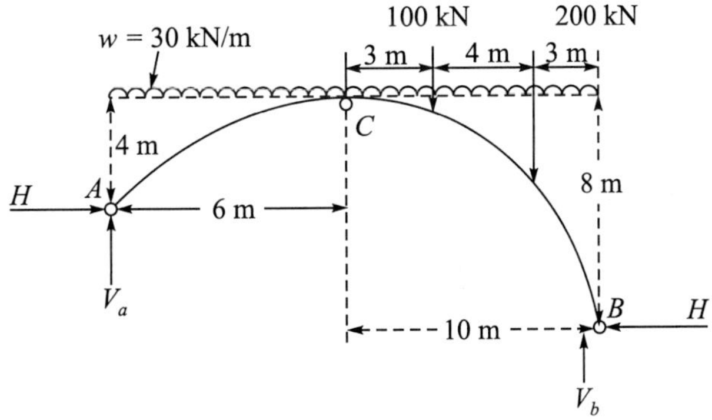

Fig. 11.16

Solutions: Taking moments about '*C*' of the forces on the *L.H.S.* of '*C*', we have,

$$6V_a = 30 \times 6 \times 3 + 4H$$

$$V_a = 90 + \frac{2}{3}H \qquad \ldots(i)$$

Similarly, taking moments about '*C*' of the forces on the *R.H.S.* of '*C*', we have,

$$10V_b = 30 \times 10 \times 5 + 100 \times 3 + 200 \times 7 + 8H$$

$$\therefore \quad V_b = 320 + \frac{4}{5}H \qquad \ldots(ii)$$

Adding eqns. (*i*) and (*ii*), we have,

$$V_a + V_b = 410 + \left(\frac{2}{3} + \frac{4}{5}\right) H = 410 + \frac{22}{15} H$$

But $\qquad V_a + V_b =$ Total load on arch

$$= 30 \times 16 + 100 + 200 = 782 \text{ kN}.$$

$$\therefore \qquad 410 + \frac{22}{15} H = 780$$

$$\therefore \qquad H = \frac{370 \times 15}{22} = 252.27 \text{ kN}$$

$$\therefore \qquad V_a = 90 + \frac{2}{3}(252.27) = \mathbf{258.18\ kN}$$

and $\qquad V_b = 780 - 258.18 = \mathbf{521.82\ kN}$

Example 11.9: *A three hinged arch ACB is supported on two columns DA and EB as shown in Fig. 11.16. The arch carries udl of 4 kN/m on the left half of span. If the lower ends of the cols are fixed, find the increase in the span of the arch. Take EI = 3 × 10^10 kNmm^2.*

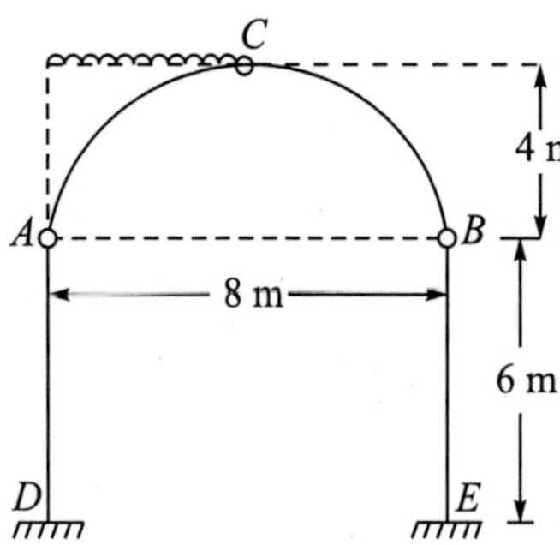

Fig. 11.17 (a)

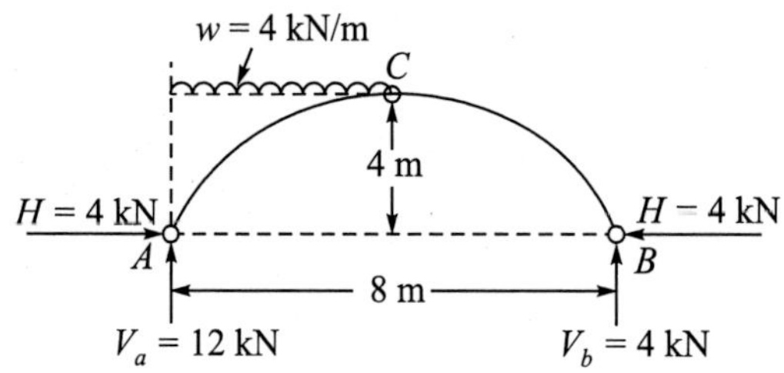

Fig. 11.17 (b)

Solution:

Considering the three hinged arch *ACB*

Taking moments about *A*,

$$V_b \times 8 = 4 \times 4 = 2$$

$$\therefore \qquad V_b = 4 \text{ kN}$$

$\therefore \qquad V_a = 16 - 4 = 12$ kN, Taking moments about *C* of thc forces on RIIS uf *C*, we have

$$H \times 4 = 4 \times 4$$

$$\therefore \qquad H = 4 \text{ kN}.$$

Now considering either column, horizontal deflection of the top of each column "δ"

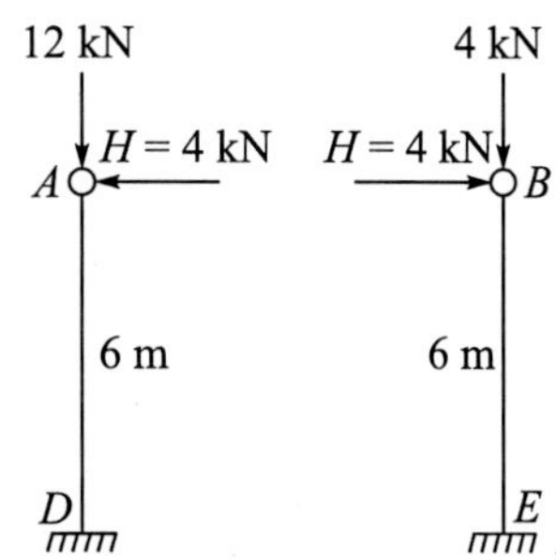

Fig. 11.17 (c)

$$\delta = \frac{Hl^3}{3\,EI} = \frac{4 \times 6^3}{3 \times 3 \times 10^{10}} \times (1000)^3$$

$$= 9.6 \text{ mm}$$

$\therefore$ Increase in span of the arch

$$= 2 \times 9.6 = 19.2 \text{ mm}.$$

SUMMARY

- A three hinged arch is a statically determinate structure, having a hingle at each abutmentor springing, and also at the crown.
- There are in all four reaction components (two at each hinge, *i.e.*, H and V). Three equations are available from the static equilibrium and one additional equation is available from the fact that the B.M. at the hinge at the crown is zero. Thus the value of 'H' can be easily calculated for any given load system.
- For larger spans, beams are very uneconomical and many a time the self weight of beams contributes to the stress in such large proportions that it is difficult to design beams for larger spans. For large spans like bridges, arches are provided instead of beams. These are nothing but curved beams that transfer loads to their plane. Arches transfer loads to abutments at springing points.
- Three hinged arch may have different shapes commonly used shapes are : Circular and parabolic.
- Horizontal thrust in an arch,

$$H = \frac{\mu_c}{y_c},$$

where μ_c = Moment at crown due to external loading

and y_c = Central rise of arch.

- If the arch is supported at different levels, then extend the arch at the upper support to a point, whose level is the same as that of a lower support. The imaginary arch may now be analysed as usual.
- When a three hinged arch is subjected to a rise in temperature, then the vertical displacement of the arch crown,

$$dy_c = \frac{l^2 + 4y_c^2}{4y_c} \times \alpha t$$

where, l = span of the arch.

y_c = central rise of the Arch.

α = co-efficient of linear expansion for the arch material.

t = change in temperature.

- When a point load W rolls over an arch, then at any section X: (*i*) the maxmimum positive bending moment takes place when the load is at the section, (*ii*) the maximum negative B.M. takes place when the load is at the crown.

SHORT ANSWER QUESTIONS

1. Define clearly the terms 'theoretical arch' and 'actual arch'.
2. What do you understand by the term Horizontal Thrust. Derive an equation for the same.
3. Derive an expression for the displacement of a three hinged arch crown, subjected to change of temperature.
4. Derive an expression for the maximum positive and negative B.M. at a section, when a udl rolls over a three hinged arch.
5. Define the terms normal thrust and radial shear force as applied in three hinged arch. Derive expression for the same.

EXERCISE FOR PRACTICE

1. A three hinged parabolic arch of span 20 m has a central rise of 5 m. Find the rise of hte arch crown if the temperature rises through 30°C. Take co-efficient of linear expansion for arch material as 12×10^{-6} per degree centigrade. **(MDU 2011)**
2. A three hinged circular arch of span 16 m and rise 4 m is subjected to two point loads of 100 kN and 80 kN at the left and right at a distance 4 m from each end. Find the reactions at the supports and resultant reaction. **(MDU 2009)**
3. A three hinged circular arch of span 21 m has a rise of 4 m. The arch is loaded with a point load of 80 kN at a horizontal distance of 6 m from the left support. Determine the horizontal thrust, two reactions and B.M. under the load. **(AMIE 2008)**
4. A three hinged parabolic arch of span 20 m and central rise of 5 m carries a point load of 200 kN at 6 m from the left hand support as shown in Fig. 11.18:
   (*a*) Find the reaction at supports *A* and *B*.
   (*b*) Draw the B.M. diagram for the arch and indicate the position of maximum B.M.

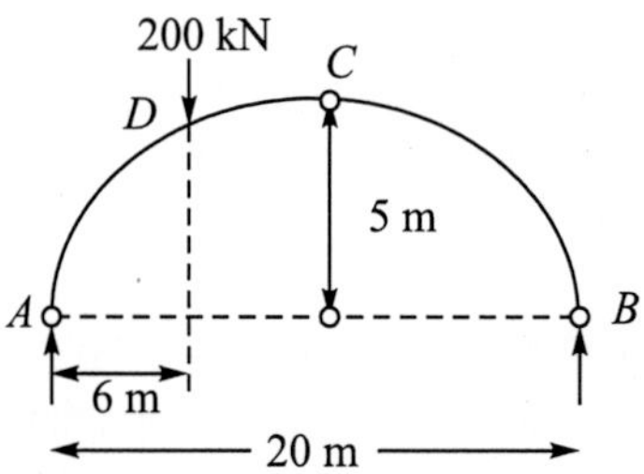

Fig. 11.18

5. A three hinged segmental arch *ACB* has a span of 30 m and a rise of 5 m. It supports a concentrated load of 150 kN at a section 7.5 m from the left support as shown in Fig. 11.19. Find the reactions at the supports and the maximum positive and negative B.M.S.

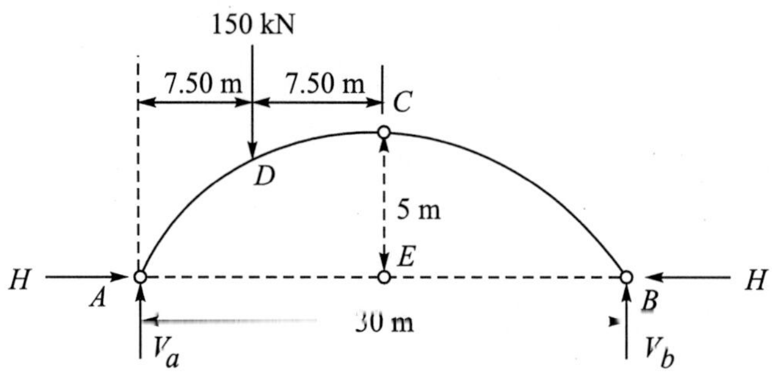

Fig. 11.19

□□□

CHAPTER

12 Shear Force and Bending Moment in Beams and Frames

Beam: A beam is a structural member subjected to a system of external forces at right angles to its axis.

Types: Depending upon the end conditions, the various types of beam are:

(*i*) Cantilever beam
(*ii*) Simply supported beam
(*iii*) Overhanging beam
(*iv*) Fixed beam
(*v*) Continuous beam

(*i*) Cantilever Beam:

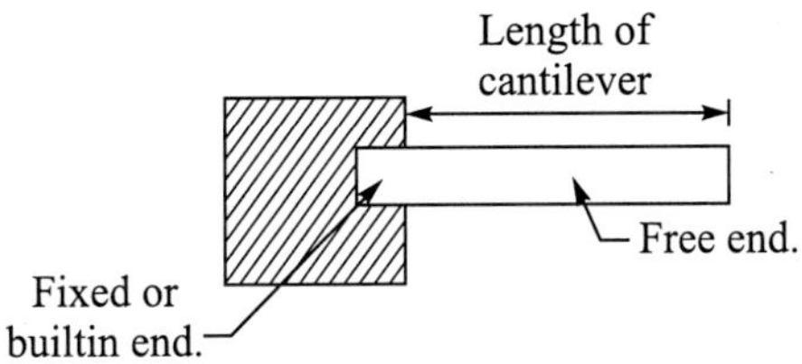

Fig. 12.1

(*ii*) Simply Supported Beam:

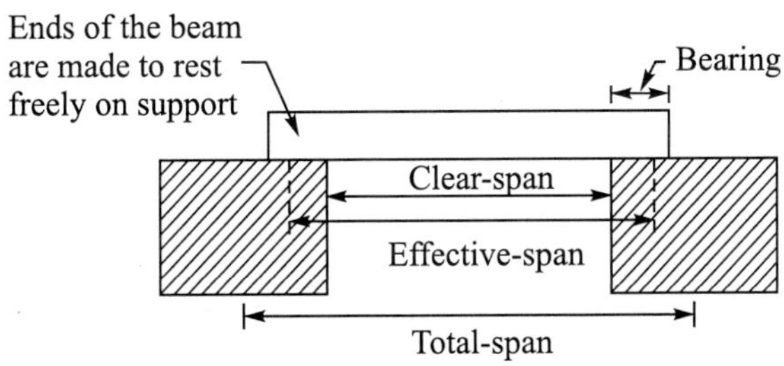

Fig. 12.2

(*iii*) Overhanging Beam:

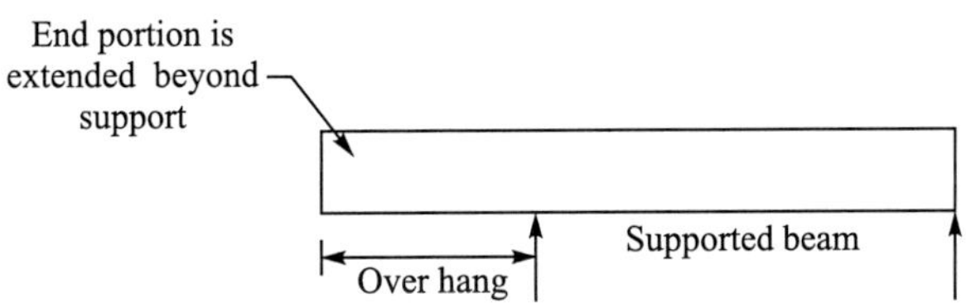

(*a*)

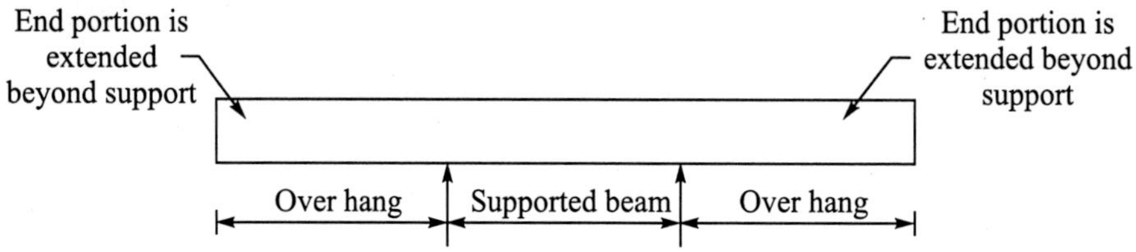

(*b*)

Fig. 12.3

(*iv*) **Fixed Beam:**

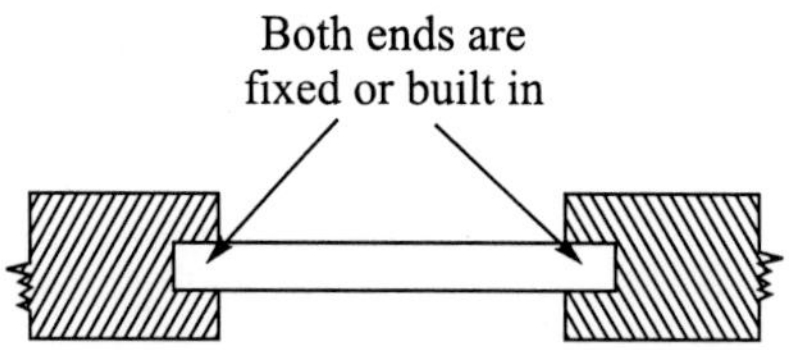

Fig. 12.4

(*v*) **Continuous Beam:** (Beam having more then two supports)

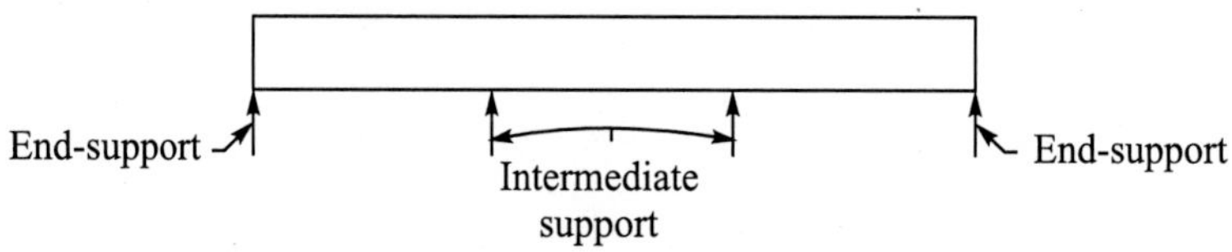

Fig. 12.5

12.1. TYPE OF END SUPPORTS OF BEAM

(*i*) **Simple or Free Support:** The supports may be in the form of walls or columns. In this case the reaction is always vertical as shown in Fig. 12.6.

Fig. 12.6

(*ii*) **Roller Support:** In this case one of the ends of the beam is supported on rollers in order to permit free movement in horizontal direction. Roller reaction is always at right angles to the roller base.

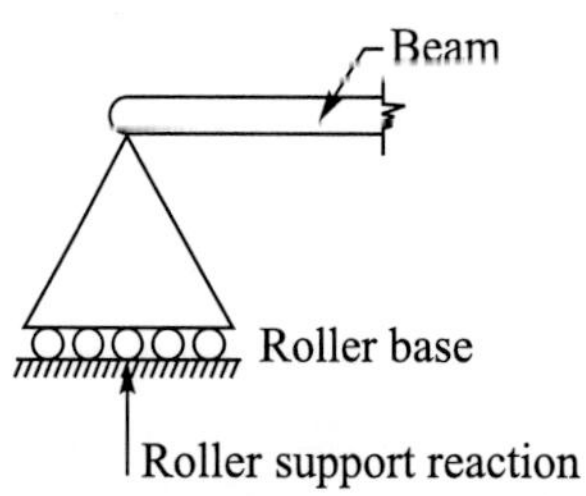

Fig. 12.7

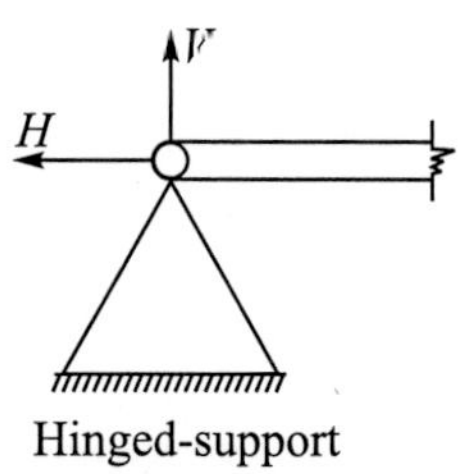

Fig. 12.8

(*iii*) **Hinged (Pinned) Support:** The reaction on such an end may be either vertical or inclined depending upon the type of loading on the beam.

(*iv*) **Fixed-Support:** Ends fixed rigidly or built in as shown in Fig. 12.9.

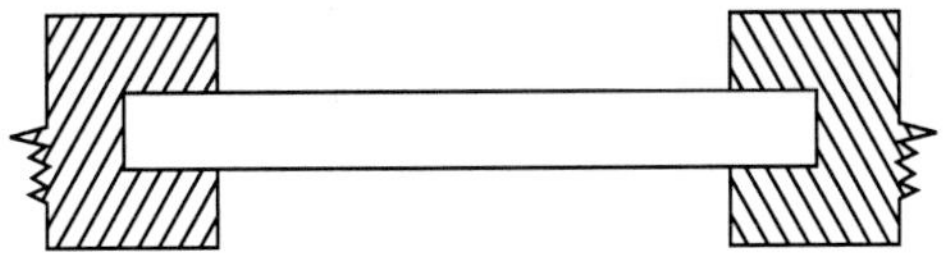

Fig. 12.9

Types of Loads:

(*a*) Concentrated or point load

(*b*) Uniformally distributed load (*udl*)

(*c*) A-couple.

(*d*) Uniformly varing load.

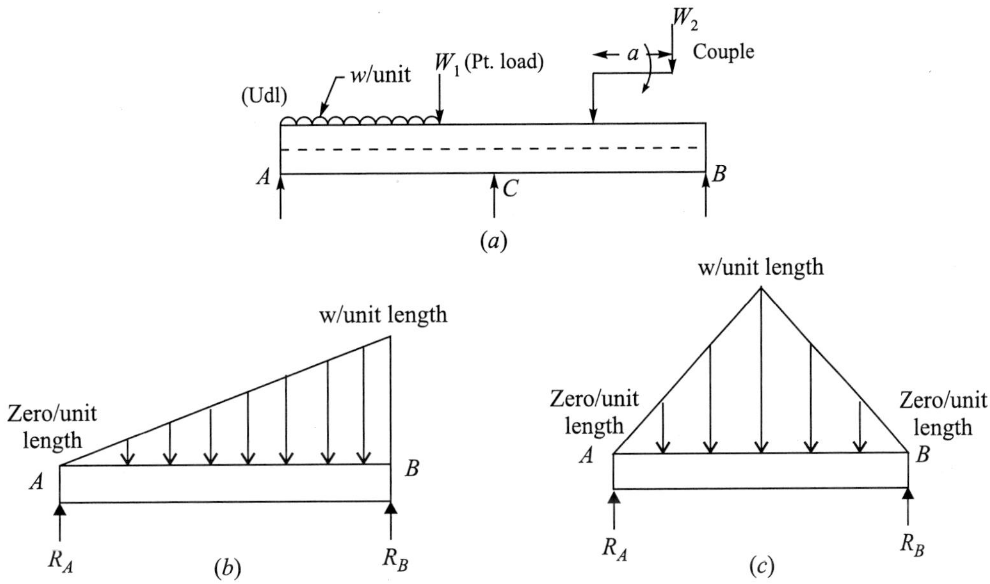

Fig. 12.10

12.2. CONCEPT OF SAGGING AND HOGGING BENDING MOMENT

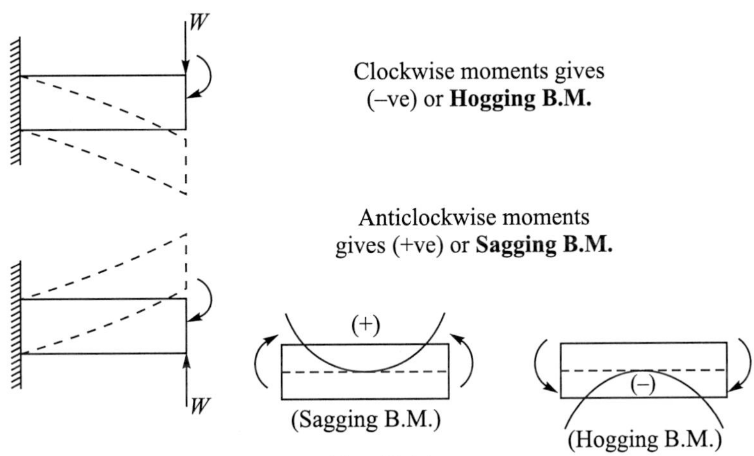

Fig. 12.11

Note :

1. Algebraic sum of moments, if clockwise, it is (–) ive *B.M.* (Hogging)
2. Algebraic sum of moments, if anticlockwise, it is (+) ive *B.M.* (Sagging)

12.3. DEFINITIONS

(*a*) *S.F.*: Shear force at **any section** can be defined as the **algebraic sum** of all the **vertical forces** acting on the beam on either side **(Left or Right) of the section.**

Sign Conventions: (*i*) Considering *RHS* of section *X-X*. If the algebraic sum of the vertical forces is downwards, it is called (+) shear, and if upwards, it is called (–) shear.

Ref. figure (*i*)

Downward vertical force $= W_1 + W_2 = 20 + 10 = 30$ kN (↓)

upward vertical forces $= W_3 = 15$ kN (↑)

Algebraic sum of total forces to the *R.H.S.* of section *X-X*,

$= 30$ kN (↓) – 15 kN (↑)

$= 15$ kN (↓), Hence (+) shear.

(*ii*) Next, consider *L.H.S.* of section *X-X*. If the algebraic sum of the vertical forces is downwards, it is called (–ive) shear, and if upwards, it is called (+) shear.

Ref. figure (*ii*)

Downward vertical forces $= W_1 + W_2 = 15 + 20 = 35$ kN (↓)

upward vertical forces $= W_3 = 20$ kN (↑)

∴ Algebraic sum of total forces to the *L.H.S.* of section *X-X*,

$= 35$ kN (↓) – 20 kN (↑) = 15 kN (↓)

Hence it is (–ive) shear.

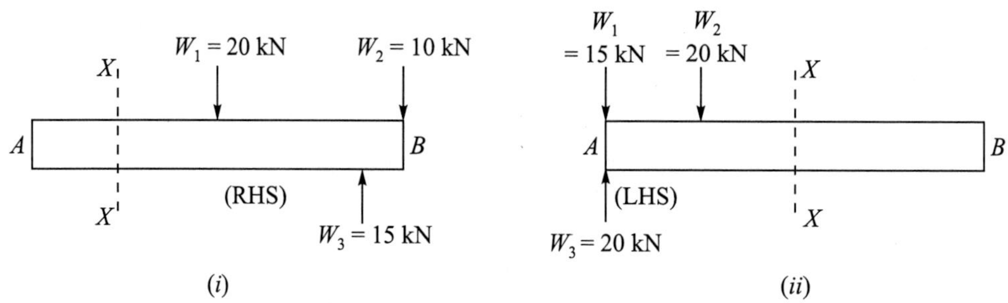

Fig. 12.12

Sign Conventions for Shear Forces

(*i*) Upwards forces to the left of a section and downward forces to the right of it will be considered as +ve. (shown in Fig. 12.13 (*a*)) *e.g.*, wristwatch clockwise movement.

Fig. 12.13

(*ii*) Downward forces to the left of a section and upward forces to the right of it will be considered as –ve. (Shown in Fig. 12.14 (*b*)) *e.g.*, wristwatch anticlockwise movement.

Bending-Moment: (*B.M.*)

B.M: The *B.M.* at any section can be defined as the algebraic sum of moments of all the vertical forces either to the left or right of the section considered.

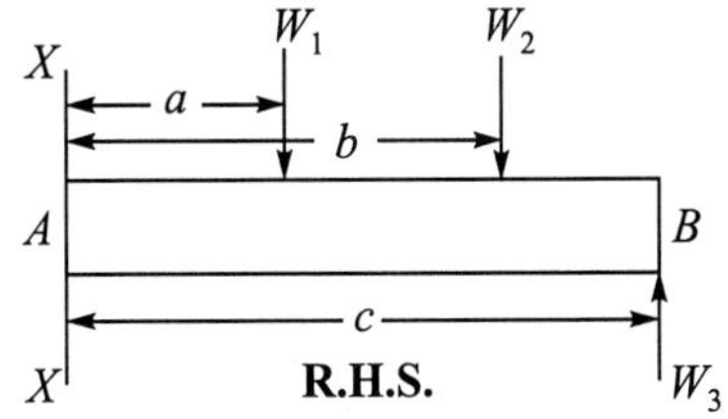

Fig. 12.14 (a)

Sign Conventions:

(*i*) Considering the *R.H.S.* of the section *X-X.* If the algebraic sum of the moments is anticlockwise ↺, *B.M.* is (+), and if clockwise ↻, *B.M.* is (–).

(*ii*) Considering the *L.H.S.* of the section *X-X.* If the algebraic sum of the moments is clockwise ↻, then the *B.M.* is (+), and if anticlockwise ↺, it is (–) ive.

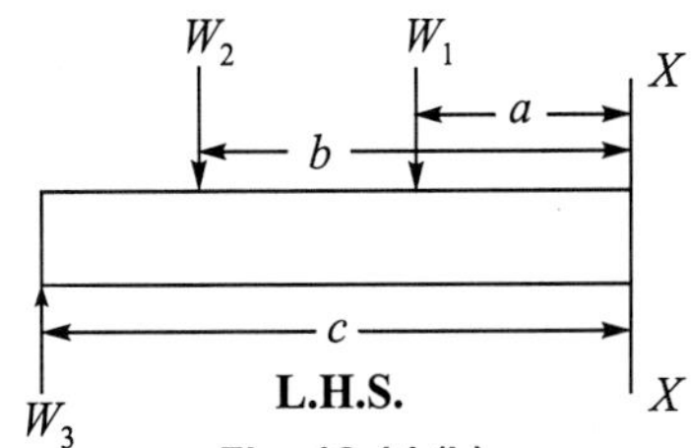

Fig. 12.14 (b)

Definitions of *S.F.* and *B.M.* Diagrams:

(*i*) *S.F.* diagram for a structural member is a diagram which shows the values of *S.F.* at various sections of the member.

(*ii*) *B.M.* diagram for a structural member is a diagram which shows the values of *B.M.* at various sections of the member.

Note: The (+) ive values of *S.F.* and *B.M.* are plotted above base line and the –ive values below the base line. For *Udl* loading it is represented by a parabolic curve. We will now analyse all types of beams for finding *S.F.* and *B.M.* values under different loading conditions.

Salient Features of Shear Force Diagram and Bending Moment Diagram

The following are the important points for drawing shear force and bending moment diagrams:

(*i*) The base line equal to the length (span) of the beam on certain scale is drawn. The positive values of shear force and bending moments are plotted above the base line, and negative values below the base line.

(*ii*) If there is a vertical load (including reaction) at a section, the shear force diagram will increase or decrease suddenly *i.e.*, by a vertical straight line at the section.

(*iii*) The shear force between any two vertical loads will be constant and hence shear force diagram between vertical loads will be horizontal. The bending moment diagram will be inclined.

(*iv*) If there is a uniformly distributed load (*udl*) between two section, the shear force diagram will be an inclined straight line, and the bending moment diagram will be a curve.

(*v*) For uniformly varying loads, shear force diagram will be a second degree parabolic curve and bending moment diagram will be cubic curve (Third degree).

(*vi*) The bending moment at the two supports of a simply supported beam and at the free end of a cantilever will be zero.

(*vii*) For a cantilever beam whole of the shear force diagram will be either above the base line (+ve) or below the base line (–ve) but whole of the bending moment diagram will be always below the base line (–ve).

(*viii*) For a simply supported beam, the shear force diagram will be above the base line for some portion of the beamed below the base line for rest of the beam. But whole of the bending moment diagram will be always above the base line (+ve).

(*ix*) For a overhanging beam, both shear force diagram and bending moment diagram each will be on both sides of the base lines.

(*a*) Cantilever:

Case I: Carring Concentrated Load "*W*" at the Free End.

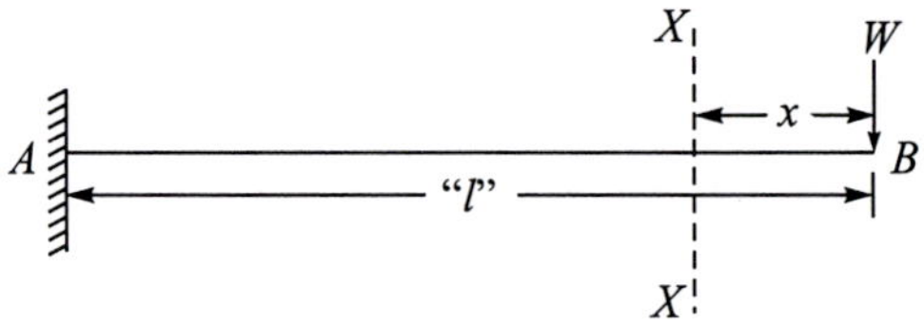

Fig. 12.15 (a)

Consider a cantilever beam *AB* of length "*l*" and carring a concentrated load *W* at the free end *B* as shown in Fig. 12.15.

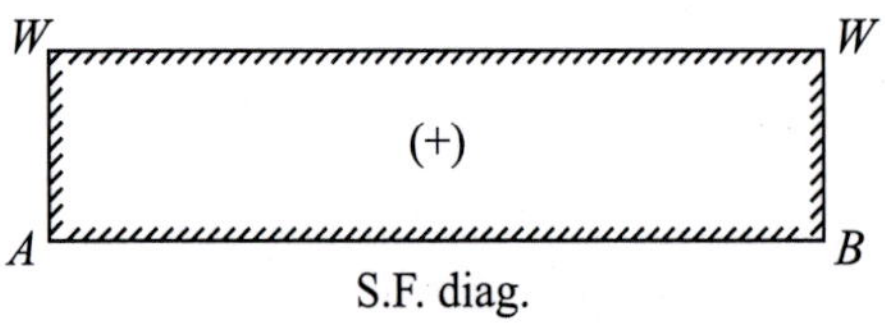

Fig. 12.15 (b)

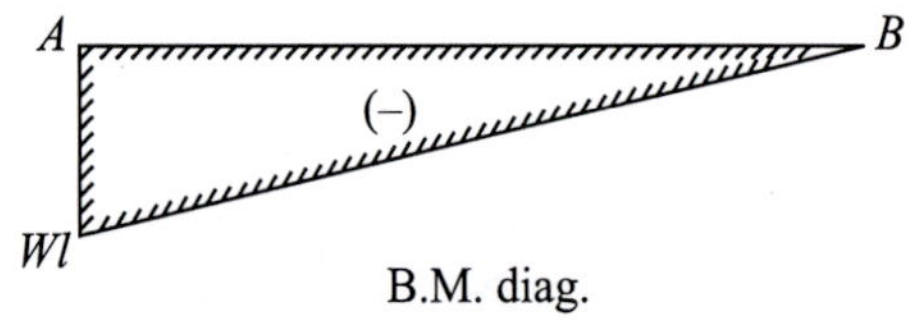

Fig. 12.15 (c)

(*a*) *B.M.* Diagram: At the see – *X-X* the B.M. = (–) $W \times x$, (since clockwise) which becomes (–) ($W \times l$), when $x = l$ (Again being clockwise) Hence the B.M. as above.

(*b*) *S.F.* Diagram: Consider section *X-X*, at a distance '*X*' from the free end.

$$S.F. \text{ at } X\text{-}X = S_X = (+)\ W \qquad \text{(Since downwards)}$$

This will remain constant between $A + B$ as there is no other load between $A + B$. Hence the *S.F.* diagram drawn above.

Case II: Cantilever Beam carring Uniformly Distributed Load (*udl*)

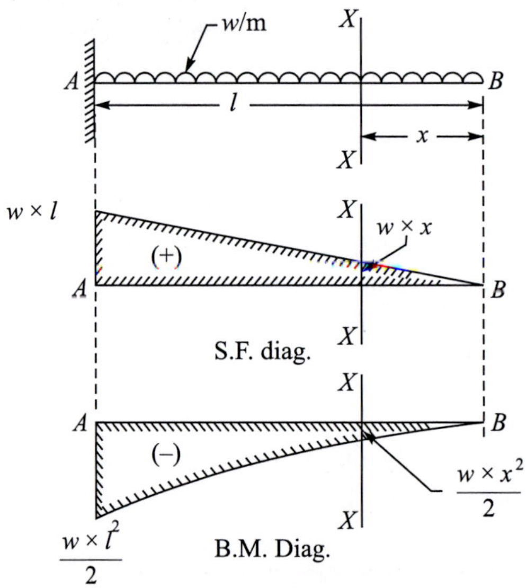

Fig. 12.16

***S.F.* diagram:** Consider any *X-X*, at a distance "*x*" from *B*.

$$S.F. \text{ at } X\text{-}X,\ S_X = (+)\ w \times x \qquad \text{(Downwards)}$$

at $x = \text{"}l\text{"},\ S_X = (+)\ w \times l$ (Downwards)

***B.M.* diagram:** *B.M.* at section *X-X*, distance "*x*" from *B*.

$$M_X = (-)\ wx \times \frac{x}{2} = (-)\ \frac{wx^2}{2} \qquad \text{(Clockwise)}$$

$\therefore$ at $x = \text{"}l\text{"},\ M_X = (-)\ \dfrac{w \times l^2}{2}$ (Clockwise)

Thus we find, in the case of *Udl*, *S.F.* is according to linear law and *B.M.* varies as per parabolic law.

Case III: Cantilever of length "*l*" carring a *Udl* w/m and a concentrated load '*W*' at the free end.

***S.F.* Diagram:** Consider any section *X-X*, at a distance "*x*" from '*B*'.

$$S.F. \text{ at } X\text{-}X,\ S_X = W \times wx$$

at $x = 0$, *i.e.* at "*B*", $S_X = (+)\ W$ (Downwards)

at $x = l$, *i.e.* at "*A*", $S_X = (+)\ (W + wl)$ (Downwards)

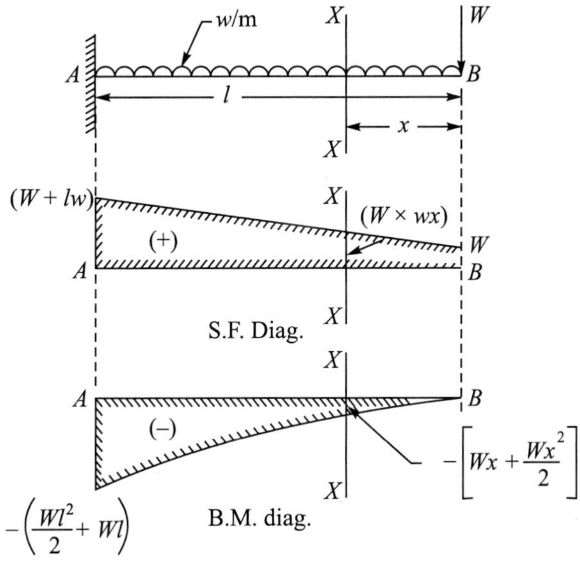

Fig. 12.17

***B.M.* Diagram:** *B.M.* at section *X-X*, distance "*x*" from *B*.

$$M_X = (-)\left[W \times x + \frac{wx^2}{2}\right] \qquad \text{(Clockwise)}$$

and at $x = l$,

$$M_X = (-)\left[W \times l + \frac{wl^2}{2}\right] \qquad \text{(Clockwise)}$$

Here also, *S.F.* follows linear law and *B.M.* follows parabolic law.

Case IV: Cantilever Carring Partial *Udl*

***S.F.* Diagram:** Consider any section *X-X*, between *D* and *B*, distance "*x*" from *B*.

$$S.F. \text{ at } X\text{-}X,\ S_X = (+)\ w \times x$$

at $x = 0,\ S_X = 0.$

At $x = a,\ S_X = (+)\ w \times a$

Next consider section *X-X*, between *D* and *A*, distance "*x*" from *B*.

$$S_X = (+)\ w \times a \quad \text{and} \quad x = \text{“}l\text{”},\ S_X = (+)\ wa$$

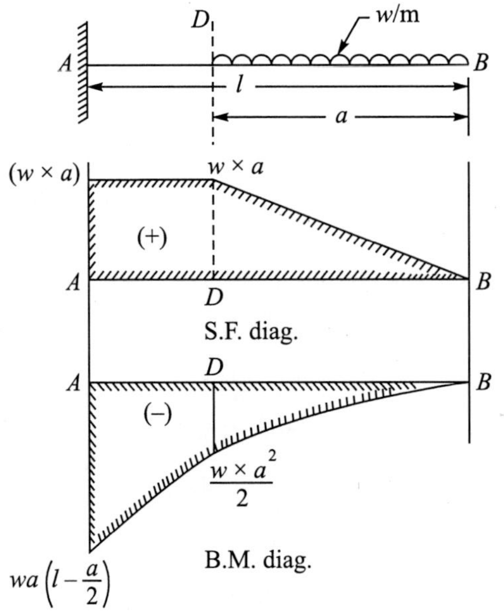

Fig. 12.18

***B.M.* Diagram:** Between position *DB*, *B.M.* at section *X-X*, distance "*x*" from *B*.

$$M_X = (-)\ \frac{w \times x^2}{2}$$

At $x = 0,\ M_X = 0$

At $x = a,$

$$M_X = (-)\ \frac{w \times a^2}{2}.$$

Between portion *D* and *A*, at a distance *x* form *B*.

$$M_X = (-)\ w \times a\left(x - \frac{a}{2}\right)$$

At $x = a,$

$$M_X = (-)\ w \times a\left(a - \frac{a}{2}\right) = (-)\ \frac{w \times a^2}{2}$$

and at $x = l,\ M_X = (-)\ w \times a\left(l - \frac{a}{2}\right)$

Example 12.1: *Cantilever carring several concentrated loads:*

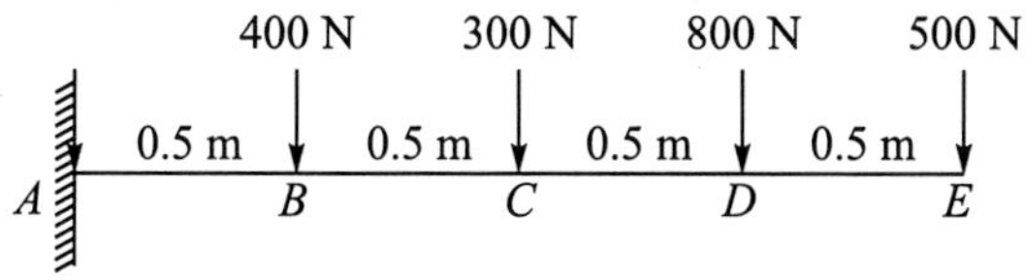

Fig. 12.19

Solution: At any section, between D and E, distance "x" from E.

$$S.F.,\ S_X = (+)\ 500\text{ N}$$
$$B.M.,\ M_X = (-)\ 500x$$

At $x = 0,\ M_X = 0.$

At $x = 0.5$ m

$$M_X = -\ 500 \times 0.5 = (-)\ 250\text{ Nm}$$

At any section, between C and D, distance "x" from E.

$$S.F.,\ S_X = (+)\ 500 + (800) = (+)\ 1300\text{ N.}$$
$$B.M., = M_X = (-)\ 500 \times x - 800\ (x - 0.5)$$
$$= -\ 1300x + 400$$

At $x = 0.5$

$$M_X = (-)\ 1300 \times 0.5 + 400$$
$$= (-)\ 250\text{ Nm.}$$

At $x = 1.0$ m,

$$M_X = -\ 1300 + 400$$
$$= (-)\ 900\text{ Nm.}$$

At any section, between B and C, distance "x" from E.

$$S.F.,\ S_X = (+)\ 500 + 800 + 300$$
$$= 1600\text{ N.}$$
$$B.M. = M_X = (-)\ 500\ x - 800\ (x - 0.5) - 300\ (x - 1)$$
$$= (-)\ 1600x + 700\text{ Nm}$$

At $x = 1$ m.

$$M_X = -\ 1600 + 700$$
$$= -\ 900\text{ Nm}$$

At $x = 1.5$ m,

$$M_X = -\ 1600 \times 1.5 + 700$$
$$= (-)\ 1700\text{ Nm}$$

At any section, between A and B, distance "x" from E.

$$S.F. = S_X = (+)\ 500 + 800 + 300 + 400$$
$$= 2000\text{ N}$$
$$B.M. = M_X = (-)\ 500\ x - 800\ (x - 0.5) - 300\ (x - 1) - 400\ (x - 1.5)$$
$$= (-)\ 2000x + 1300\text{ Nm}$$

At $x = 1.5$ m,

$$M_X = (-)\ 2000 \times 1.5 + 1300$$
$$= (-)\ 1700\text{ Nm}$$

At $x = 2$ m,

$$M_X = (-)\ 2000 \times 2 + 1300 = (-)\ 2700 \text{ Nm}$$

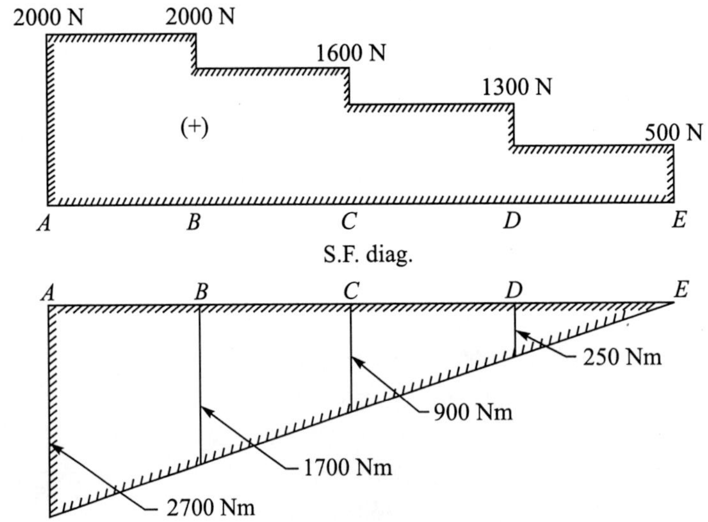

Fig. 12.20

Example 12.2: *A cantilever beam 4 m long carries a gradually varying load, zero at the free end to 3 kN/m at the fixed end. Draw B.M. and S.F. diagram for the beam.*

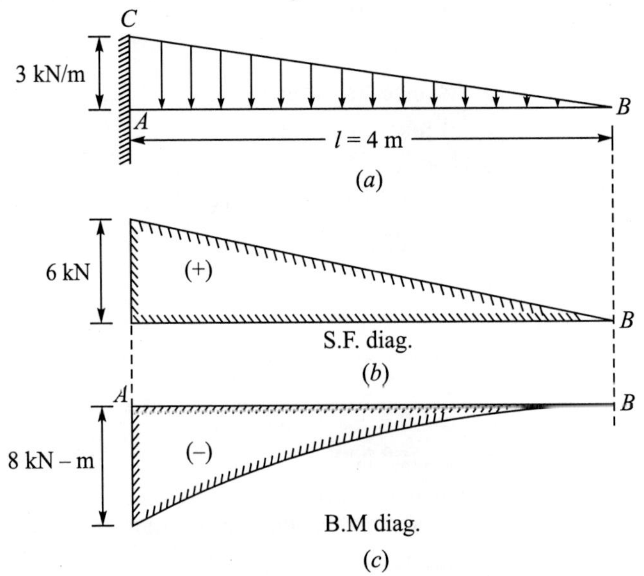

Fig. 12.21

Solution: Given, span "l" = 4 m and gradually varing load at A (w) = 3 kN/m.

***S.F.* diagram:**

S.F. at free end $B = F_B = 0$

S.F. at *A* = Total load on beam, $F_A = (+)\ \frac{1}{2} \times 3 \times 4 = +6$ kN = Area of $\Delta\ ABC$

***B.M.* diagram:**

B.M. at free end *B* = 0, $M_B = 0$, B.M. at *A* = – (Area of ΔABC × Distance of *C.G.* of load Δ from *A*)

i.e., $$M_A = (-)\left(6 \times \frac{1}{3} \times 4\right) = (-)\ 8 \text{ kN-m.} \qquad (-\text{ sign due to hogging B.M.})$$

The *S.F.* and *B.M.* diagram are drawn thus.

Example 12.3: *A cantilever beam of 2 m span is subjected to a gradually varing load from 2 kN/m to 5kN/m as shown in Fig. 12.22. Draw the S.F. and B.M. diagram for the beam.*

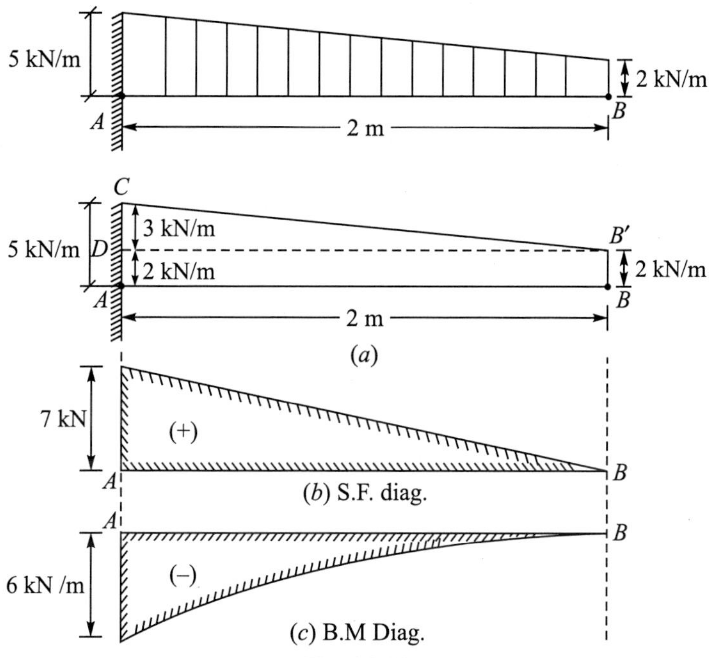

Fig. 12.22

Solution: Given, span "*l*" = 2 m, Gradually varing load at *A* (*wa*) = 5 kN/m and varing to 2 kN/m at *B*.

The load may be assumed to be split up into: (*i*) a *Udl* w_1 of 2kN/m over the entire span and (*ii*) a gradually varing load w_1 from zero at *B* to 3kN/m at *A*.

***S.F.* diagram:** $F_B = 0$

F_A = Area of Load Diagram

$$= +\left[\underbrace{(2 \times 2)}_{\text{Area of rectangle } BB'DA} + \underbrace{\left(\frac{3 \times 2}{2}\right)}_{\text{Area of } \Delta B'DC}\right] = (+)\ 7 \text{ kN.}$$

***B.M.* diagram:** $M_B = 0$

M_A = – [(Area of Loaded rectangle *BB′DA*) × Distance of its C.G. from *A*]
+ [(Area of Δ *B′DC*) × Distance of its C.G. from *A*]

$$= -\left[(2\times 2)\times(1)+\left(\frac{1}{2}\times 3\times 2\right)\times\left(\frac{1}{3}\times 2\right)\right]$$

$$= (-)\ 6\text{ kN}\quad (-\text{ sign due to Hogging})$$

(*b*) "Simply Supported Beam"

Case I: Having concentrated load at mid-span.

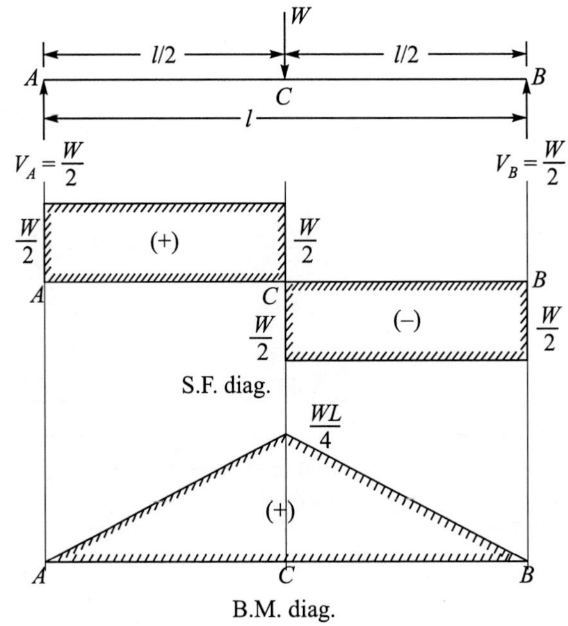

Fig. 12.23

Solution: The vertical reactions at A and B viz.,

$$V_A = V_B = \frac{W}{2}.$$

For any section between A and C,

S.F. $\qquad S_X = (+)\ \frac{W}{2}$

and for any section between C and B,

S.F. $\qquad S_X = (\ )\ \frac{W}{2}$

At pt. "C", the *S.F.* changes from $(+)\ \frac{W}{2}$ to $(-)\ \frac{W}{2}$.

At any section, between A and C, distance "x" from A.

$$B.M.\ M_X = (+)\ \frac{W}{2}\times x \qquad \text{(Sagging)}$$

At $\qquad x = 0,$

$\qquad M_X = 0,$

At $\qquad x = \frac{l}{2},\ M_X = W\times\frac{l}{4}$

Hence the *B.M.* increases uniformally from zero at "*A*" to $\frac{W \times l}{4}$ at "*C*".

Similarly, the *B.M.* decreases uniformally from $\frac{W \times l}{4}$ at "*C*" to zero at "*B*".

Case II: Having concentrated load placed eccentrically on the span.

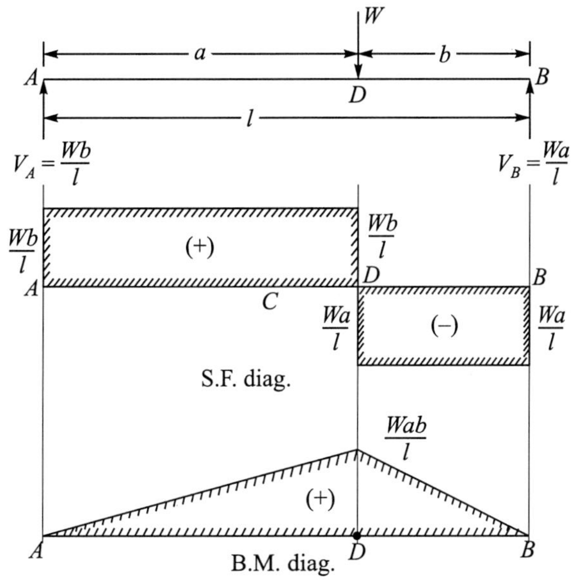

Fig. 12.24

Taking moments about *A*,

M^A,

$$V_B \times l = W_a$$

$\therefore$
$$V_B = \frac{Wa}{l}$$

and
$$V_A = W - \frac{Wa}{l} = \frac{W(l-a)}{l} = \frac{Wb}{l}$$

For any section between *A* and *D*.

S.F.,
$$S_X = V_A = (+)\ \frac{Wb}{l}$$

For any section between *D* and *B*.

S.F.,
$$S_X = (-)\ V_B = (-)\ \frac{Wa}{l}$$

For any section between *A* and *D*, distant "*x*" from "*A*"

B.M.,
$$M_X = (+)\ \frac{Wb}{l} \times x \qquad \text{(Sagging)}$$

At
$$x = 0,\ M_X = 0$$

At
$$x = a,\ M_X = \frac{Wab}{l}$$

It may be observed from the *S.F.* and *B.M.* diagrams that max. *B.M.* occurs at '*D*', where *S.F.* changes its sign. This is called **point of contraflexure.**

Case III: Simply supported beam carring no. of concentrated loads.

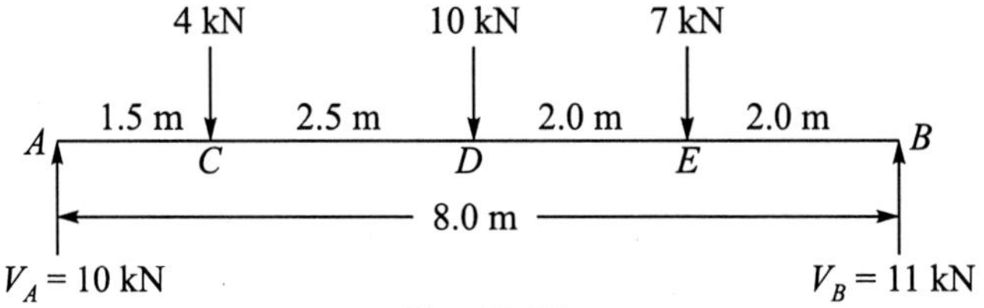

Fig. 12.25

Solution: For equilibrium of system, taking moments about *A*,

$$V_B \times 8 = 4 \times 1.5 + 10 \times 4 + 7 \times 6 = 88 \text{ kN}$$

$$\therefore \quad V_B = \frac{88}{8} = 11 \text{ kN.}$$

$$\therefore \quad V_A = 21 - 11 = 10 \text{ kN}$$

S.F. between *A* and *C* $= 10$ kN

S.F. between *C* and *D* $= 10 - 4 = 6$ kN

S.F. between *D* and *E* $= 10 - 4 - 10 = (-)\ 4$ kN

S.F. between *E* and *B* $= 10 - 4 - 10 - 7 = (-)\ 11$ kN

Now, *B.M.* at *A* $= 0$

B.M. at *C* $= 10 \times 1.5 = (+)\ 15$ kNm (Sagging)

B.M. at *D* $= 10 \times 4 - 4 \times 2.5 = (+)\ 30$ kNm (Sagging)

B.M. at *E* $= 11 \times 2 = (+)\ 22$ kNm (Sagging)

Again note that, max. *B.M.* occurs where the *S.F.* changes its sign.

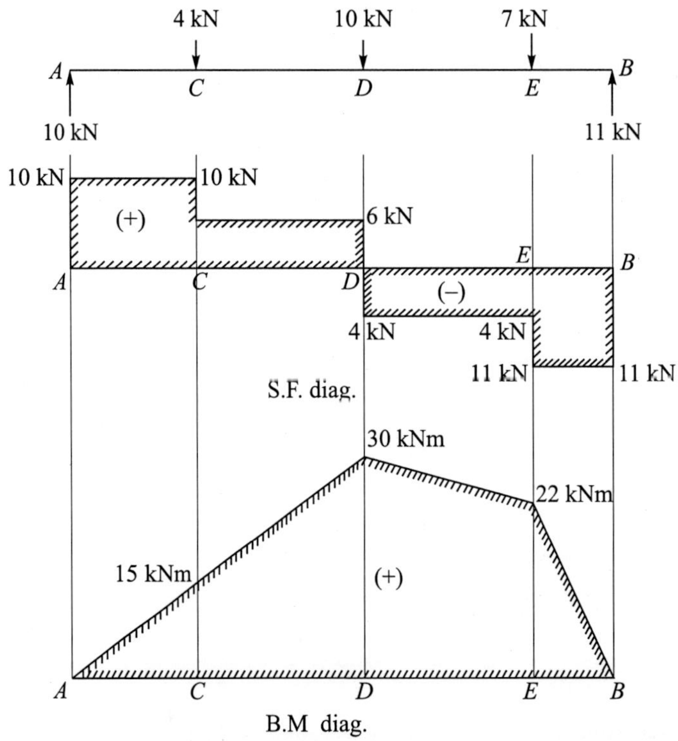

Fig. 12.26

Case IV: *S.S.* beam carrying *Udl* over the whole span.

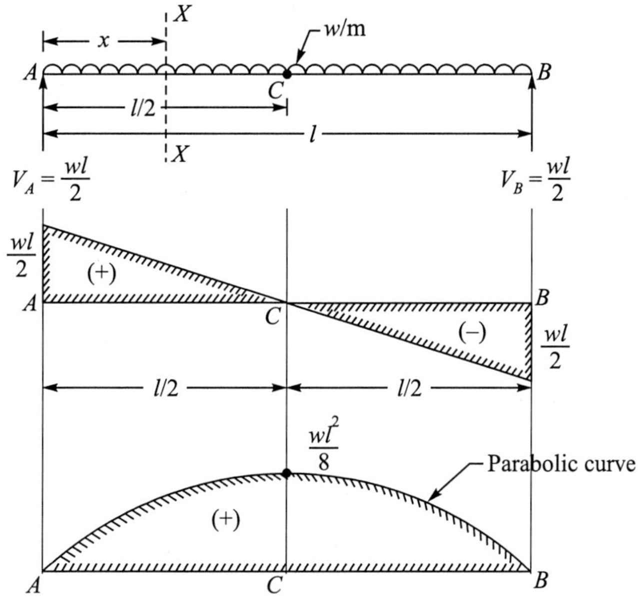

Fig. 12.27

Solution: $$V_A = V_B = \frac{wl}{2}$$ (Since loading is symmetrical)

Consider section- *X-X*, distant "*x*" from *A*,

S.F., $$S_X = (+)\ V_A - wx = (+)\ \frac{wl}{2} - wx$$

and *B.M.*, $$M_X = (+)\ V_A x - \frac{wx^2}{2} = \frac{wlx}{2} - \frac{wx^2}{2}$$

∴ $$M_X = (+)\ \frac{w}{2} x\,(l - x)$$

At $$x = 0,\ S_X = \frac{wl}{2} \quad \text{and} \quad M_X = 0$$

At $$x = l,\ S_X = \frac{wl}{2} - wl = (-)\ \frac{wl}{2} \quad \text{and} \quad M_X = 0$$

At $$x = \frac{l}{2},\ S_X = \frac{wl}{2} - \frac{wl}{2} = 0,$$

$$M_X = (+)\ \frac{w}{2} \times \frac{l}{2}\left(l - \frac{l}{2}\right) = (+)\ \frac{wl^2}{8}$$

The *S.F.* diagram is a st. line *S.F.* uniformally changes from (+) $\frac{wl}{2}$ at *A* (–) $\frac{wl}{2}$ at *B* and *S.F.* at mid pt. of span = 0. However, *B.M.* diagram is parabola.

Case V: *S.S.* beam carring *udl* on an intermediate part of the span.

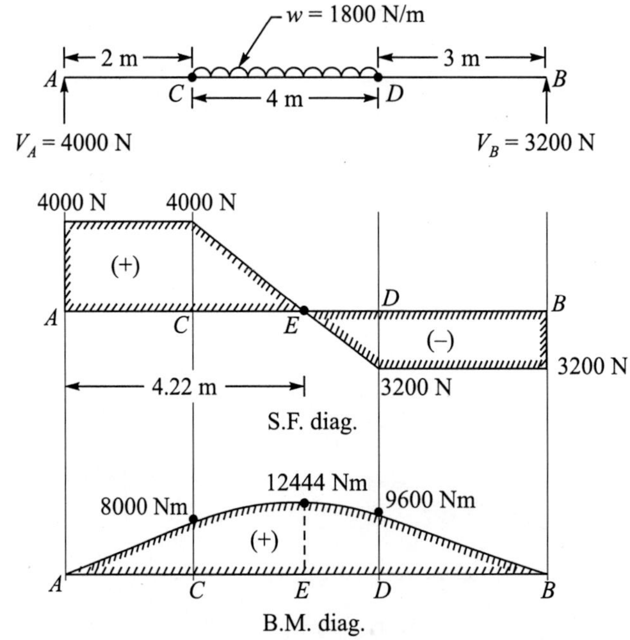

Fig. 12.28

Solution: For equilibrium of the system, taking moments about *A*, M^A,

$$V_B \times 9 = 1800 \times 4\ (2 + 2)$$

$\therefore$ $V_B = 3200$ N

$\therefore$ $V_A = 1800 \times 4 - 3200 = 4000$ N.

At any section between *A* and *C*,

S.F., $S_X = (+)\ 4000$ N.

Section between *D* and *B*,

S.F., $S_X = (-)\ 3200$ N

Section between *C* and *D*, distant "*x*" from "*A*",

S.F., $S_X = (+)\ 4000 - 1800\ (x - 2)$

At $x = 2$ m, $S_X = (+)\ 4000$ N.

$x = 6$ m, $S_X = (+)\ 4000 - 1800 \times 4$

$= (-)\ 3200$ N.

Let *S.F.* be zero at "*x*" from *A*,

$\therefore$ $4000 - 1800\ (x - 2) = 0$

$\therefore$ $x = 4.22$ m.

Next, *B.M.* at $A = 0$

B.M. at $B = 0$

B.M. at $C = (+)\ 4000 \times 2 = (+)\ 8000$ Nm

B.M. at $D = (+)\ 3200 \times 3 = (+)\ 9600$ Nm

At any section, in *CD*, distant "*x*" from *A*,

B.M., $$M_X = 4000x - 1800\,\frac{(x-2)^2}{2}$$

$$= 4000x - 900\,(x-2)^2, \text{ since } S.F. \text{ is zero at } x = 4.22 \text{ mm}$$

∴ Substituting, $$x = \frac{38}{9}\text{ m} = 4.22 \text{ m. to max. } B.M.$$

∴ *B.M.*, $$M_X = (+)\,4000 \times \frac{38}{9} - 900\left(\frac{38}{9} - 2\right)^2 = 12444.4 \text{ N-m.}$$

Example 12.4: *A simply supported beam of 5 m span carries a triangular load of 30 kN. Draw S.F. and B.M. diagram for the beam.*

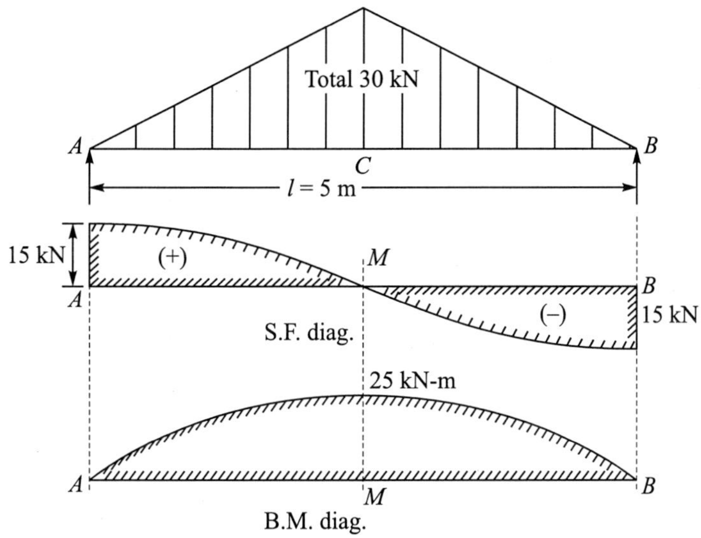

Fig. 12.29

Solution: Given, span "*l*" = 5 m and total triangular load (*w*) = 30 kN.

By symmetry, $$R_A = R_B = \frac{30}{2} = 15 \text{ kN.}$$

***S.F.* diagram:** The values at supports are:

$$F_A = (+)\,R_A = (+)\,15 \text{ kN}$$
$$F_B = (-)\,R_B = (-)\,15 \text{ kN}$$

***B.M.* diagram:** It is zero at *A* and *B* also the max. *B.M.* will occur at the centre *i.e.*, at *M*, where the *S.F.* changes sign.

$$M_{max} = \frac{wl}{6} = \frac{30 \times 5}{6} = 25 \text{ kN-m}$$

The diagrams are drawn thus.

Example 12.5: *The intensity of loading on a simply supported beam of 6 m span increase gradually from 800 N/m at one end to 2000 N/m run at the other as shown in Fig. 12.30. Find the position and amount of max B.M. Also draw the S.F. and B.M. diagram.*

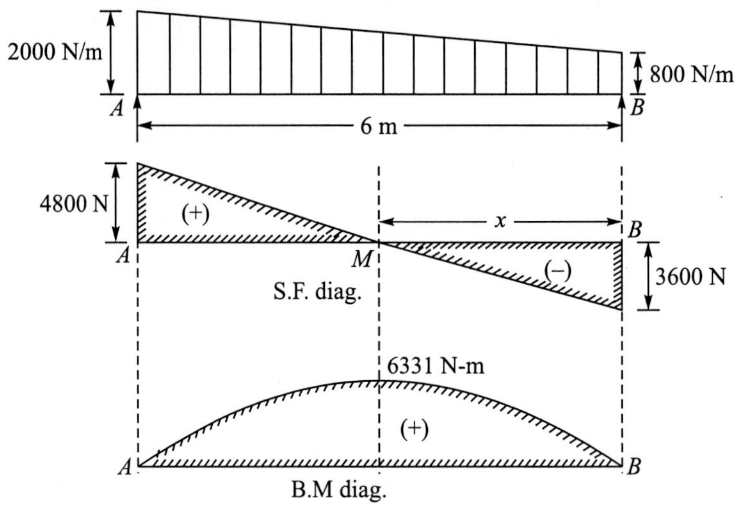

Fig. 12.30

Solution: Given span (l) = 6 m, gradually varing load at A (w_A) = 2000 N/m to 800 N/m (wb) at B. The weight may be assumed to be split up with: (i) a *Udl* (w_1) of 800 N/m over the entire span and (ii) a gradually varing load of zero at B to 1200 N/m at A.

∴ Total *Udl*, $w_1 = 800 \times 6 = 4800$ N

$$R_B = \frac{4800}{2} + \frac{3600}{3} = 3600 \text{ N}$$

and $$R_A = \frac{4800}{2} + \frac{2 \times 3600}{3} = 4800 \text{ N}$$

***S.F.* diagram:** $F_A = (+)\ R_A = 4800$ N

$F_B = (–)\ R_B = (–)\ 3600$ N

***B.M.* diagram:** The *B.M.* is zero at A and B and the max. *B.M.* will occur at M, where the *S.F.* changes sign.

Max. *B.M.*: We know that the max. *B.M.* will occur at M, where the *S.F.* changes sign. Let "x" be the distance between B and M. We also know that *S.F.* at a distance 'x' from M,

$$= (–)\ 3600 + 800x + \frac{1}{2} \times 1200x \times \frac{x}{6}$$

$$= –3600 + 800x + 100x^2 – 100x^2 + 800x – 3600$$

Now to find the position of M, left us equate the above equation to zero, *i.e.*,

$$100x^2 + 800x – 3600 = 0$$

or $$x^2 + 8x – 36 = 0$$

∴ $$x = \frac{-8 \pm \sqrt{(8)^2 + (4 \times 36)}}{2} = 3.21 \text{ m}$$

and $$B.M. \text{ at } M = M_X = 3600x - \left[800 \times \frac{x^2}{2}\right] - \left[\frac{1}{2} \times 1200x \times \frac{x}{6} \times \frac{x}{3}\right]$$

$$= 3600x - 400x^2 - \frac{100}{3}x^3$$

$$= (3600 \times 3.21) - 400 \times (3.21)^2 - \frac{100}{3} \times (3.21)^3 \text{ N-m}$$

$$= 11556 - 4122 - 1102 = \mathbf{6332\ N\text{-}m}$$

Overhang Beam Cases

Example 12.6:

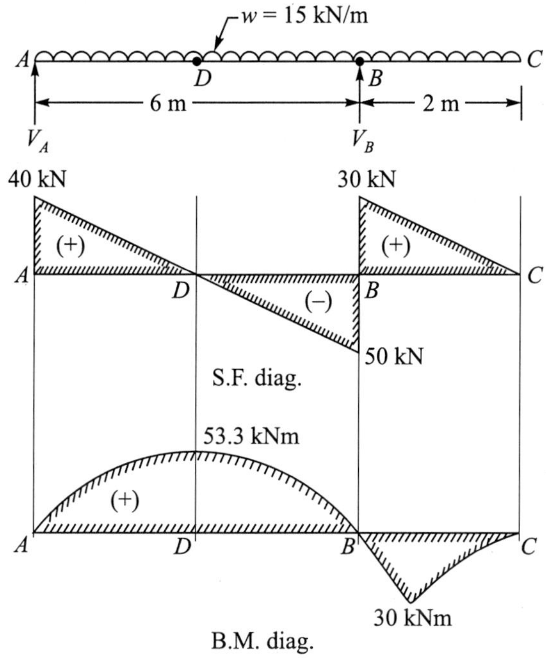

Fig. 12.31

Solution: Taking moments about A, M^A,

$$V_B \times 6 = 15 \times 8 \times \frac{8}{2}$$

$\therefore \quad V_B = 80$ kN

$\therefore \quad V_A = (15 \times 8) - 80 = \mathbf{40\ kN}$

Consider section AB, and a see at a distance "x" from "A"

S.F. $\quad S_X = 40 - 15x$, general expression of *S.F.*

At $\quad x = 0$, *S.F.* = (+) 40 kN

$\quad x = 6$, *S.F.* = $40 - 15 \times 6$ = (–) 50 kN.

Section of zero shear,

Equating the general expression of shear force = 0.

i.e., $\quad 40 - 15x = 0.$

$\quad x = 2.67$ m

Take any section XX, in segment BC, distant "x" from "C"

S.F. $\quad S_X = (+)\ 15x$

At $\quad x = 0$, *S.F.* = 0

At $\quad x = 2$ m, *S.F.* = 15×2 = (+) 30 kN

For *B.M.*, consider any section in *AB*, distant "*x*" from "*A*",

$$M_X = 40x - 15 \times \frac{x^2}{2}$$

$\therefore$

$$\boxed{M_X = 40x - 7.5x^2}$$

At $x = 0,\ M_X = 0$

At $x = 6$ m, $M_X = 40 \times 6 - 7.5 \times 6^2 = (-)\ 30$ kNm

Max. *B.M.* occurs at the section of zero shear,

i.e., $$M_X = 40 \times \frac{8}{3} - 7.5\left(\frac{8}{3}\right)^2 = (+)\ \frac{160}{3} = (+)\ 53.33 \text{ kN-m.}$$

At any section in *CB*, distant "*x*" from "*C*"

$$M_X = (-)\ \frac{15x^2}{2} = (-)\ 7.5x^2$$

At $x = 0,\ M_X = 0$, At $x = 2$m, $M_X = -\ 7.5 \times 2^2 = (-)\ 30$ kNm

Example 12.7:

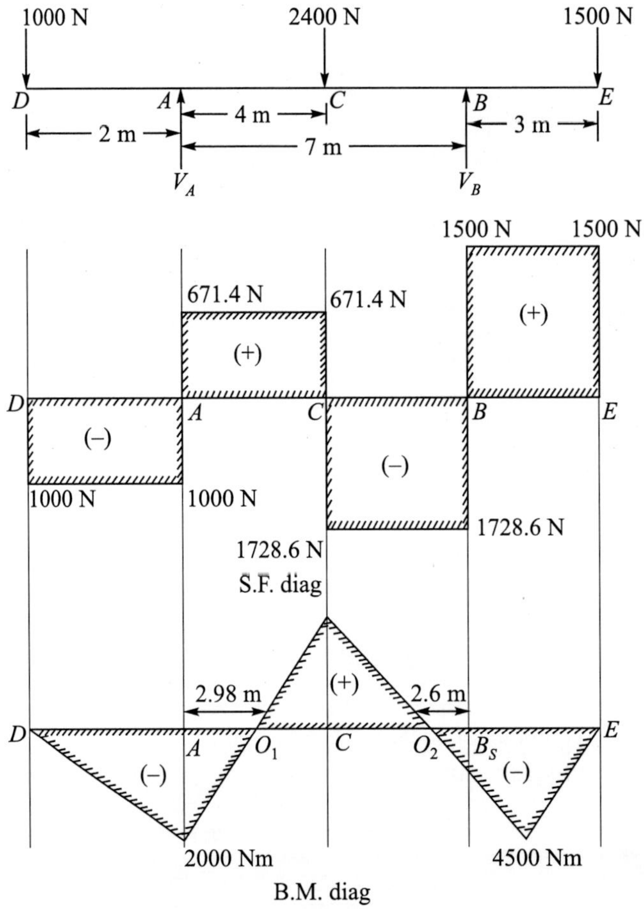

Fig. 12.32

Solution: Taking moments about A, M^A,

$$V_B \times 7 + 1000 \times 2 = 2400 \times 4 + 1500 \times 10$$

$\therefore$ $V_B = (+)\ 3228.6$ N

and $V_A = 4900 - 3228.6 = (+)\ 1671.4$ N.

***S.F.* Calculations:**

S.F. between D and $A = (-)\ 1000$ N

S.F. between A and $C = (-)\ 1000 + 1671.4 = (+)\ 671.4$ N

S.F. between C and $B = (+)\ 1500 - 3228.6 = (-)\ 1728.6$ N

S.F. between B and $E = (+)\ 1500$ N.

***B.M.* Calculations:**

B.M. at D, $M_D = 0$

B.M. at A, $M_A = (-)\ 1000 \times 2 = (-)\ 2000$ Nm

B.M. at C, $M_C = (+)\ 1671.4 \times 4 - 1000 \times 6 = (+)\ 685.6$ Nm

B.M. at B, $M_B = (-)\ 1500 \times 3 = (-)\ 4500$ Nm

B.M.E, $M_E = 0$

Point of contraflexure "0_1" between segment A and C,

Let it be "x" mtr. from "A", equating *B.M.* = 0,

$$1671.4 \times x = 1000\ (x + 2) = 0$$

$\therefore$ $x = 2.98$ m from "A"

Point of contraflexure "0_2" between segment C and B,

Let it be "x" mtr. from "B", equating *B.M.* = 0,

$$3228.6 \times x - 1500\ (x - 3) = 0$$

$$x = 2.6 \text{ m from "}B\text{"}$$

Example 12.8:

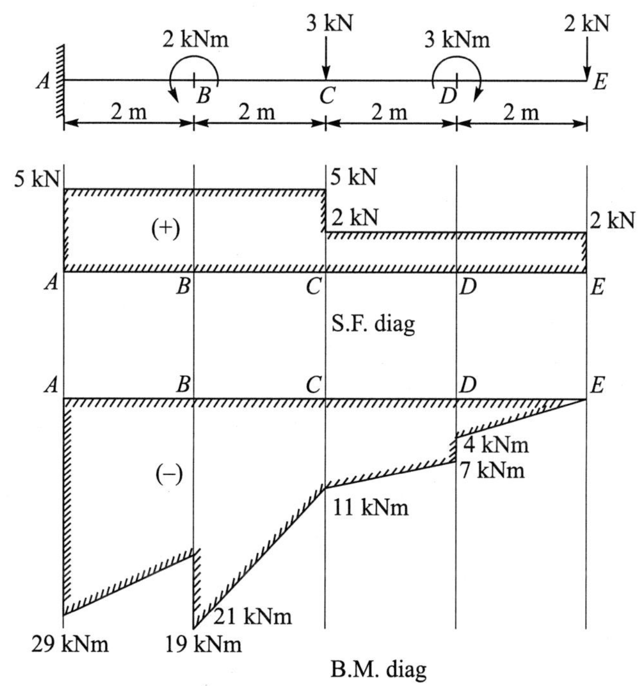

Fig. 12.33

Solution: For equilibrium,

V_B = 5 kN (upwards), for fixed support "*A*",

Total forces acting,

(*i*) Couple at "*B*" = 2.0 kNm (↺)

(*ii*) Moment due to = 3 × 4 = 12 kNm (↻)
3 kN load

(*iii*) Couple at *D* = 3.0 kNm (↻)

(*iv*) Moment due to load 2 kN = 2 × 8 = 16 kNm (↻)

Hence, at "*A*", the fixed support will provide a balancing moment of 29 kNm (Anticlockwise). Hence, the reaction at "*A*" will consist of an upward reacting force of 5 kN and an anticlockwise moment of 29 kN-m

S.F. between *A* and *C* = (+) 5 kN

S.F. between *C* and *E* = (+) 2 kN:

***B.M.* Calculations:**

B.M. at *E* = 0

B.M. just on the *R.H.S.* of *D* = −2 × 2 = (−) 4 kNm

B.M. just on the *L.H.S.* of *D* = −4 − 3 = (−) 7 kNm

B.M. at "*C*" = −2 × 4 − 3 = (−) 11 kNm

B.M. at *R.H.S.* of *B* = −2 × 6 − 3 − 3 × 2 = (−) 21 kNm

B.M. at *L.H.S.* of *B* = −21 + 2 = (−) 19 kNm

B.M. at "*A*" = −2 × 8 − 3 − 3 × 4 + 2 = (−) 29 kNm
= Reacting moment at "*A*".

Example 12.9:

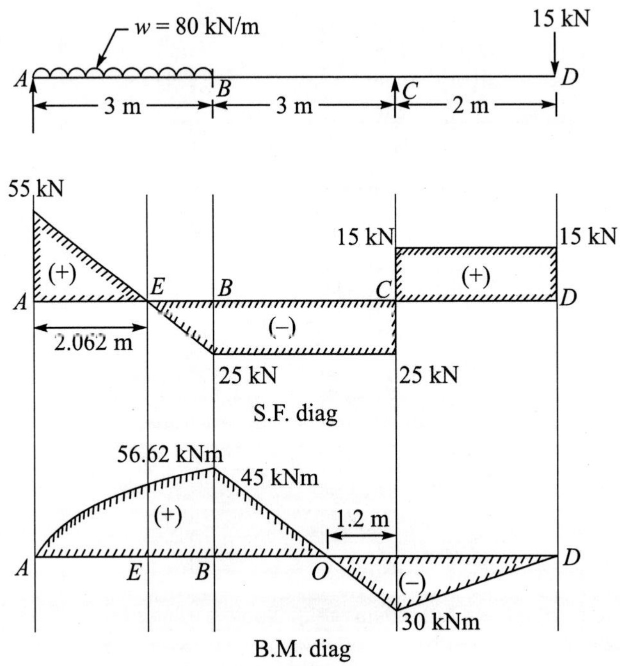

Fig. 12.34

Solution: Taking moments about *A*, M^A, we have,

$$V_C \times 6 = 80 \times \frac{3}{2} + 15 \times 8$$

$$V_C = 40 \text{ kN}$$

$$\therefore \quad V_A = 80 + 15 - 40 = 55 \text{ kN}$$

***S.F.* Calculations:**

S.F. at $A =$ (+) 55 N

S.F. at $B = 55 - 80 =$ (–) 25 kN

S.F. between *B* and $C =$ (–) 25 kN

S.F. between *C* and $D =$ (+) 15 kN

Let *S.F.* be zero at "*x*" mtr. from "*A*",

$$55 - \frac{80}{3} x = 0, \quad \therefore \quad x = 2.062 \text{ m from "}A\text{"},$$

***B.M.* Calculations:**

Next *B.M.* at "*A*" $= M_A = 0$

B.M. at "*B*" $= M_B = 55 \times 3 - 80 \times \frac{3}{2} =$ (+) 45 kNm

B.M. at *E, i.e.,* 2.062 mtr. from "*A*", $= 55 \times 2.062 - \frac{80}{3} \times \frac{(2.062)^2}{2} =$ (+) 56.62 kN-m

B.M. at "*C*" $= M_C = -15 \times 2 =$ (–) 30 kN-m

B.M. at "*D*" $= M_D = 0$

Points of contraflexure, this will be between *B* and *C* segment.

Let it be "*x*" mtrs from "*C*"

$$40x - 15(2 + x) = 0$$

$$\therefore \quad \mathbf{x = 1.2 \text{ m}}$$

***S.F.* and *B.M.* diagram for frames:**

Example 12.10: *Draw B.M. for the frame shown in Fig. 12.35.*

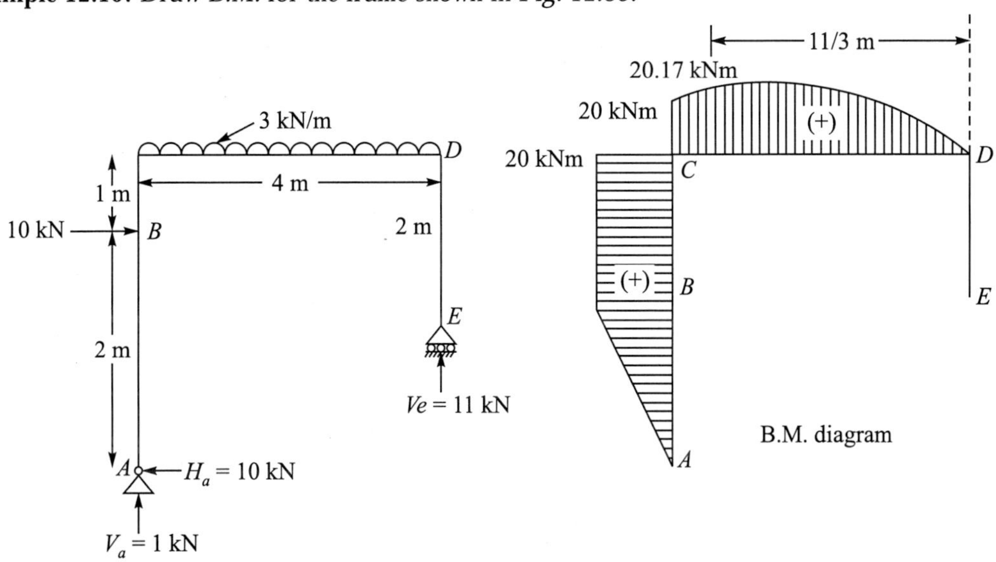

Fig. 12.35

Solution: Reactions at support, taking moments about A,

$$V_e \times 4 = (10 \times 2) + (3 \times 4 \times 2)$$

$$\therefore \quad V_e = 11 \text{ kN} \uparrow$$

$$\therefore \quad V_a = (3 \times 4) - 11 = 1 \text{ kN} \uparrow, \; H_a = 10 \text{ kN} \leftarrow$$

Member *DE*: The *B.M.* at every section of this member = 0.

Member *CD*: At any section distant x from D, the *B.M.* is given by,

$$M = 11x - 3 \times \frac{x^2}{2} = 11x - 1.5x^2$$

At $x = 0$, $M = 0$, At $x = 4$, $M = 11 \times 4 - 1.5 \times 4^2 = (+)\ 20$ kNm

For M to be max, $\dfrac{dM}{dx} = 0,$

$$\therefore \quad \frac{dM}{dx} = 11 - 3x = 0, \; \therefore \; x = \frac{11}{3} \text{ m.}$$

Putting $x = \dfrac{11}{3}$ m in the expression for *B.M.*

$$M_{max} = 11 \times \frac{11}{3} - 1.5\left(\frac{11}{3}\right)^2 = 20.17 \text{ kN-m}$$

Member *AC*: *B.M.* at $A = 0$, *B.M.* at $B = (+)\ 10 \times 2$

$= (+)\ 20$ kNm

B.M. at $C = (+)\ 10 \times 3 - 10 \times 1 = (+)\ 20$ kNm

The *B.M.* diagram is drawn as shown.

Example 12.11: *Draw B.M. diagram for the frame shown in the Fig. 12.36.*

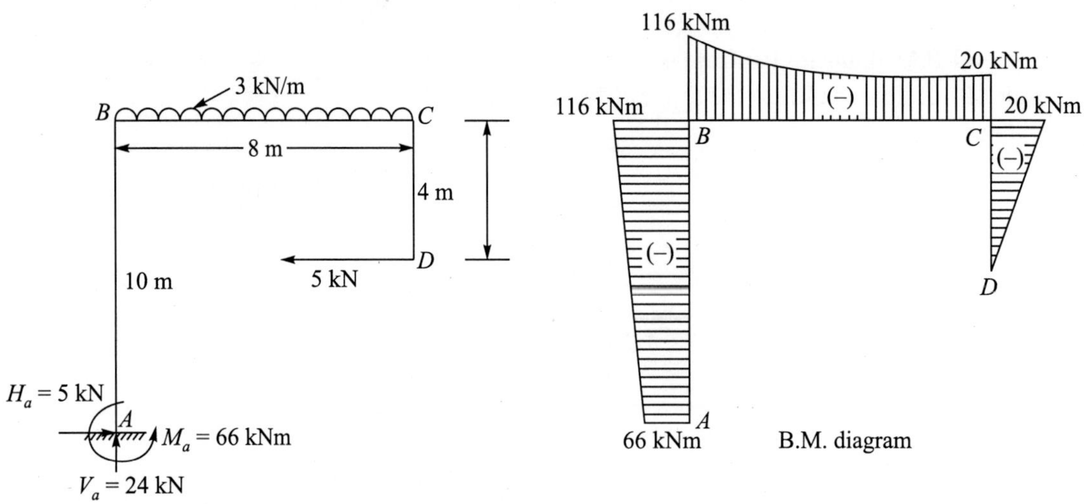

Fig. 12.36

Solution: At the fixed support A,

Reacting force $\quad H_a = 5$ kN $\rightarrow$

Reacting moments $\quad M_a = (3 \times 8 \times 4) - (5 \times 6) = 66$ kNm (Anticlockwise)

Member *DC*: At any section in *DC* distant "*y*" from *D*, *B.M.* is given by, $M = -5y$

At $y = 0, M = 0$, At $y = 4, M = -20$ kNm

Member *BC*: At any section distant *x* from *C*, the *B.M.* is given by,

$$M = -5 \times 4 - 3 \times \frac{x^2}{2}, \quad \therefore \quad M = -(20 + 1.5x^2)$$

At $x = 0, M = -20$ kNm, At $x = 4, M = -44$ kNm

and At $x = 8, M = -116$ kNm.

Member *AB*: At any section distant *y* from *A*, the *B.M.* is given by, $M = -(66 + 5y)$

At $y = 0, M = -66$ kNm, At $y = 10, M = -116$ kNm

The *B.M.* diagram is drawn thus.

Example 12.12: *Draw the B.M. diagram for the frame shown in Fig. 12.37.*

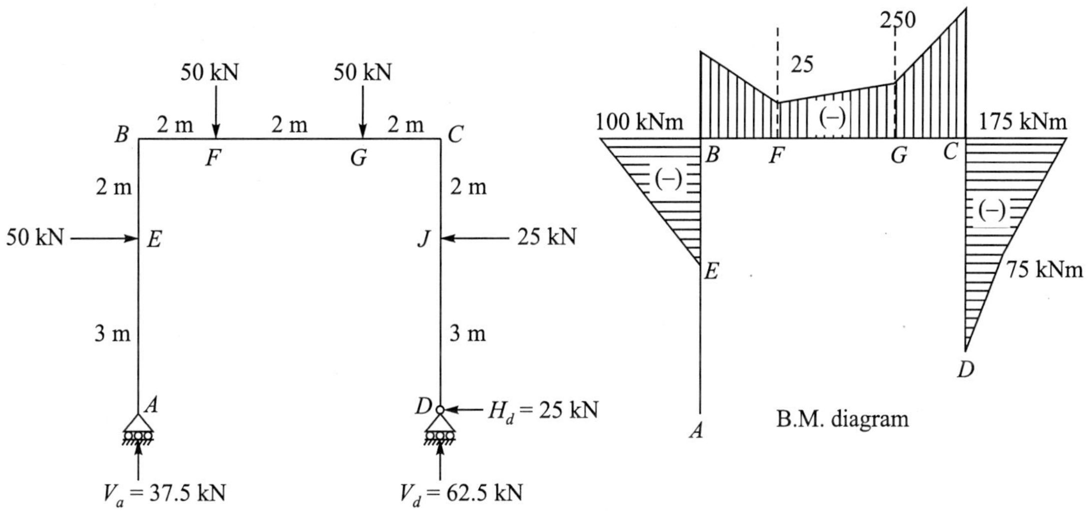

Fig. 12.37

Solution: Let V_a and V_d be the vertical reactions at *A* and *D*. Let H_d be the horizontal reaction at *D*. Resolving horizontally,

$$H_d + 25 = 50, \quad \therefore \quad H_d = 25 \text{ kN} \leftarrow, H_a = 0$$

Taking moments about *D*,

$$V_a \times 6 + 50 \times 3 = 50 \times 4 + 50 \times 2 + 25 \times 3$$

$$\therefore \quad V_a = 37.5 \text{ kN} \uparrow$$

$$V_d = 100 - 37.5 = 62.5 \text{ kN} \uparrow$$

***B.M.* Calculations:**

B.M. at $A = 0$

B.M. at $E = 0$

B.M. at $B = -50 \times 2 = -100$ kNm

B.M. at $F = 37.5 \times 2 - 50 \times 2 = -25$ kNm

B.M. at $G = 37.5 \times 4 - 50 \times 2 - 50 \times 2 = -50$ kNm

$$B.M. \text{ at } C = -25 \times 2 - 25 \times 5 = -175 \text{ kNm}$$

$$B.M. \text{ at } J = -25 \times 3 = -75 \text{ kNm}$$

∴ The *B.M.* diagram is drawn thus.

Example 12.13: *Draw the B.M. diagram for the frame shown in Fig. 12.38.*

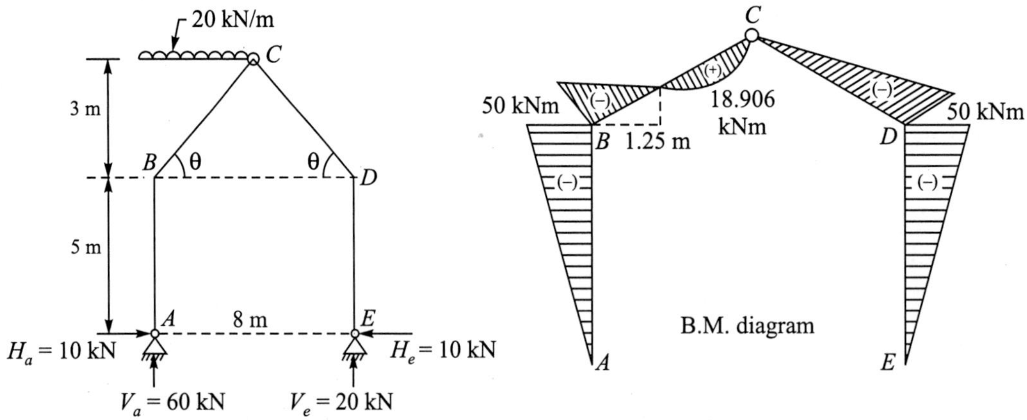

Fig. 12.38

Solution: Taking moments about *A*, $V_b \times 8 = 20 \times 4 \times 2$

$$\therefore \quad V_b = 20 \text{ kN} \uparrow,$$

$$\therefore \quad V_a = (20 \times 4) - 20 = 60 \text{ kN} \uparrow$$

Since hinge is provided at *C*, the *B.M.* at *C* = 0.

Taking moments about *C*, from right,

$$H_e \times 8 = 20 \times 4, \quad \therefore \quad H_e = 10 \text{ kN}$$

Resolving horizontally,

$$H_a = H_e, \quad \therefore \quad H_a = H_e = 10 \text{ kN}$$

Members *AB* and *ED*, *B.M.* at *A* = *B.M.* at *E* = 0

B.M. at *B* = *B.M.* at *D* = –10 × 5 = –50 kNm

Member *BC*:

$$\tan\theta = \frac{3}{4}, \sin\theta = \frac{3}{5}, \cos\theta = \frac{4}{5}$$

At any section in *BC*, the *B.M.* is given by,

$$M = 60x - 10\,(5 + y) - 10x^2$$

$$= 60x - 10\left(5 + \frac{3}{4}x\right) - 10x^2$$

$$\therefore \quad M = 60x - 50 - 7.5x - 10x^2$$

$$= -10x^2 + 52.5x - 50$$

At, $x = 0$, $M = -50$ kNm, At $x = 4$,

$$M = -10 \times 4^2 + 52.5 \times 4 - 50 = 0.$$

For M to be max, $$\frac{dM}{dx} = 0$$

$$\therefore \quad \frac{dM}{dx} = -20x + 52.5 = 0.$$

$$\therefore \quad x = 2.625 \text{ m}$$

$$M_{max} = -10\,(2.625)^2 + 52.5\,(2.625) - 50 = (+)\ 18.906 \text{ kNm}$$

Point of contraflexure in BC,

Equating the general expression for $B.M.$ to zero

$$-10x^2 + 52.5x - 50 = 0, \quad \therefore \quad x^2 = 5.25x + 5 = 0$$

$$x = \frac{5.25 \pm 2.75}{2} \text{ m}$$

$$x = 1.25 \text{ m and } 4 \text{ m}.$$

SUMMARY

- Conditions for equilibrium of a beam are :
  (*a*) Algebraic sum of all forces in any direction is zero.
  (*b*) Algebraic sum of the moments of all forces (including the support reactions) about any point is zero.)

  Thus, the sum of all downward forces is equal to the sum of all upward forces and the sum of all clockwise moments about any point is equal to the sum of all anticlockwise moments about that point.
- S.F. at a section is the algebraic sum of all vertical forces to any one side of the section. B.M. at a section is the algebric sum of the moments of all forces to one side of the section.
- In case of inclined loads, resolve them into vertical and horizontal components. Horizontal components cause only thrust in the beam. S.F. and B.M. are caused only by the vertical components.
- At a hingle in a beam there is no B.M. but there is S.F.
- B.M. is maximum at the point where the S.F. is zero or where it changes direction from positive to negative or vice-versa.
- The point of contraflexure or the point of inflexion is the point where the B.M. changes its nature or is zero.
- A load acting at a point is known as concentrated or point load.
- S.F. for a U.D.L. varies according to straight line law where as B.M. varies according to a parabolic law.
- B.M. is zero at hinges and at simple supports.
- The rate of change of shear force is equal to the rate of loading and the rate of change of bending moment is equal to the shear force at the section.

OBJECTIVE TYPE QUESTIONS

1. If a cantilever beam is subjected to a point load at its free end, then the S.F. under the point load is:
   (*a*) Zero (*b*) less than the load
   (*c*) equal to load (*d*) more than the load
2. When S.F. at a pont is zero, then the B.M. at that point will be
   (*a*) zero (*b*) minimum (*c*) maximum (*d*) infinity
3. The point of contrafluxure is the point where
   (*a*) S.F. changes sign (*b*) B.M. change sign
   (*c*) S.F. is maximum (*d*) B.M. is maximum
4. Maximum B.M. is a S.S. beam carring a Udl is
   (*a*) $\frac{wl^2}{8}$ (*b*) $\frac{wl^2}{4}$ (*c*) $\frac{wl^2}{a\sqrt{3}}$
5. B.M. at the supports in a S.S. beam is always
   (*a*) Zero (*b*) Maximum (*c*) Minimum (*d*) None of these

Answers: 1. (*c*), 2. (*c*), 3. (*b*), 4. (*a*), 5. (*a*)

EXERCISE FOR PRACTICE

1. Draw the S.F. and B.M. diagram for the beam shown in the Fig. 12.39. Also determine the location of the maximum B.M. and point of contrafluxure. **(MDU 2011)**

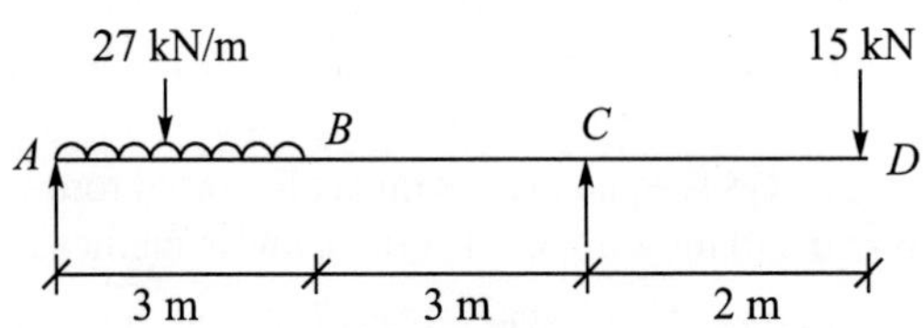

Fig. 12.39.

2. Analyse the frames shown in Fig. 12.40 and draw B.M. of S.F. Diagrams. **(MDU 2009)**

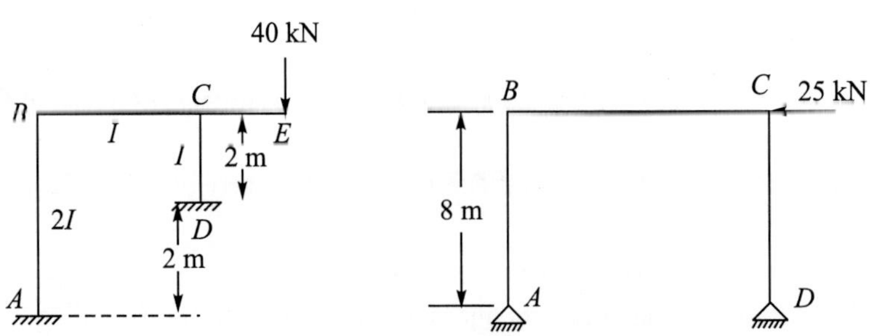

Fig. 12.40.

3. Draw B.M. and S.F. diagram for the portal frame shown below after completely analying it. (MDU 2010)

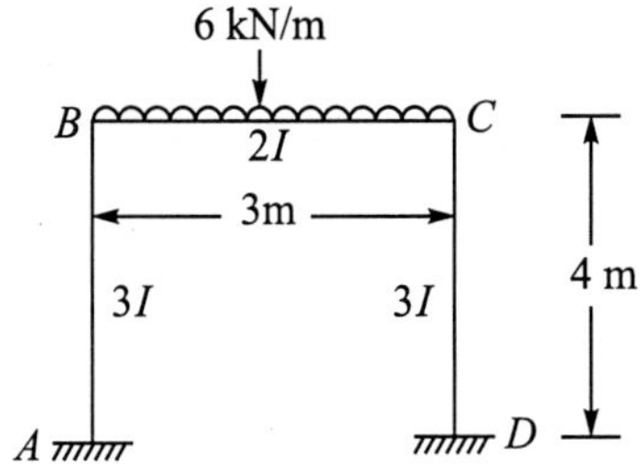

Fig. 12.41.

4. Draw the S.F. and B.M. diagram for the beam shown below: (MDU 2010)

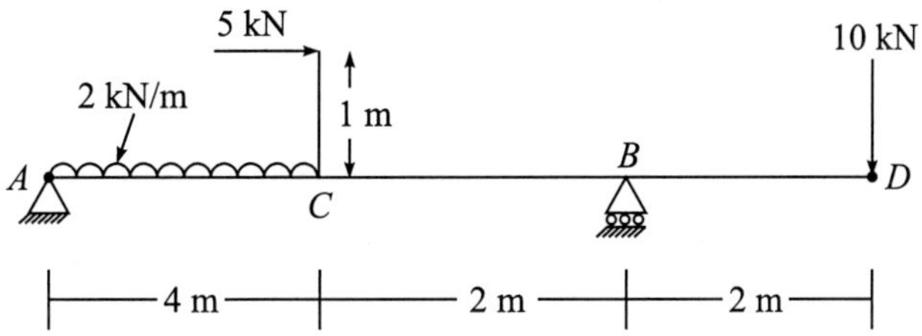

Fig. 12.42

5. Draw the S.F. and B.M. diagram for the beam shown in Fig. 12.43 below: (MDU 2009)

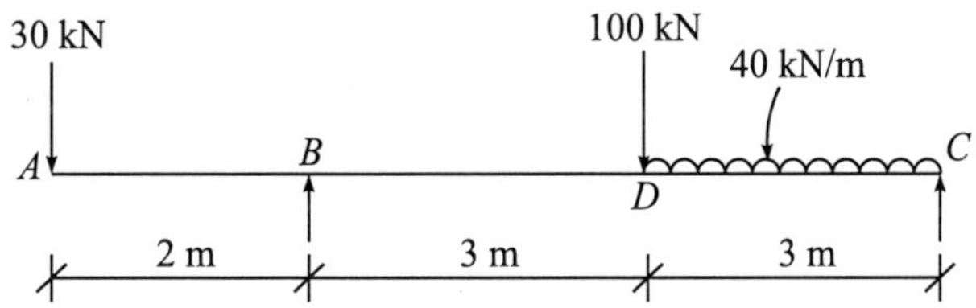

Fig. 12.43

6. Draw the axial force, S.F. and B.M. diagram for the Gable frame shown below. (MDU 2011)

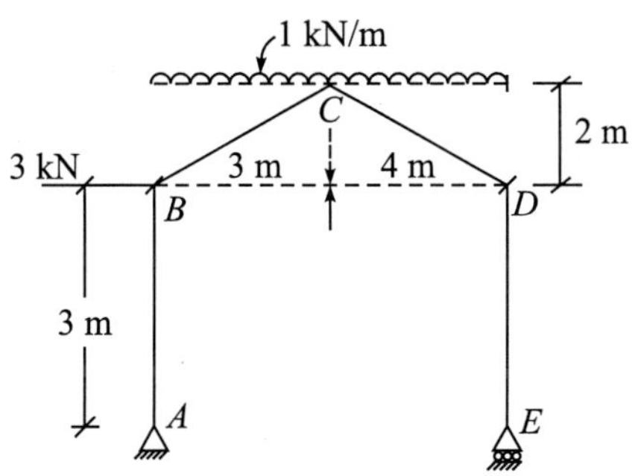

Fig. 12.44

CHAPTER

13 Unsymmetrical Bending

Definition: In the simple theory of bending using the well known flexure formula $\frac{M}{I} = \frac{f}{y}$, it is assumed that the N-axis of the beam is perpendicular to the plane of loading. This condition applies that the plane of loading or the plane of bending is co-incident with or parallel to a plane containing a principle centroidal axis of inertia of the *X*-section of the beam. If however the plane of loading or that of bending, does not lie in or parallel to a plane that contains the principle centroidal axes of the *X*-section, the bending is called **"Unsymmetrical Bending"** Fig. 13. (*a*) and (*b*) shows cases of unsymmetrical bending, in which the plane of load "*W*" is vertical and does not coincide with the principal centroidal axis *UU* and *VV*. In case of unsymmetrical bending the direction of the *N*. Axis will not be 1 for to the plane of bending.

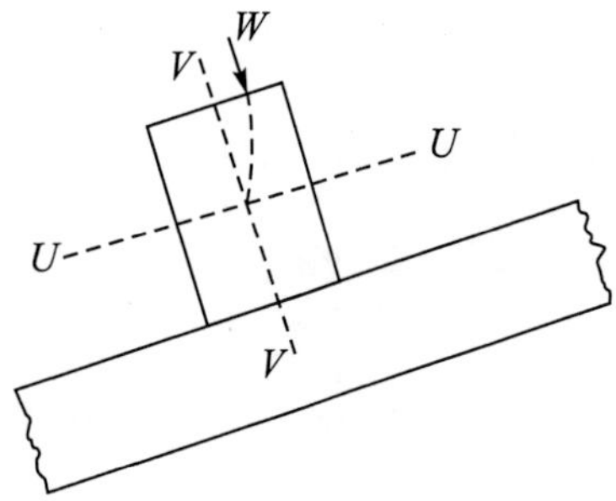

Fig. 13.1 (a)

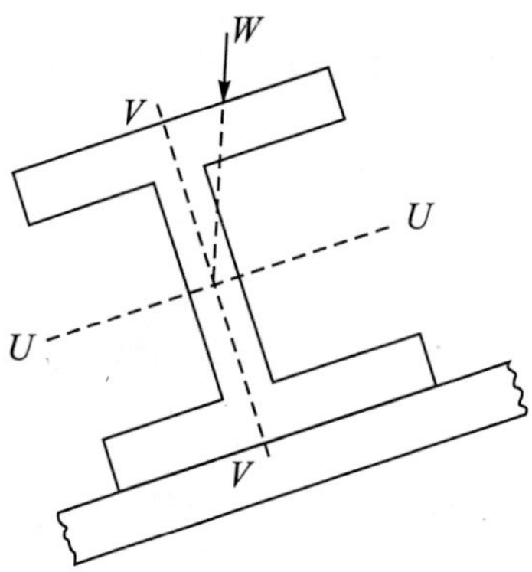

Fig. 13.1 (b)

13.1. CENTROIDAL PRINCIPAL AXES OF A SECTION

The centroidal principal axes of a section are defined as a pair rectangular axes through the *C.G.* of plane area such that the product of inertia is zero.

Let *U-U*, *V-V* be the principal centroidal axes.

X-X, *Y-Y* be any pair of centroidal rectangular axes. "α" be the angle between *U-U* and *X-X* axes.

If *U-U*, *V-V* are the principal axes, the product of inertia $\Sigma\, u \cdot v \cdot \delta\, a = 0$, where δa is an elementary area with co-ordinates *u* and *v* referred to the principal axes. If a plane area has an axis of symmetry, it is obviously a principal axis, since the axes of symmetry has to satisfy the condition $\Sigma\, u\, v\, \delta\, a = 0$ about it.

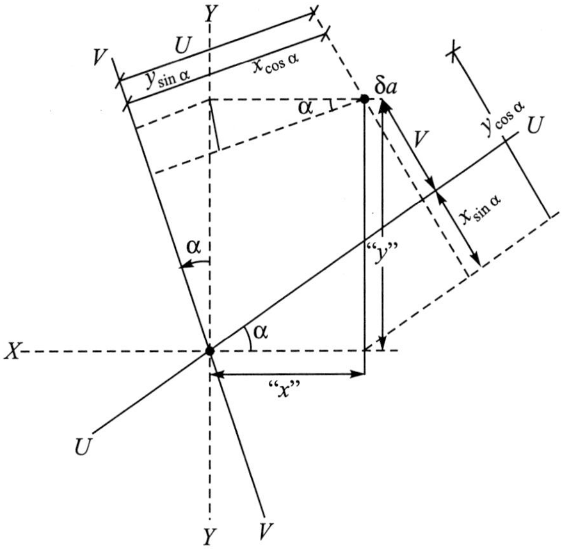

Fig. 13.2

In general however a plane area may not have any axis of symmetry. In that case the principal axes may be located provided its properties about any pair of rectangular axes *X-X, Y-Y* are known.

Let *x, y* be the co-ordinates of an elementary area δ*a*, with respect to the *X-Y* axis, and *u,v* be the corresponding co-ordinates with respect to the principal axes *U-V*.

By definition,

$$I_{XX} = \Sigma\, y^2 \cdot \delta a;\ I_{YY} = \Sigma\, x^2\, \delta a;\ I_{XY} = \Sigma\, xy^2\, \delta a$$

Similarly, $$I_{UU} = \Sigma\, v^2\, \delta a;\ I_{VV} = \Sigma\, u^2\, \delta a;\ I_{UV} = \Sigma\, uv^2\, \delta a$$

The relationships between *x, y* and *u, v* co-ordinates are:

$$u = x\cos\alpha + y\sin\alpha$$

$$v = y\cos\alpha - x\sin\alpha$$

Hence,

$$\begin{aligned} I_{UU} &= \Sigma\, v^2 \cdot \delta a = \Sigma\,(y\cos\alpha - x\sin\alpha)^2\, \delta a \\ &= \cos^2\alpha\, \Sigma\, y^2\, \delta a + \sin^2\alpha\, \Sigma\, x^2\, \delta a - 2\sin\alpha \cdot \cos\alpha \cdot \Sigma\, xy\, \delta a \\ &= I_{XX}\cos^2\alpha + I_{YY}\sin^2\alpha - I_{XY}\sin^2\alpha \qquad \ldots(13.1) \end{aligned}$$

$$\begin{aligned} I_{VV} &= \Sigma\, u^2\, \delta a = \Sigma\,(x\cos\alpha + y\cos\alpha)^2\, \delta a \\ &= \sin^2\alpha\, \Sigma\, y^2\, \delta a + \cos^2\alpha\, \Sigma\, x^2\, \delta a + 2\sin\alpha\cos\alpha\, \Sigma\, xy\, \delta a \\ &= I_{XX}\sin^2\alpha + I_{YY}\cos^2\alpha + I_{XY}\sin 2\alpha \qquad \ldots(13.2) \end{aligned}$$

and

$$\begin{aligned} I_{UV} &= \Sigma\, uv\, \delta a = \Sigma\,(x\cos\alpha + y\sin\alpha)\,(y\cos\alpha - x\sin\alpha)\, \delta a \\ &= \cos^2\alpha\, \Sigma\, xy\, \delta a - \sin^2\alpha\, \Sigma\, xy\, \delta a + \sin\alpha \cdot \cos\alpha\, (\Sigma\, y^2\, \delta a - \Sigma\, x^2\, \delta a) \\ &= \cos^2\alpha \cdot I_{XY} - \sin^2\alpha \cdot I_{XY} + \sin\alpha\cos\alpha\, (I_{XX} - I_{YY}) \\ &= \left(\frac{I_{XX} - I_{YY}}{2}\right)\sin 2\alpha + I_{XY}\cos^2\alpha \qquad \ldots(13.3) \end{aligned}$$

Since *U-U* and *V-V* are the principal axis,

$$I_{UV} = 0 = \left(\frac{I_{XX} - I_{YY}}{2}\right)\sin 2\alpha + I_{XY}\cos^2\alpha$$

or $$\tan 2\alpha = (-)\frac{2 I_{XY}}{I_{XX} - I_{YY}} \qquad ...(13.4)$$

Knowing I_{XX}, I_{YY} and I_{XY}, the angle α can be calculated from = *n*. (13.4)

Substituting α in = *n* (13.1) and (13.2), the M.I. about principal axes can be determined.

13.2. BENDING STRESS IN BEAMS SUBJECTED TO UNSYMMETRICAL BENDING

In the case of simple bending, where the plane of loading (or bending) coincides with one of the pricipal plane, the *N*-axis is perpendicular to the principal plane and passes through the centroid of the section. In the case of unsymmetrical bending, N-axis is not perpendicular to the plane of bending. The bending stress at any point in the beam subjected to unsymmetrical bending can be determined by the following methods:

1. Resolution of B.M. into two components along principal axes.
2. Resolution of B.M. into two components along any rectangular axes through the centroid.
3. Locating *N*-axis of the section.

Resolution of B.M. into two Components along the Principal Axes: Let the plane of bending (*M*) be inclined at an angle θ with one of the principal planes. The B.M. '*M*' can be resolved into two components viz. $M \cos\theta$ along plane *VV* and *M* and $\sin\theta$ along the plane *UU*. Having resolved the B.M. in the two components, the simple theory of bending can be applied to bending occurring in the principal planes. The intensity of bending stress at any point $P(u, v)$ will be the algebraic sum of the stress due to the component bending moments.

For the component $M \cos\theta$, bending takes place about *V-V* for which *U-U* becomes the *N*-axis. For the component $M \sin\theta$, bending takes place about *U-U* for which *V-V* becomes the *N*-axis.

Hence, the final bending stress at $P = f_b$ $$= \frac{M\cos\theta}{I_{UU}}v + \frac{M\sin\theta}{I_{VV}}u \qquad ...(13.5)$$

The co-ordinates *u, v* will be positive in that quadrant of *U-V* planes in which B.M. is applied. Thus *u* is positive in quadrants *I* and *II*, while *v* is positive in quadrants *I* and *IV*. Since the co-ordinates of any extreme point of the section are known, the bending stress can be calculated.

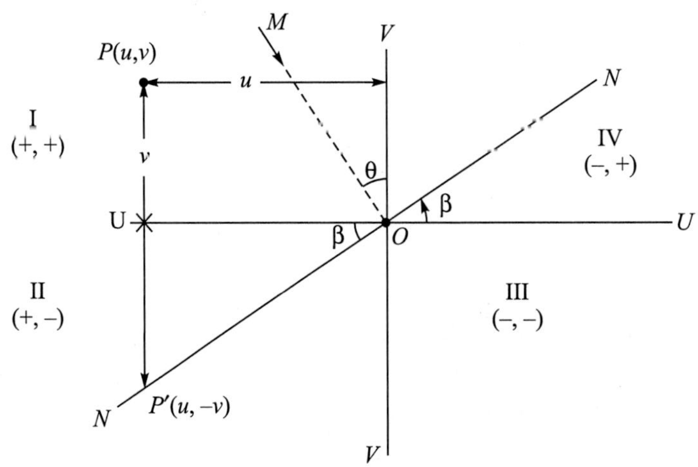

Fig. 13.3

13.3. RESOLUTION OF B.M. INTO ANY TWO RECTANGULAR AXES THROUGH THE CENTROIDS

The most general method of finding the bending stress at any point is to resolve it along any two rectangular axes passing through the centroids of the section. Let *X-X* and *Y-Y* be the centroidal axes.

The resolved component of *M* along the *Y-Y* axis is designated as M_{XX} and is equal to $M \cos \theta$. Similarly resolved component of *M* along *X-X* axis is designated as M_{YY} and is equal to $M \sin \theta$.

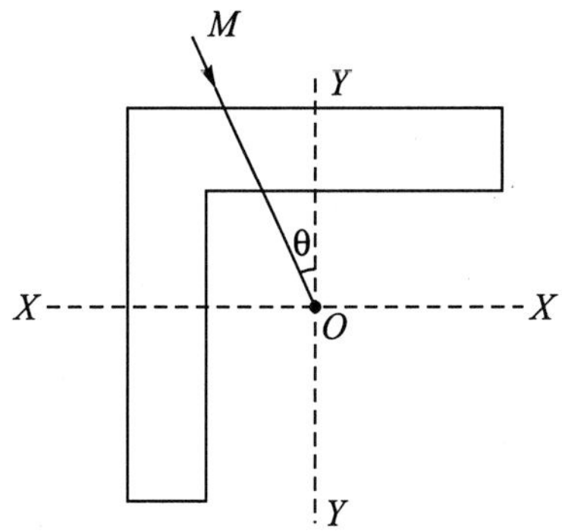

Fig. 13.4

The bending stress f_b at any point *P* (*x*, *y*) can be expressed as

$$f_b = a_1 x + b_1 y \quad \ldots(13.6)$$

Since the variation of bending stress is linear, a_1 and b_1 are constant to be determined.

Now M_{XX} = B.M. about *X*-axis.

$$= \int f_b \cdot \delta a \cdot y = \int (a_1 x + b_1 y)\, y \cdot \delta a$$

$$= a_1 \int xy\, \delta a + b_1 \int y^2\, \delta a$$

$$= a_1 I_{XY} + b_1 I_{XX} \quad \ldots(13.7)$$

M_{YY} = B.M. about *Y*-axis.

$$= \int f_b \cdot \delta a \cdot x = \int (a_1 x + b_1 y)\, x \cdot \delta a$$

$$= a_1 \int x^2\, \delta a + b_1 \int xy\, \delta a$$

$$= a_1 I_{YY} + b_1 I_{XY} \quad \ldots(13.8)$$

Solving (13.7) and (13.8), we get

$$a_1 = \frac{M_{YY} \cdot I_{XX} - M_{XX} \cdot I_{XY}}{I_{XX} \cdot I_{YY} - I^2\ XY} \quad \ldots(13.9)$$

and

$$b_1 = \frac{M_{XX} \cdot I_{YY} - M_{YY} \cdot I_{XY}}{I_{XX} \cdot I_{YY} - I^2\ XY} \quad \ldots(13.10)$$

Substituting the values of a_1 and b_1 in eq. (13.5)

$f_b = a_1 x + b_1 y$, we get

$$f_b = \frac{M_{YY} \cdot I_{XX} - M_{XX} \cdot I_{XY}}{I_{XX} \cdot I_{YY} - IXY^2} \cdot x + \frac{M_{XX} \cdot I_{YY} - M_{YY} \cdot I_{XY}}{I_{XX} \cdot I_{YY} - IXY^2} \cdot y$$

Thus, the bending stress "f_b" can be calculated at any point whose co-ordinate (*x*, *y*) are known.

13.4. LOCATION OF NEUTRAL AXIS

As stated earlier, in the case of unsymmetrical bending, the *N*-axis is neither perpendicular to the plane of bending, nor perpendicular to any of the principal planes.

Let θ = Inclination of the plane of bending to the *V-V* axis.

β = Inclination of the *N*-axis with the *U-U* axis.

The *N*-axis can be located as under. Fig. 13.3 (*i*) shows the *N*-axis *N-N*, inclined at an angle β with the *U-U* axis. At any point (such as *P*) on it, the bending stress is equal to zero.

Hence equation,

$$f_b = \frac{M\cos\theta}{I_{UU}}v + \frac{M\sin\theta}{I_{VV}}u = 0$$

or
$$v = -u \times \frac{I_{UU}}{I_{VV}}\tan\theta$$

This is the equation of the *N*-axis which is a st. line.

It is clear that when $v = 0$, $u = 0$, hence the *N*-axis passes through the centroid of the section, also

$$\tan\beta = -\frac{v}{u}, \quad \text{But} \quad -\frac{v}{u} = \frac{I_{UU}}{I_{VV}}\tan\theta$$

Hence,
$$\tan\beta = \frac{I_{UU}}{I_{VV}}\tan\theta$$

Thus, the *N*-axis can be located from above equation.

13.4.1. Shear Centre

The shear centre (for any transverse section of the beam) is the point of intersection of the bending axis and the plane of the transverse section.

Shear centre of a section can be defined as a point about which the applied force is balanced by the set of shear forces obtained by summing the shear stresses over the section (for unsymmetrical sections such as angle section and channel section, summetion of shear stresses in each leg gives a set of forces which should be in equilibrium with the applied shear force). Shear centre is also known as "Centre of twist".

In case of a beam having two axes of symmetry the shear centre coincides with the centroid. In case of a section having one axis of symmetry the shear centre does not coincide with the centroid but lies on the axis of symmetry.

Also when the load passes through the shear-centre then there will be only bending in the cross-section but no twisting.

The principle involved in locating the shear centre for a cross-section of a beam is that the loads acting on the beam must lie in a plane which contains the resultant shear force on each cross-section of the beam as computed from the shearing stresses produced in the beam when it is loaded so that it does not twist at its ends.

13.5. SHEAR CENTRE FOR CHANNEL SECTION

Figure 13.5 shows a channel section (flanges $b \times t_1$ and web $h \times t_2$) with *X-X* as the horizontal symmetrical axis.

Let S = Applied shear force (vertical downwards) and (then S is the shear force in the web in the upward direction)

S_1 = Shear force in the top flange (there will be equal and opposite shear force in the bottom flange as shown)

Now shear stress (τ) in the flange at a distance of x from the right hand edge (of the top flange)

$$\tau = \frac{SA\bar{y}}{I_{XX}t}$$

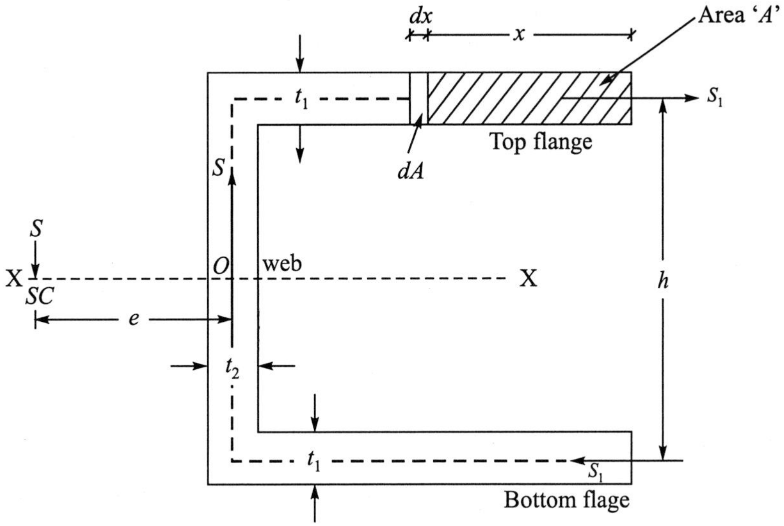

Fig. 13.5

$$A\bar{y} = (t_1 \cdot x)\ \frac{h}{2} \qquad \text{(where } t = t_1\text{, thickness of the flange)}$$

$$\therefore \qquad \tau = \frac{St_1 \cdot x}{I_{XX} \cdot t_1} \times \frac{h}{2} = \frac{Sxh}{2\,I_{XX}}$$

Shear force in elementary area,

$$(dA = t_1 \cdot dx) = \tau \cdot dA = \tau \cdot t_1 \cdot dx$$

Total shear force in top flange

$$= \int_0^b \tau \cdot t_1 \cdot dx \qquad \text{(where } b = \text{breadth of the flange)}$$

$$S_1 = \int_0^b \frac{Sxh}{2\,I_{XX}} \cdot t_1 \cdot dx = \frac{Sht_1}{2\,I_{XX}} \int_0^b x \cdot dx$$

or

$$S_1 = \frac{Sht_1}{I_{XX}} \cdot \frac{b^2}{4}$$

Let e = Distance of shear centre (SC) from the web along the symmetric axis XX

Taking moments of the shear forces about the centre 0 of the web, we get,

$$S \cdot e = S_1 \cdot h = \frac{Sht_1}{I_{XX}} \cdot \frac{b^2}{4} \cdot h = \frac{S \cdot t_1\, h^2\, b^2}{4\,I_{XX}}$$

$$\therefore \qquad e = \frac{b^2\, h^2\, t_1}{4\,I_{XX}} \qquad \ldots(13.11)$$

Now,

$$I_{XX} = 2\left[\frac{b \times t_1^3}{12} + b \cdot t_1 \left(\frac{h}{2}\right)^2\right] + \frac{t_2\, h^3}{12}$$

$$= \frac{b\,t_1^3}{6} + \frac{b\,t_1\,h^2}{2} + \frac{t_2\,h^3}{12} = \frac{b\,t_1\,h^2}{2} + \frac{t_2\,h^3}{12}$$

(Neglecting the term $\frac{b t_1^3}{6}$, being negligible in comparison to other terms)

or $$I_{XX} = \frac{h^2}{12}(t_2 h + 6\, bt_1)$$

Substituting the value of I_{XX} in eq. (13.10), we get

$$e = \frac{b^2 h^2 t_1}{4} \times \frac{12}{h^2 (t_2 h + 6\, bt_1)} = \frac{3b^2 t_1}{(t_2 h + 6\, bt_1)}$$

Let $bt_1 = Af$ (area of flange)

and $ht_2 = Aw$ (area of web)

Then, $$e = \frac{3\, b\, Af}{Aw + 6\, Af} = \frac{3\, b}{6 + \frac{Aw}{Af}}$$

i.e., $$\text{“}e\text{”} = \frac{3\, b}{6 + \frac{Aw}{Af}}$$

13.6 SHEAR CENTRE OF AN EQUAL ANGLE SECTION

An equal angle section of size $b \times b \times t$ is subjected to a shear force F at point 0, the inter-section of centre lines of its two legs. Axis of symmetry of the equal angle Fig. 13.6 shall be its neutral axis too. Resultant F_R of the shear forces F_1 and F_2 induced in the two legs of the angle has to be equal and opposite to the applied shear force F. Point 0 is thus the shear centre.

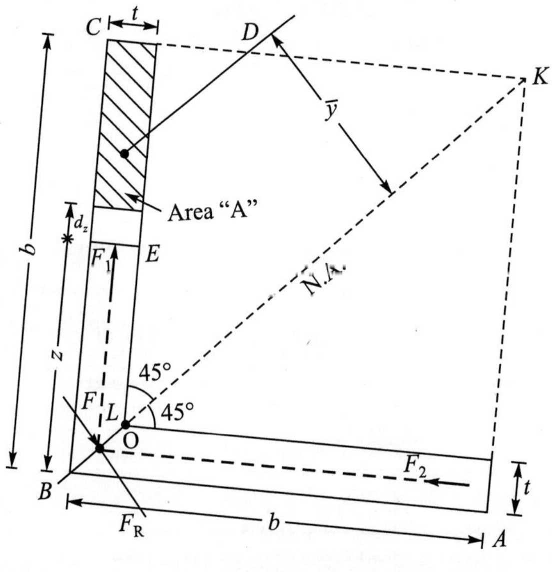

Fig. 13.6

Thus $F_R = F.$

Let τ be the shear stress at an element of width d_z at a distance z from B in leg BC. If A is the area of the portion of leg between the element under consideration and the end of the leg and $\bar{y}$ is the distance of its centroid from the neutral axis, then,

$$A = t \times (b - z)$$

and
$$\bar{y} = \left[b - \left(\frac{b-2}{2}\right)\right] \sin 45^\circ = \frac{b+2}{2\sqrt{2}}$$

New I for the whole section is:

$I = 2$ (I for Δ BCK – I for Δ LDK) about BK.

$$= \frac{b^4 - (b-t)^4}{12}$$

Since t is very small as compared to b so on neglecting values of t^2 and t^3 in the above equation, we have,

$$I = \frac{b^3 t}{3}$$

Now,
$$\text{“}\tau\text{”} = \frac{F(A\bar{y})}{It}$$

$\therefore$ Shear force F_1 in leg BC of the section is:

$$F_1 = \int \tau \, dA = \int_0^b \frac{F(A\bar{y})}{It} \, dA$$

$$= \int_0^b \frac{F}{It}\left[\left\{t(b-z) \times \frac{b+2}{2\sqrt{2}}\right\}\right] \times t \cdot d_z$$

$$= \frac{Ft}{2\sqrt{2}\,I} \int_0^b (b^2 - z^2)\, d_z.$$

$$= \frac{Ft}{2\sqrt{2}\,I}\left[b^2 z - \frac{z^3}{3}\right]_0^b = \frac{Ftb^3}{3\sqrt{2}\,I} = \frac{Ftb^3}{3\sqrt{2}} \times \frac{3}{b^3 t} = \frac{F}{\sqrt{2}}$$

Since by symmetry
$$F_2 = F_1 = \frac{F}{\sqrt{2}}$$

and F_1 and F_2 are at right angles to each other so resultant $S.F.$ is

$$F_R = \sqrt{F_1^2 + F_2^2} = F$$

So "O" is the shear centre and F_1 and F_2 are components of the resisting force F_R.

SUMMARY

- Reasons of unsymmetrical bending are :
  (*i*) The section is symmetrical but the load line is inclined to both the principal axes.
  (*ii*) The section itself is unsymmetrical and the load line is along any centroidal axis.

- In the simple theory of bending, using the well known flexure formula = $\frac{M}{I} = \frac{f}{y}$, it is assumed that the neutral axis of the *X*-sec of the beam is perpendicular to the plane of loading. This condition implies that the plane of loading, or the plane of bending is coincident with, or parallel to, a plane containing a principal centroidal axes of inertia of the cross-section of the beam. If however the plane of loading or that of bending, does not lie in a plane that contains the principal centroidal axes of the cross-section, the bending is called "Unsymmetrical bending".

- Product of inertia $I_{XY} = \int xy\,dA$. Product of inertia of a section about its principal axis is zero.

- If the section is symmetrical, the principal axes are along the axes of symmetry.

- Stresses due to unsymmetrical bending:
  Resultant bending stress at a point,

$$\sigma_b = M\left[\frac{v\cos\theta}{I_{UU}} + \frac{u\sin\theta}{I_{VV}}\right]$$

where, M = B.M.

u, v = co-ordinates of the point

θ = Angle of inclination of axes of M with respect to the principal axis VV.

- Angle of inclination of N.A. with respect of principal axis UU,

$$\alpha = \tan^{-1}\left(\frac{I_{UU}}{I_{VV}}\tan\theta\right)$$

- Deflection of a beam under load W causing unsymmetrical bending,

$$\alpha = \frac{kWl^3}{E}\sqrt{\frac{\sin^2\theta}{I_{VV}^2} + \frac{\cos^2\theta}{I_{UU}^2}}$$

where k = A constant depending on the end condition of the beam and position of the load along the beam.

l = Length of the beam.

θ = Angle of inclination of load W with respect to VV principal axis.

- The shear centre for any transverse section of the beam is the point of intersection of the bending axis and the plane of hte transverse section. Shear centre is also known as centre of twist.

- In case of a beam having two axes of symmetry, the shear centre coincides with the centroid.

EXERCISE FOR PRACTICE

1. Determine the principal moment of inertia of the section shown in Fig. 13.7, below. All dimensions are in mm.

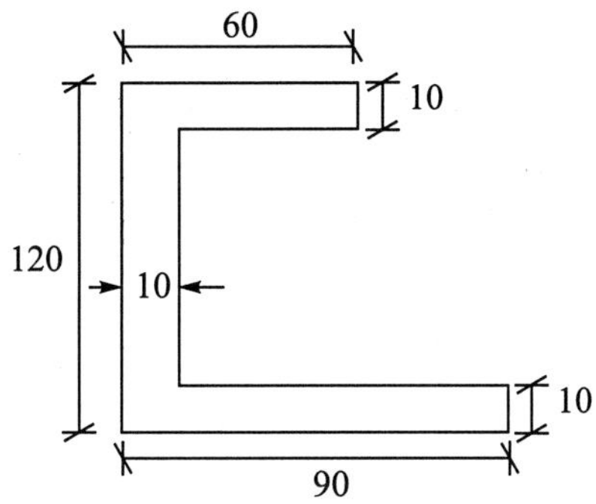

Fig. 13.7.

2. Determine the centroidal principal moments of inertia of an equal angle 30 mm × 30 mm × 10 mm.
3. A timber beam 250 mm wide by 300 mm deep is used as S.S. beam on a span of 5 m. It is subjected to a concentrated load of 30 N at the mid-section of the span. If the plane of the load makes an angle of 45° with the vertical plane of symmetry. Find the direction of N.A. and maximum stress.

CHAPTER

14 Cables and Suspension Bridges

Suspension bridges are used for highways, where the span of a bridge is more than 200 m. These are very light and used for pedestrian traffic only. Essentially a suspension bridge consists of the following elements:

(*i*) The cable

(*ii*) Suspenders

(*iii*) Decking including the stiffening girders

(*iv*) Supporting towers and

(*v*) Anchorages

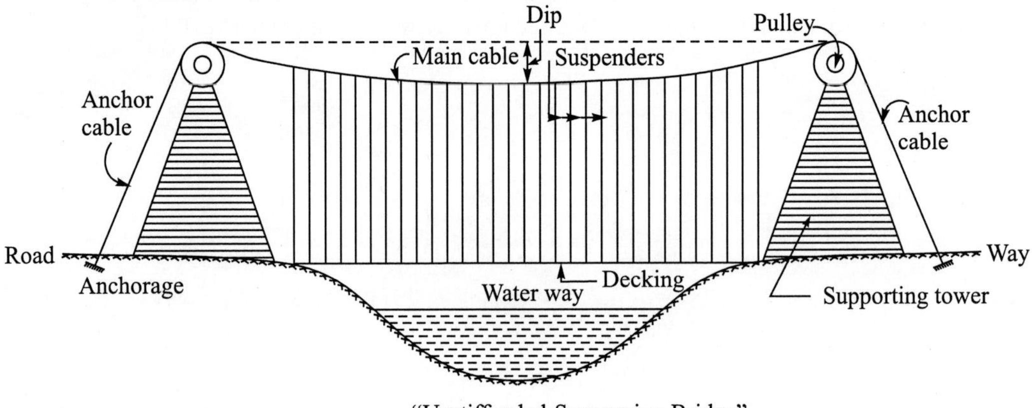

"Unstiffended-Suspension Bridge"

Fig. 14.1 (a)

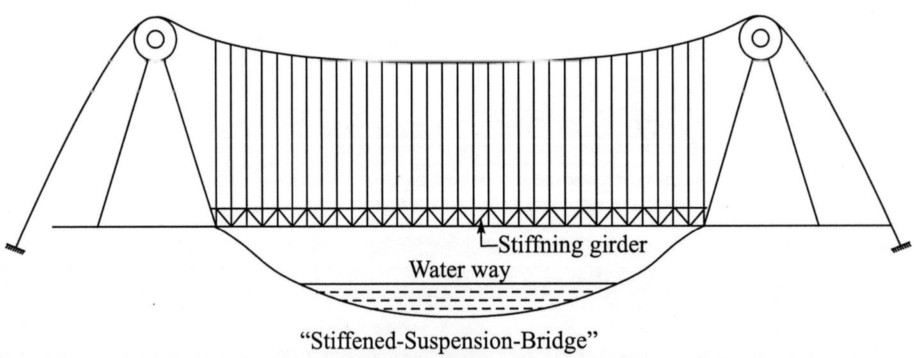

"Stiffened-Suspension-Bridge"

Fig. 14.1 (b)

The suspension bridge consists of two cables, which are stretched over the spans to be bridged. Each cable, passing over two towers, anchored by backways to a firm foundation as shown in Fig. 14.2. As the cable is flexible throughout, therefore it cannot resist any moment and can adopt any shape under the loads, that is why the bending moment at every point of the cable is taken as zero. The roadway is suspended from the cables by means of hangers or suspenders. Since the hangers are large in number, therefore the load transmitted by the hangers, is taken as a uniformally distributed load. It may be noted that since the cable is in tension throughout, this type of construction is most economical.

The traffic load of the decking is transferred to main cable through the suspenders. Since cable is the main load bearing member, the curvature of the cable of an unstiffened bridge changes as the load moves on the decking. To avoid this, the decking is stiffened by provision of either a three hinged or two hinged stiffening girder. The cable consists of either wire rope or parallel wires joined with clips or eye bar links. The suspenders consists of round rods with turn buckles so that adjustment in their lengths may be done as required.

14.1. CABLE CARRYING UNIFORMLY DISTRIBUTED LOAD (*UDL*)

Figure 14.2(*a*) shows a light cable of span "*l*" and dip "*h*" subjected to a *Udl* of w/m on the horizontal span.

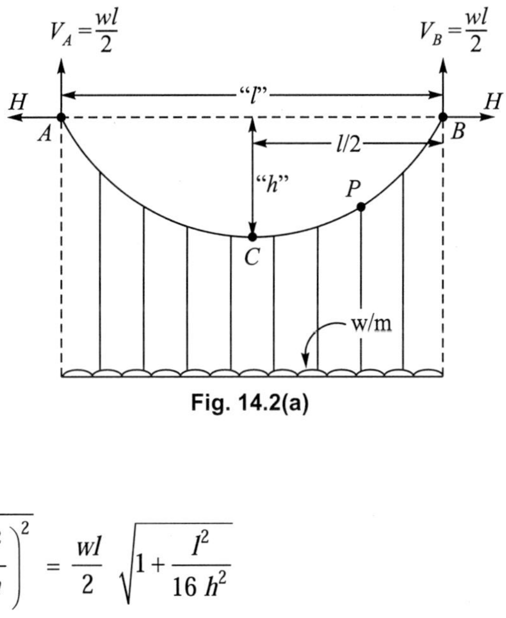

Fig. 14.2(a)

Vertical Reactions

V_a and V_b are equal

$$= \frac{wl}{2}.$$

Taking moments about *C* of all the forces *R.H.S.* of *C*,

$$M^C, H \times h + w \times \frac{l}{2} \times \frac{l}{4} - \frac{wl}{2} \times \frac{l}{2}$$

or $$'H' = \frac{wl^2}{8h}$$

Max tension T_{max} = Resultant reaction at *A* or *B*,

$$= \sqrt{\left(\frac{wl}{2}\right)^2 + \left(\frac{wl^2}{8h}\right)^2} = \frac{wl}{2}\sqrt{1 + \frac{l^2}{16h^2}}$$

Let *T* be the tension at any point *P* see Fig. 14.2 (*b*) of the cable. Consider the equilibrium of the part *PB* of the cable.

Let θ be the inclination of the tangent at *P* with the horizontal.

Resolving the forces acting on this part of the cable horizontally, we get, $T \cos \theta = H$

Hence the horizontal component of the tension at any section equals the horizontal reaction *h* obviously at the lowest point *C* of the cable, the tension $T_C = H$.

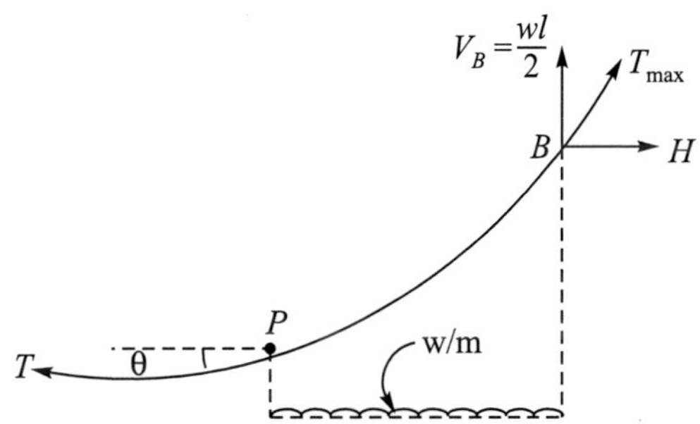

Fig. 14.2 (b)

Hence for the cable,

Minimum tension $= T_C = H$ (occurring at mid. point of the cable)

Maximum tension, $T_{max} = \frac{wl}{2}\sqrt{1+\frac{l^2}{16\,h^2}}$ (occurring near each support)

14.2. THE SHAPE OF THE LOADED CORD

Let the co-ordinates of any point *P* of the cable be (x, y) with respect to *C* the lowest point of the cable as origin.

Consider the equilibrium of the part *CP* of the cable.

The forces keeping the part *CP* in equilibrium are the following:

(*i*) External load "*wx*" acting vertically downwards.

(*ii*) Tension $T_C = H$, (acting horizontally at *C*)

(*iii*) Tension *T* at *P*, (acting along the tangent at *P*)

Let θ be the inclination of the tangent at *P* with the horizontal.

Resolving the forces on this part vertically and horizontally, we get,

$$T\sin\theta = wx,$$

$$T\cos\theta = T_C = H$$

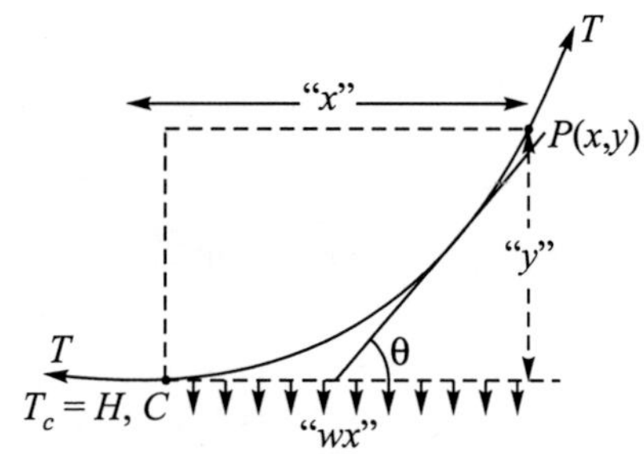

Fig. 14.3

$\therefore$ $$\tan\theta = \frac{wx}{H}$$

But $$\tan\theta = \frac{dy}{dx}$$

$\therefore$ $$\frac{dy}{dx} = \frac{wx}{H}$$

$\therefore$ $$dy = \frac{wx}{H} \times dx$$

Integrating, we get

$$y = \frac{wx^2}{2\,H} + k, \text{ where '}k\text{' is a constant.}$$

At "*C*", *i.e.,* $x = 0, y = 0$

$\therefore$ $k = 0$

$\therefore$ $$\text{"}y\text{"} = \frac{wx^2}{2\,H}$$

This is the equation of a parabola,

Since $$H = \frac{wl^2}{8\,h}, y = \frac{wx^2}{2} \times \frac{8\,h}{wl^2}$$

$\therefore$ $$\boxed{\text{"}y\text{"} = \frac{4\,h}{l^2} \times x^2}$$

with either support as origin, the equation to the cable will be $\text{"}y\text{"} = \frac{4\,h}{l^2}\,x\,(l - x)$

14.3. LENGTH OF THE CABLE

Let 'l' be the span of the cable. Let 'h' be the dip of the cable.

The equation of the cable with the lowest point 'C' as origin is given by,

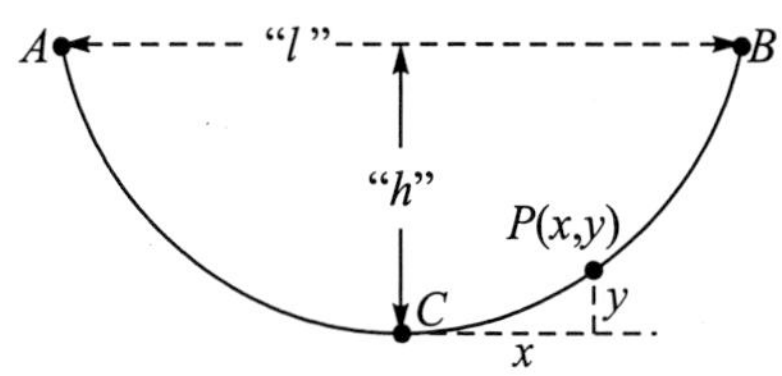

Fig. 14.4

$$\text{“}y\text{”} = \frac{4h}{l^2} \times x^2$$

$$\therefore \quad \frac{dy}{dx} = \frac{8hx}{l^2}$$

we know that,

$$\frac{ds}{dx} = \sqrt{1+\left(\frac{dy}{dx}\right)^2}$$

$$\therefore \quad \frac{ds}{dx} = \left[1+64\frac{h^2x^2}{l^4}\right]^{\frac{1}{2}}$$

$$\therefore \quad \frac{ds}{dx} = 1+\frac{1}{2}\times 64\,\frac{h^2x^2}{l^4}$$

(neglecting the other forms of the expression which are very small)

$$\therefore \quad ds = \left[1+32\frac{h^2x^2}{l^4}\right]dx$$

$\therefore$ Total length of the cable,

$$\text{“}L\text{”} = 2\int_0^{l/2}\left[1+32\frac{h^2x^2}{l^4}\right]dx$$

$$\text{“}L\text{”} = \left[\frac{l}{2}+32\frac{h^2}{l^4}\times\frac{1}{3}\times\frac{l^3}{8}\right] \quad \therefore \quad \text{“}L\text{”} = l+\frac{8}{3}\times\frac{h^2}{l}$$

14.4. TEMPERATURE STRESSES IN CABLE

Let the span and dip of a cable be "l" and "h" respectively.

The length of the cable is given by

$$\text{“}L\text{”} = l+\frac{8}{3}\times\frac{h^2}{l}$$

Suppose due to rise in temperature, the length of the cable increase by "dL" so that dip of the cable increase by "dh". The relation between dL and dh can be obtained by determining the above relation,

$$dL = \frac{16h}{3l}\times dh \quad \therefore \quad dh = \frac{3}{16}\times\frac{l}{h}dL$$

Let the rise in temperature be t°C.

$\therefore$ New length of cable $= L' = L(1+\alpha t)$, here α = co-efficient of linear expansion of the cable.

Increase in length of cable $= dL = L' - L = \alpha\, t\, L$

$$= \alpha\, t\left[l+\frac{8}{3}\frac{h^2}{l}\right] = \alpha\, t\, l + \frac{8}{3}\frac{h^2}{l}\alpha\, t$$

Since the quantity $\frac{8}{3}\frac{h^2}{l}\alpha t$ is very small, it can be ignored.

$$\therefore \quad dL = \alpha t l, \text{ But } dh = \frac{3}{16}\frac{l}{h}dL = \frac{3}{16}\frac{l}{h}\alpha t l$$

$$\therefore \quad dh = \frac{3}{16}\alpha t \frac{l^2}{h}$$

$\therefore$ Max. Tension in cable = $T_{max} = \sqrt{V^2 + H^2}$

When the central dip is small, we can make the approximation, $T_{max} = H$. Suppose the loading on the cable is w/m, we know that,

$$\text{"}H\text{"} = \frac{wl^2}{8h}$$

As 'h' increase, H decreases.

When 'h' increase to $h + dh$,

'H' decreases to $H - dH$

Differentiating the above equation,

$$dH = -\frac{wl^2}{8h^2}dh$$

$$\therefore \quad \frac{dH}{H} = (-)\frac{dh}{h}$$

$$\text{Stress in the cable} = f = \frac{T_{max}}{A} = \frac{H}{A}$$

$\therefore$ Let df be the change in stress in cable,

$$\therefore \quad \frac{df}{f} = \frac{dH}{H} = (-)\frac{dh}{h}$$

$$\therefore \quad \frac{df}{f} = -\frac{3}{16}\alpha t \frac{l^2}{h^2}$$

(Minus sign indicates that when the temperature rises the central dip increases, as a result of which the horizontal thrust decreases).

14.5. CABLE WITH TWO HINGED STIFFENING GIRDER

In this case the girders supporting the roadway are hinged only at ends. The structure in this case is statically indeterminate. An approximate solution is possible by the strain energy method. But by making the assumption that the girder is so rigid that the load on the girder is transmitted to the cable as *Udl*, the analysis can be made simple.

For instance, consider the point loads W_1 and W_2 on the stiffening girder. *Udl* transmitted to the cable

$$= w_e = \frac{W_1 + W_2}{\text{Span}}$$

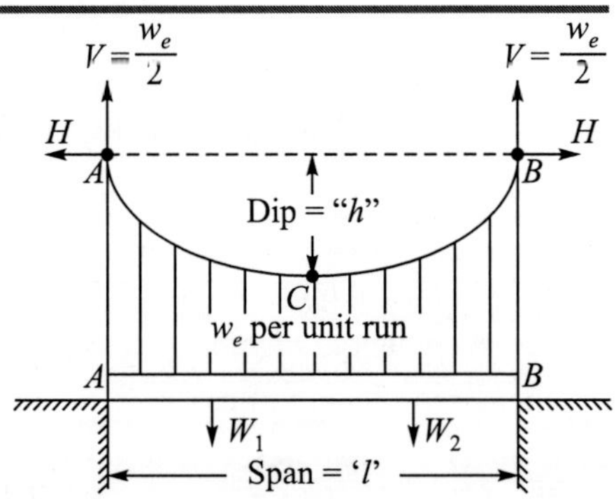

Fig. 14.5

In general, whatever may be the loading on the girder, $w_e = \dfrac{\text{Total load on the span of the girder}}{\text{Span}}$

Horizontal reaction at each end of the cable, $= H = \dfrac{w_e l^2}{8h}$

Vertical reaction at each end of the $= V = \dfrac{w_e l}{2}$

$\therefore \qquad T_{max} = \sqrt{V^2 + H^2}$

For the girder,

B.M. at any section = Beam moment − *H* moment

S.F. at any section = Bear shear − $H \tan \theta$

as in the case of the 3-hinged stiffening girder.

14.6. SUSPENSION BRIDGE WITH THREE HINGED STIFFENING GIRDER

In suspension cable bridges, it is necessary to maintain the parabolic shape of the cable. To satisfy this condition, the load transmitted to the cable should be a *Udl.* This is achieved as follows:

The external loads are carried out by two main three hinged or two hinged girders while the girders are supported not only at the ends but also by closely spaced hangers connected to the cables.

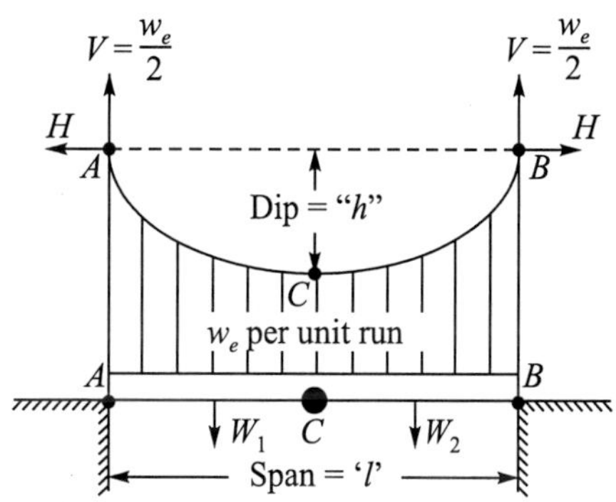

Fig. 14.6(a)

Figure 14.6 (*a*) shows three hinged stiffening girder supported at the ends and connected to the cable. Let the girder be subjected to any external load system W_1, W_2.

Let w_e per unit run be the *Udl* transferred to the cable. Obviously the girder will be subjected to an upward *Udl* "w_e" per unit run.

Now let us consider, the equilibrium of the cable and the stiffening girder separately.

(*i*) Consider the cable: This is subjected to *Udl* of w_e per unit run transmitted by the hangers. The equation to the cable with say the end '*A*' as origin is

$$y = \frac{4h}{l^2}\, x\,(l - x).$$

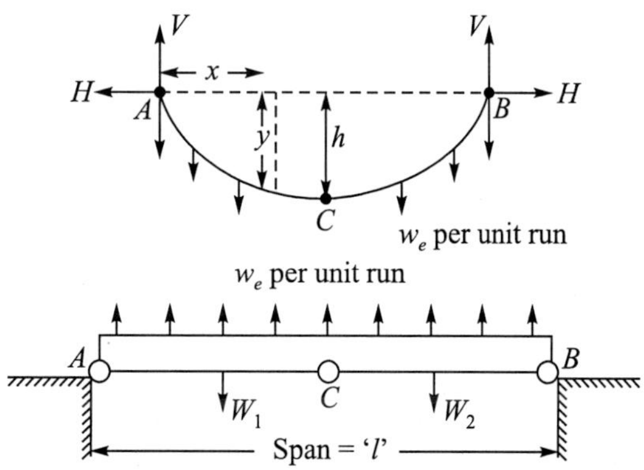

Fig. 14.6(b)

Vertical reaction at each end support

$$= V = \frac{w_e l}{2}$$

Horizontal reaction at each end support.

$$= H = \frac{w_e l^2}{8h}$$

Max. Tension, $T_{max} = \sqrt{V^2 + H^2}$.

***(ii)* Now consider the girder:** This is subjected to two load systems, namely,

(*i*) The applied external load system W_1 and W_2.

(*ii*) Upward *Udl* load of w_e per unit run. Consider any section *X-X* of the girder distant '*x*' from the end *A*.

The actual *B.M.* at the section of the girder is due to the above two load systems.

Let *M* be the actual *B.M.* at the section,

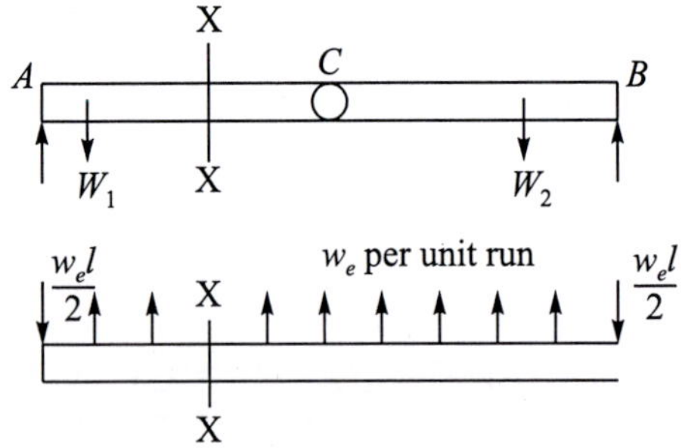

Fig. 14.7

$$M = M_x - \left[\frac{w_e l}{2} x - \frac{w_e x^2}{2}\right]$$

$$\therefore \quad M = M_x - \frac{w_e}{2} x(l - x) \qquad \ldots(14.1)$$

where M_x is the *B.M.* at the section for the first of the above load systems.

***(iii)* Now consider the cable:** Considering the corresponding pt. '*x*' of the cable,

we have,

$$H_y = \frac{w_e l^2}{8h} \times \frac{4h}{l^2} x(l - x)$$

$$\therefore \quad H_y = \frac{w_e}{2} x(l - x) \qquad \ldots(14.2)$$

From equations (14.1) and (14.2), we get

$$M = M_X - hy$$

The quantity *Hy* is called the "*H*-moment".

The actual *B.M.* at any section

$$= M = \text{Beam moment} - H \text{ moment}$$

Applying this condition to the hinge '*C*' of the girder, where *B.M.* is zero, we get

$M = 0 = M_C - H \times h$, where M_C is beam-moment at *C*.

$$\therefore \quad \text{"}H\text{"} = \frac{M_C}{h}$$

Now it is easy to find the *Udl* transmitted to the cable from the relation,

$$\text{"}H\text{"} = \frac{w_e l^2}{8h}$$

$$\therefore \quad w_e = \frac{8hH}{l^2}$$

$$\therefore \qquad V = \frac{w_e l}{2}$$

$$T_{max} = \sqrt{V^2 + H^2}$$

Next, shear force at any section of the girder,

$$= S = S_X - \frac{w_e l}{2} + w_e x = S_x - \frac{w_e}{2}(l - 2x) \qquad \text{...(14.3)}$$

where S_x is the $S.F.$ at the section due to the applied external load system considering the span as that of a simply supported girder.

Now for the cable with the left end as origin, the equation to the cable is

$$y = \frac{4h}{l^2} x (l - x)$$

$$\therefore \qquad \text{Slope at any section} = \tan\theta = \frac{dy}{dx} = \frac{4h}{l^2} x (l - 2x);$$

$$H = \frac{w_e l^2}{8h}$$

$$H \tan\theta = \frac{w_e l^2}{8h} \times \frac{4h}{l^2} x (l - 2x)$$

$$\therefore \qquad H \tan\theta = \frac{w_e}{2}(l - 2x) \qquad \text{...(14.4)}$$

From equations (14.3) and (14.4), we get,

$$S = S_x - H \tan\theta$$

Thus the actual $S.F.$ "S" at any section is given by S = Beam shear – $H \tan\theta$.

14.7. PROBLEMS ON CABLE SUSPENSION BRIDGES

Example 14.1: *A cable of span 'l' has its ends at heights h_1 and h_2 above the lowest point of the cable. It carries a Udl of w/m of span. Determine the vertical and Horizontal reactions at each end.*

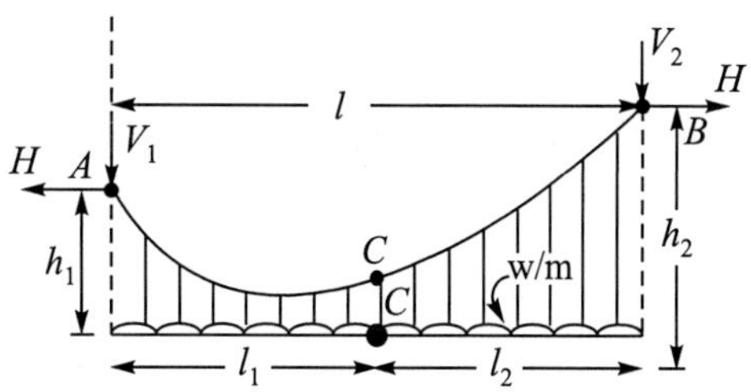

Fig. 14.8

Solution: Let 'C' be the lowest point of the cable. Let A and B be at heights h_1 and h_2 above "C".

Let the horizontal distance between A and C be l_1 and between B and C be l_2.

$$\therefore \qquad l_1 + l_2 = \text{'}l\text{'}$$

Since ACB is parabolic, we have

$$\frac{x^2}{y} = \text{constant, with 'C' as origin.}$$

$$\frac{l_1^2}{h_1} = \frac{l_2^2}{h_2}$$

$$\therefore \qquad \frac{l_1}{\sqrt{h_1}} = \frac{l_2}{\sqrt{h_2}} = \frac{l_1 + l_2}{\sqrt{h_1} + \sqrt{h_2}}$$

$$= \frac{l}{\sqrt{h_1}+\sqrt{h_2}}, \quad \therefore \quad l_1 = \frac{l\sqrt{h_1}}{\sqrt{h_1}+\sqrt{h_2}}$$

and
$$l_2 = \frac{l\sqrt{h_2}}{\sqrt{h_1}+\sqrt{h_2}}$$

Let V_1 and V_2 be the vertical reactions at A and B. Let 'H' be the horizontal reaction at each support. Taking moments about 'C' of the forces on the *L.H.S.* of 'C', we have,

$$V_1 l_1 = H h_1 + \frac{w l_1^2}{2}, \quad \therefore \quad V_1 = H \times \frac{h_1}{l_1} + \frac{w l_1}{2} \qquad \ldots(i)$$

11^{ly}, M^C, *R.H.S* of "C",

$$V_2 l_2 = H h_2 + \frac{w l_2^2}{2}, \quad \therefore \quad V_2 = H \times \frac{h_2}{l_2} + \frac{w l_2}{2} \qquad \ldots(ii)$$

Adding (*i*) and (*ii*), we get,

$$V_1 + V_2 = H\left[\frac{h_1}{l_1}+\frac{h_2}{l_2}\right] + \frac{w}{2}(l_1 + l_2)$$

$$\therefore \quad V_1 + V_2 = H\left[\frac{h_1}{l_1}+\frac{h_2}{l_2}\right] + \frac{wl}{2}$$

But
$$V_1 + V_2 = w \times l$$

$$\therefore \quad wl = H\left[\frac{h_1}{l_1}+\frac{h_2}{l_2}\right] + \frac{wl}{2}$$

$$\therefore \quad 'H' = \frac{\frac{wl}{2}}{\frac{h_1}{l_1}+\frac{h_2}{l_2}}$$

Putting,
$$"l_1" = \frac{l\sqrt{h_1}}{\sqrt{h_1}+\sqrt{h_2}} \quad \text{and} \quad "l_2" = \frac{l\sqrt{h_2}}{\sqrt{h_1}+\sqrt{h_2}}$$

$$\therefore \quad "H" = \frac{\frac{wl}{2}}{\frac{h_1(\sqrt{h_1}+\sqrt{h_2})}{l\sqrt{h_1}} + \frac{h_2(\sqrt{h_1}+\sqrt{h_2})}{l\sqrt{h_2}}}$$

$$\therefore \quad "H" = \frac{\frac{wl^2}{2}}{\sqrt{h_1}(\sqrt{h_1}+\sqrt{h_2}) + \sqrt{h_2}(\sqrt{h_1}+\sqrt{h_2})}$$

$$\therefore \quad \boxed{"H" = \frac{wl^2}{2(\sqrt{h_1}+\sqrt{h_2})^2}}$$

After finding "H", we can find the values of V_1 and V_2 from Eqns. (*i*) and (*ii*) above.

Example 14.2: *A cable carring a load of 10 kN per mtr. run of horizontal span, is stretched between supports 100 m apart. The supports are at the same level and the central dip is 8 m. Find the greatest and the least tension in the cable.*

Solution: Each vertical reaction,

$$'V' = \frac{wl}{2} = \frac{10\times100}{2} = 500 \text{ kN.}$$

Horizontal reaction,
$$'H' = \frac{wl^2}{8\,h} = \frac{10\times100^2}{8\times8} = 1562.5 \text{ kN}$$

Max. tension,
$$T_{max} = \sqrt{V^2 + H^2} = \sqrt{(500)^2 + (1562.5)^2}$$
$$= 1640.55 \text{ kN.}$$

Min. tension,
$$T_{min} = \text{“}H\text{”} = 1562.5 \text{ kN.}$$

Example 14.3: *A cable of span 'l' and dip 'h' is subjected to a Udl of w/m of horizontal span. If the dip be considered as small compared with the span, show that the difference between the greatest and least tension is appx. "wh".*

Solution: Each vertical reaction $V = \frac{wl}{2}$

Horizontal reaction
$$= H = \frac{wl^2}{8\,H}$$

Max. tension,
$$T_{max} = \sqrt{V^2 + H^2} = \sqrt{\left(\frac{wl}{2}\right)^2 + \left(\frac{wl^2}{8\,h}\right)^2}$$

$$= \frac{wl^2}{8\,h}\left[1 + \frac{16\,h^2}{l^2}\right]^{1/2} = \frac{wl^2}{8\,h}\left[1 + \frac{1}{2}\times\frac{16\,h^2}{l^2}\right], \text{ ignoring the other terms.}$$

$$= \frac{wl^2}{8\,h}\left[1 + \frac{8\,h^2}{l^2}\right] = \frac{wl^2}{8\,h} + wh$$

Min. tension
$$T_{min} = H = \frac{wl^2}{8\,h}$$

$$\therefore \quad T_{max} - T_{min} = \frac{wl^2}{8\,h} + wh - \frac{wl^2}{8\,h} = \text{“}\mathbf{wh}\text{”}$$

Example 14.4: *A cable of span 100 m and dip of 5 m is subjected to a rise of temp. of 15°C. Find the increase in dip due to rise in temp.*

Take $\alpha = 12 \times 10^6/0°C$

Solution:
Span $'l' = 100$ m, $h = 5$ m

$'t' = 15°C$, $\alpha = 12 \times 10^{-6}/0°C$.

Increase in dip
$$= dh = \frac{3}{16}\,\alpha\, t \times \frac{l^2}{h}$$

$$= \frac{3}{16} \times 12 \times 10^{-6} \times 15 \times \frac{100^2}{5} = 0.0675 \text{ m} = 67 \text{ mm.}$$

Example 14.5: *A cable having span of 100 m, dip "h" = 10 m is subjected to a rise is temp. of 10°C. The cable supports total load of 25 kN/m. Find the change in tension due to rise in temperature.*

Take $\alpha = 12 \times 10^{16}/0°C$.

Solution:
$$H = \frac{wl^2}{8\,h} = \frac{25 \times 100^2}{8 \times 10} = 3125 \text{ kN.}$$

Let '*dH*' be the change in horizontal reaction due to rise in temp.

$$\therefore \qquad \frac{dH}{H} = -\frac{dh}{h} = (-)\,\frac{3}{16}\,\alpha\,t\,\frac{l^2}{h^2}$$

$$\therefore \qquad \frac{dH}{H} = -\frac{3}{16} \times (12 \times 10^{-6}) \times 10 \times 10^2 = -\,0.00225$$

$$\therefore \qquad dH = 0.00225 \times 3125 = (-)\ 7.03 \text{ kN.}$$

(–) sign shows decrease in horizontal reaction.

Example 14.6: *A suspension bridge of 100 m span and roadway 6 m has two-hinged stiffening girders supported by two cables, has dip 10 m. The D.L. of the bridge is 5 kN/m² and the L.L. is 10 kN/m², which covers the left half of the span. Determine the S.F. and B.M. for the girder at 25 m from the left end. Find also the max. tension in the cable for this position of load.*

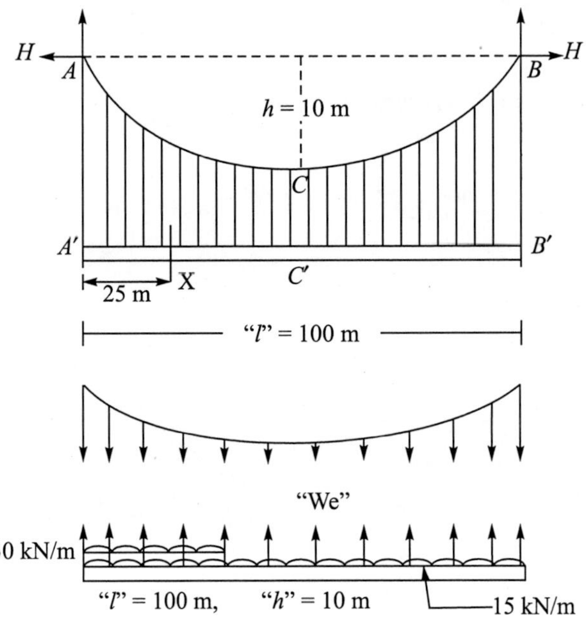

Fig. 14.9 (a)

Solution: *D.L.* per mtr. run of girder

$$= \frac{6 \times 5}{2} = 15 \text{ kN/m}$$

L.L. per mtr. span of girder

$$= \frac{6 \times 10}{2} = 30 \text{ kN/m}$$

Now consider analysis of the girder.

There will be no *S.F.* and *B.M.* for the girder due to *D.L.*

Total *L.L.* = 30 × 50 = 1500 kN.

Udl transmitted to the cable due to

$$L.L. = \frac{1500}{100} = 15 \text{ kN/m}$$

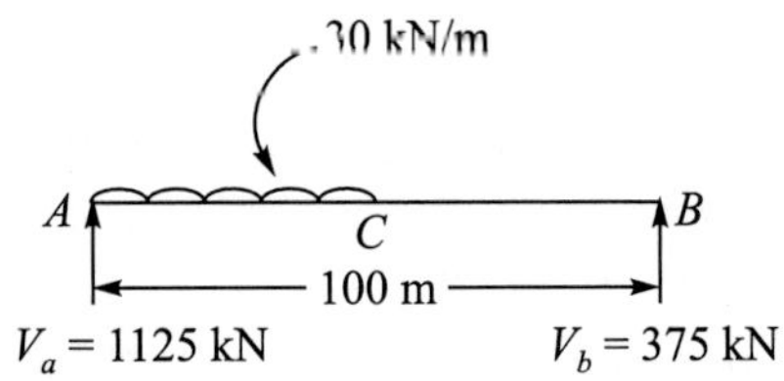

Fig. 14.9 (b)

Taking the girder as a *S.S.* girder carrying the *L.L*, the reactions at the supports are,

$$V_a = 1125 \text{ kN} \quad \text{and} \quad V_b = 375 \text{ kN}$$

Beam moment at 25 m from *A*, $= 1125 \times 25 - 30 \times \dfrac{25^2}{2} = 18750$ kN-m

Horizontal reaction for the cable due to *L.L.*

$$= H = \frac{wl^2}{8h} = \frac{15 \times 100^2}{8 \times 10} = 1875 \text{ kN}$$

The dip of the cable 25 mtr. from the *L.H.* end

$$= y = \frac{4h}{l^2} x(l - x) = \frac{4 \times 10}{100 \times 100} \times 25\,(100 - 25) = 7.5 \text{ m}$$

B.M. at 25 m from *A* = Beam moment – *H* moment

= Beam moment – *Hy*

= 18750 – 1875 × 7.5 = 4687.5 kNm

At 25 m from *A*, slope of the cable

$$= \tan\theta = y = \frac{4h}{l^2}(l - 2x) = \frac{4 \times 10}{(100)^2}(100 - 50) = 0.2$$

Actual *S.F.* for the girder at 25 m from the left end = Beam shear – $H \tan\theta$

= (1125 – 30 × 25) – 1875 × 0.2 = 0

Now consider the cable,

D.L. transmitted to cable = 15 kN/m

and *L.L.* Transmitted to cable = 15 kN/m

∴ Total load on the cable = 30 kN/m

Vertical reaction at the end of the cable = *V*

$$= \frac{w_e l}{2} = \frac{30 \times 100}{2} = 1500 \text{ kN}$$

Horizontal reaction at the end of cable = *V*

$$= \frac{w_e l}{2} = \frac{30 \times 100}{2} = 1500 \text{ kN}$$

Horizontal reaction $\text{“}H\text{”} = \frac{w_e l^2}{8h} = \frac{30 \times 100^2}{8 \times 10} = 3750 \text{ kN.}$

∴ Max. tension, $T_{max} = \sqrt{V^2 + H^2} = \sqrt{1500^2 + 3750^2} = 4038.9 \text{ kN.}$

Example 14.7: *The 3 hinged stiffening girder of a suspension bridge of span 120 m is subjected to two point loads of 240 kN and 300 kN at distances 25 M and 80 M from L end. Find the S.F. and B.M. for the girder at a distance 40 m. from left end. dip is 12 m. Find also T_{max} and draw B.M. diagram.*

Solution: Considering the stiffening girder as a simply supported beam supporting the given external load system, we will first calculate the reactions.

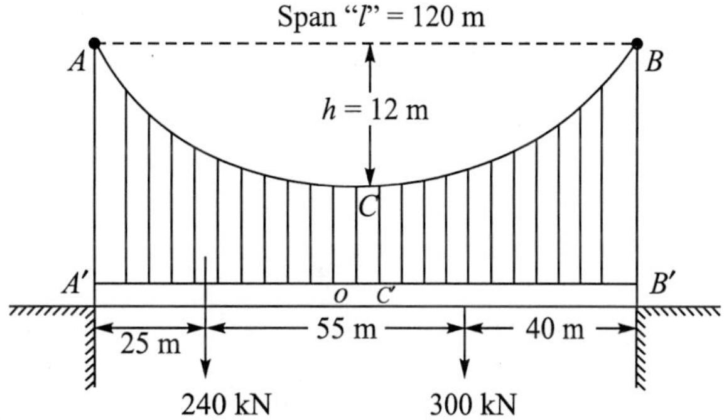

Fig. 14.10

$$M^A,\ V_b \times 120 = 240 \times 25 + 300 \times 80$$

$$\therefore \quad V_b = 250 \text{ kN},\ V_a = 240 + 300 - 250 = 290 \text{ kN}$$

Beam moment at '*C*', the middle of girder

$$M^C = 290 \times 60 - 240 \times 35 = 9000 \text{ kNm}$$

Beam moment under 240 kN load = 290 × 25 = 7250 kN-m

Beam moment under 300 kN load = 250 × 40 = 10000 kN-m

Horizontal reaction at each end of the cable

$$H = \frac{M^C}{h} = \frac{9000}{12} = 750 \text{ kN.}$$

Let the *Udl* transferred to the cable be we per unit run.

$$\therefore \quad H = \frac{w_e l^2}{8\,h} = \frac{w_e \times 120^2}{8 \times 12} = 750 \text{ kN}$$

$$\therefore \quad \text{“}w_e\text{”} = 5 \text{ kN/m}$$

Each vertical reaction for cable,

$$V = \frac{w_e l}{2} = \frac{5 \times 120}{2} = 300 \text{ kN}$$

$$\therefore \quad T_{max} \text{ in cable} = \sqrt{V^2 + H^2}$$

$$= \sqrt{300^2 + 750^2} = 807.8 \text{ kN}$$

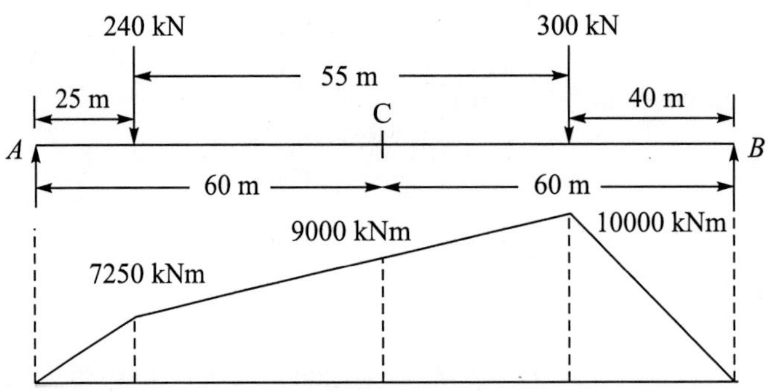

B.M. diagram

Fig. 14.11

For the girder: *S.F.* at any section is give by

$$S = \text{Beam shear} - H \tan \theta$$

For the cable, at any point,

$$\tan \theta = \frac{4\,h}{l^2}\,(l - 2x)$$

∴ At 40 m from *L.H.* end,

$$\tan \theta = \frac{4 \times 12}{120 \times 120}\,(120 - 20 \times 40) = \frac{2}{15}$$

Beam shear at 40 m from left end = 290 – 240 = (+) 50 kN

∴ Actual *S.F.* at 40 m from *L.H.* end

$$= 50 - 750 \times \frac{2}{15} = (-)\ 50 \text{ kN}$$

For the girder, the *B.M.* at any see is given by,

$$M = \text{Beam moment} - H \text{ moment}$$
$$= \text{Beam moment} - Hy$$

∴ *B.M.* at 40 m from end

$$= 290 \times 40 - 240 \times 15 = 8000 \text{ kN-m}$$

At 40 m from left hand end for cable,

$$y = \frac{4h}{l^2} \times (l - x)$$

$$= \frac{4 \times 12}{(120)^2} \times 40 \times 80 = \frac{32}{3} \text{ m.}$$

Actual *B.M.* at 40 m form *L.H.* end = Beam moment – *H* moment

$$= 8000 - 750 \times \frac{32}{3} = 0$$

Actual *B.M.* diagram for girder,

Dip of the cable 25 m from *L.H.*end

$$= \frac{4h}{l^2} \times (l - x)$$

$$= \frac{4 \times 12}{120^2} \times 25 \times 95 = 7.92 \text{ m.}$$

Actual *B.M.* at 25 m from *L.H.* end = Beam moment – H moment

$$= 7250 - 750 \times 7.92 = 1310 \text{ kN-m}$$

Also '*h*' for 80 m from *L.H.* end,

$$= \frac{4 \times 12}{(120)^2} \times 80 \times 40$$

$$= \frac{32}{3} \text{ m}$$

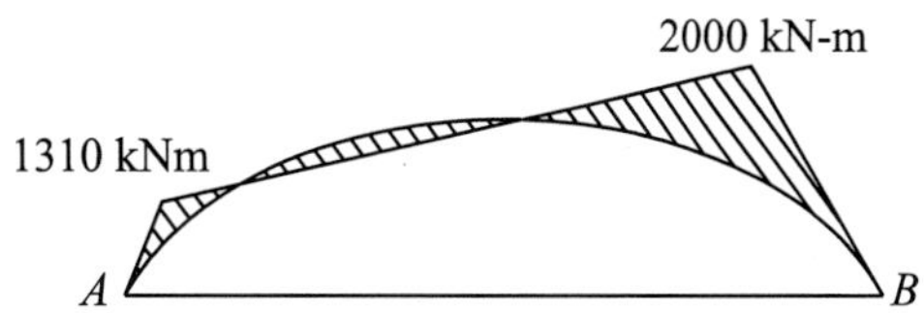

Actual B.M. for girder

Fig. 14.12

∴ Actual *B.M.* at 80 m from *L.H.* end

$$= 10000 - 750 \times \frac{32}{3} = 200.0 \text{ kN-m}$$

Example 14.8: *A cable of span 100 m has its ends at heights 8 m, above the lowest point of the cable. It carries a Udl of 10 kN/m per unit horizontal run of the span. Determine the horizontal and vertical reaction at the support. What is the length of the cable?*

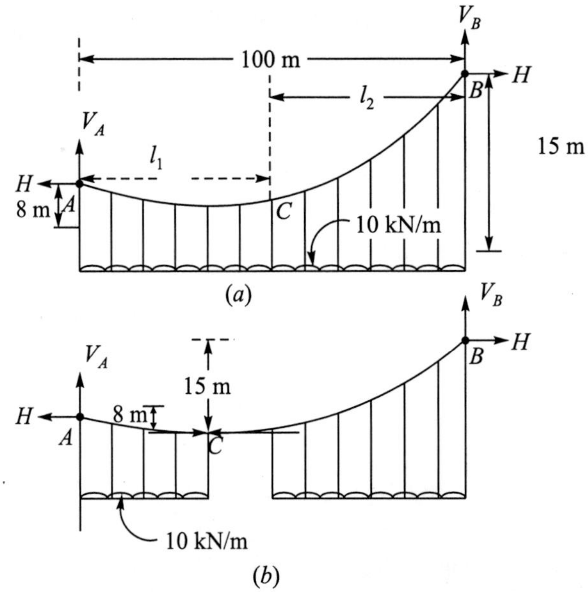

Fig. 14.13

Solution: Referring to Fig. 14.13 above, $h_1 = 8$ m, $h_2 = 15$ m, $l = 100$ m

Moment about A and B when calculated from mid section gives,

$$H \times 8 - 10 \times l_1 \times \frac{l_1}{2} = 0,$$

$$H \times (15 - 10)\ l_2 \times \frac{l_2}{2} = 0$$

$$8\ H = 5\ l_1^2;\ 15\ H = 5\ l_2^2$$

$$\therefore \quad \frac{l_1^2}{l_2^2} = \frac{8}{15}, \quad \text{or} \quad \frac{l_1}{l_2} = \frac{\sqrt{8}}{\sqrt{15}} = 0.730$$

$$\frac{l_1}{l_1 + l_2} = \frac{0.730}{1.730} \quad \text{or} \quad l_1 = \frac{0.730}{1.73}\ (l_1 + l_2) = \frac{0.730}{1.73} \times 100 = 42.214 \text{ m}$$

$$l_2 = 100 - 42.214 = 57.786 \text{ m}$$

$$V_A = wl_1 = 10 \times 42.214 = 422.14 \text{ kN}$$

$$8\ H = 10 \times \frac{l_1^2}{2}$$

$$H = \frac{10 \times 42.214^2}{8 \times 2} = 1113.76 \text{ kN}$$

$$V_B = wl_2 = 10 \times 57.786 = 577.86 \text{ kN}$$

Length of the cable,

$$L = l + \frac{2}{3} \times \frac{h_1^2}{l_1} + \frac{2}{3} \times \frac{h_2^2}{l_1} = 100 + \frac{2}{3} \times \frac{8^2}{42.214} + \frac{2}{3} \times \frac{15^2}{57.786}$$

$$= 103.607 \text{ m}$$

SUMMARY

- Cables and arches are efficient forms of structure to span large openings. Suspension bridges are used for highways, where the span of a bridge is more than 200 m.
- As the cable is flexible throughout, therefore it can not resist any moment and can adopt any shape under the loads, that is why B.M. at every point of the cable is taken as zero.
- A suspension bridge essentially consists of: (*a*) cable, (*b*) suspenders, (*c*) decking including stiffening girders, (*d*) supporting towers and Anchorages.
- The cable consists of either SWR (Steel Wire Rope) or parallel wires joined with clips or eye bar links. The cable can take direct tension only.
- Due to oscillations and sag, under loads these types of bridges are not suitable for railways or very heavy traffic.
- The suspension cable is supported on either side. It passes over smooth friction less pulley and anchored into the solid rock.
- The roadway is suspended from the cables by means of hangers or suspenders and the load taken by the hangers is taken as uniformly distributed load.

SHORT ANSWER QUESTIONS

1. Derive the equation for the cable and show that it hangs in the form of a parabola.
2. Obtain a relation for the: (*i*) horizontal thrust and (*ii*) tension in the cable.
3. Derive an expression for the length of the cable.
4. What are stiffening girders ? Discuss their uses and types.
5. Obtain relation for the temperature stresses in suspension cable.

EXERCISE FOR PRACTICE

1. A suspension cable bridge of span 100 m and central dip 10 m is suspended from the same level at two towers. The bridge cable is stiffened by a three hinged stiffened girder which carries a hinge concentrated of 50 kN at a point 30 m from one end. Find out the vertical and horizontal thrust, tension in cable and inclination of tension with horizontal axis. Sketch the S.F. diagram for the girder. **(MDU 2009)**
2. A cable supported at 120 m apart at the same level has a central dip of 12 m, carries a load of 20 kN/m of horizontal length. Calculate change in horizontal tension when temperature rises through 20°F. Take $\alpha = 6 \times 10^{-6}$ per°F. Also explain do stiffening girders transfer Udl to each suspender. **(MDU 2011)**
3. A suspension cable of 100 m span has 10 m dip. It is stiffened by a three hinged girder, whose D.L. is 20 kN/m. Determine the maximum tension in the cable, if a point load of 500 kN rolls over the girder. **(AMIE 2006)**
4. A suspension cable of 60 m span and central dip of 6 m is strengthened by two hinged stiffening-girders. The D.L. of stiffening girder is 20 kN/m. The girder is also subjected to a L.L. of 30 kN/m over the left half of the span. Find the maximum tension in the cable. **(MDU 2010)**
5. A cable supported at 120 m apart at the same level, has a central dip of 10 m. Find the increase in dip due to rise of temperature of 20°C. Take $\alpha = 12 \times 10^{-6}$ °C.
6. A suspension bridge of 40 m span and 3 m wide platform is subjected to a load of 64 kN/m^2. The bridge is supported by a pair of cables having central dip of 4.5 m. Find the necessary cross sectional area of thecable, if the maximum permissible stress in the cable material not to exceed 1.2×10^5 kN/m^2. **(Madras Univ. 2008)**
7. A cable of span 50 m is suspended from two pegs 6 m and 1.25 m above the lowest point of the cable. Find the length of the cable between the two pegs. **(KU 2010)**

❑❑❑

CHAPTER

15 Analysis of Statically Determinate Trusses/Frames

Statically determinate means–Any structural system which can be analysed by equation of statics under equilibrium condition, *viz.*,

$$\left.\begin{aligned}\Sigma V &= 0\\ \Sigma H &= 0\\ \Sigma M &= 0\end{aligned}\right\}$$

Plane Truss: The plane truss consists of a number of bars jointed together, such that they lie in one plane and form a frame-work which is stable against any type of loading acting in the same plane.

Framed Structure: A framed structure consists of a number of members connected to each other so as to form a frame to support an external load system. The simplest frame is a triangle. This frame is called the basic perfect frame and can be easily analysed by the conditions of equilibrium.

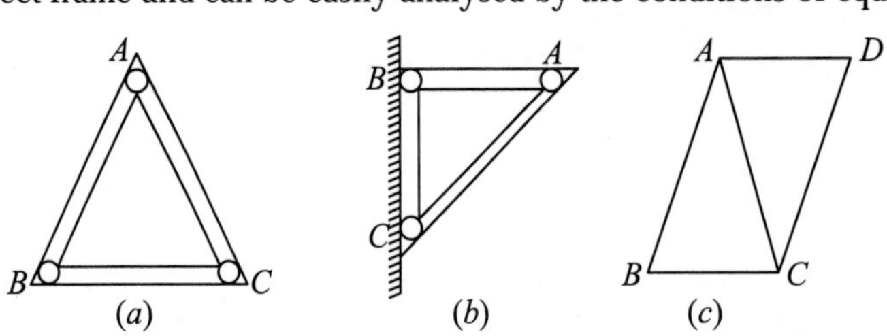

Fig. 15.1

Types of Plane Trusses:

(*a*) Simple trusses

(*b*) Compound trusses

(*c*) Complex trusses

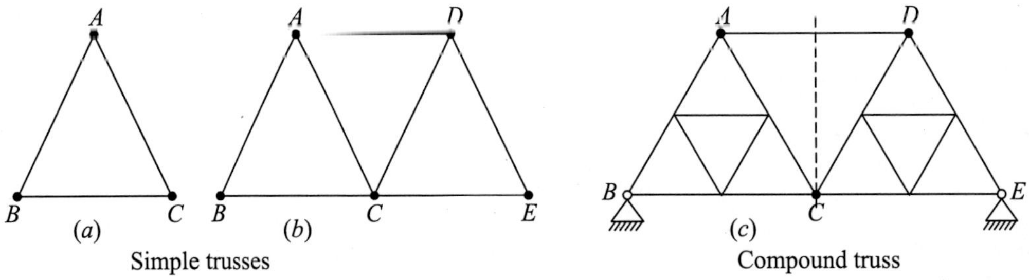

Fig. 15.2

A simple truss is formed by interconnecting simple rigid frames such that complete constraint is provided. A compound truss is cut by a section through the common hinge.

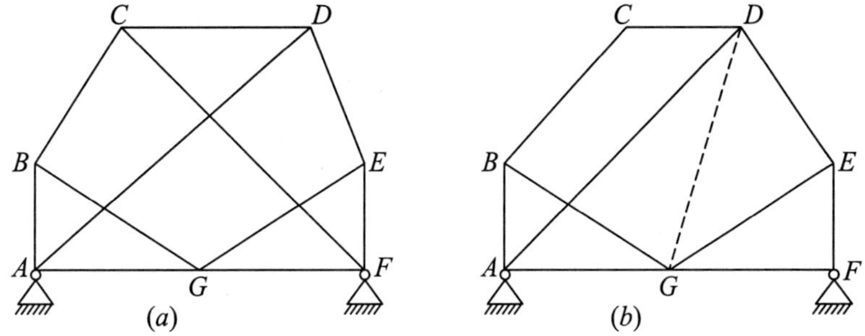

In case of complex trusses, the member *CF* is removed and replaced by *DG*.

15.1. RELATION BETWEEN THE NUMBER OF JOINTS AND THE NUMBER OF MEMBERS

(In a perfect frame)

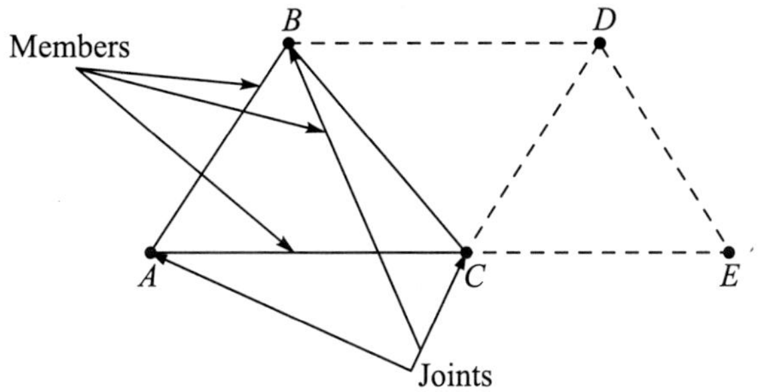

(Perfect-frame)

Fig. 15.3

Let there be '*n*' members and '*j*' joints in a perfect frame.

Suppose we remove 3 members *AB*, *BC* and *CA* and three joints *A*, *B* and *C*.

We are now left with $(n - 3)$ members and $(j - 3)$ joints.

Now from the figure we find that, for each joint, there are two members.

Hence, for a perfect and stable frame, the number of members can be found from the equation,

$$(n - 3) = 2\,(j - 3)$$

$$\therefore \qquad \boxed{n = 2j - 3}$$

i.e., **No. of members required = Twice the No. of joints – 3**

For example, in case of a square frame,

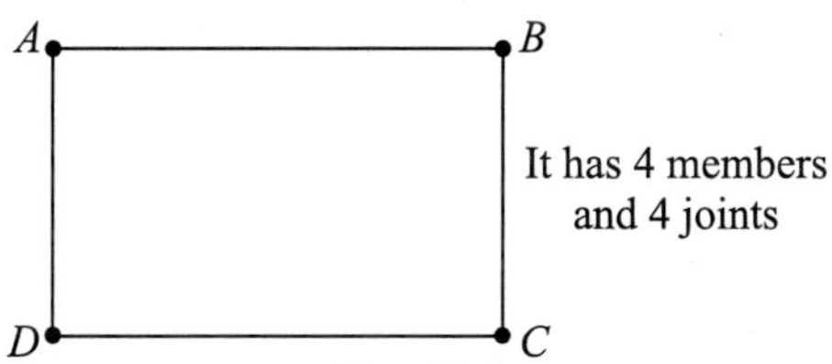

Fig. 15.4

For stability, we need $2 \times 4 - 3 = 5$ members. Such a frame is called a 'Deficient' frame.

For making it stable, join any of the diagonals. If we join other diagonal *BD*, even then the frame would become stable. It has members more than required, hence called redundant frame.

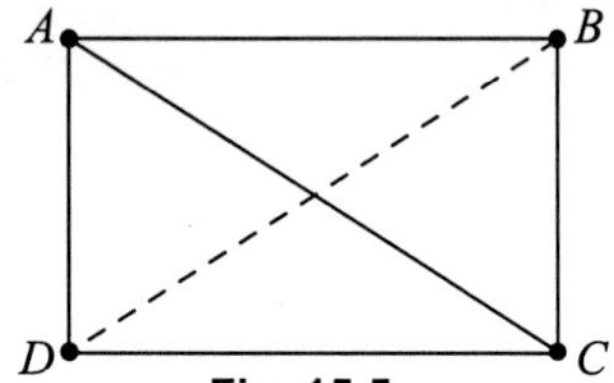

Fig. 15.5

In general, let the frame has "n" members and "j" joints.

If $n = 2j - 3 \rightarrow$ It is called perfect frame.

If $n < 2j - 3 \rightarrow$ It is called deficient frame.

If $n > 2j - 3 \rightarrow$ It is called redundant frame.

Important Note:

To determine which members of a truss do not carry forces:

(*a*) A single force cannot form a system in equilibrium. Hence if there is only one force acting at a joint, then for the equilibrium of the joint. This force = 0

(*b*) If 2 forces act at a joint, then for the equilibrium of the joint, these two forces should act along the same straight line. The two forces will be equal and opposite. If the two forces acting at a joint are not along the same straight line, then for the equilibrium of the joint, each force = 0

(*c*) If 3 forces act a joint and 2 of them are along the same straight line, then for the equilibrium of the joint, the 3rd force should be = 0.

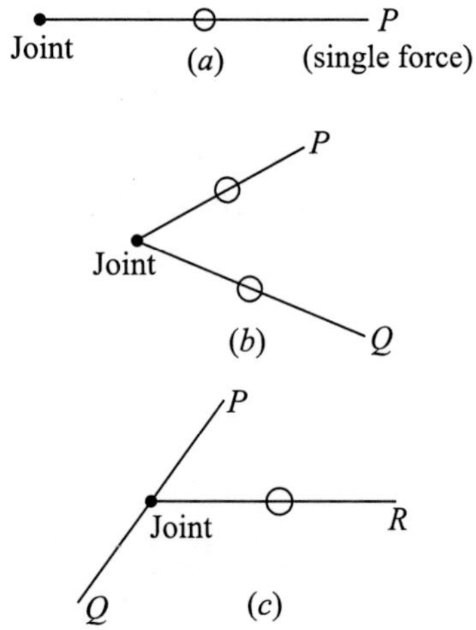

Fig. 15.6

Now consider the truss as shown below,

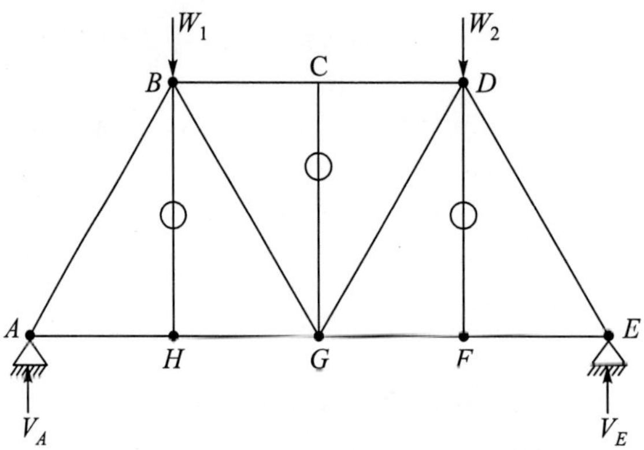

Fig. 15.7

Consider joint "H", forces at the joint 'H' are:

P_{ha} in the member HA

P_{hg} in the member HG

P_{hb} in the member HB

Since P_{ha} and P_{hg} are in the same straight line force $P_{hb} = 0$, similarly, $P_{cg} = 0$ and $P_{fd} = 0$

15.2. ANALYSIS OF PLANE TRUSS: (IT INVOLVES DETERMINATION OF REACTIONS AT SUPPORT AND FORCES IN THE MEMBERS OF THE TRUSS)

Methods:

(*i*) Method of joints
(*ii*) Method of sections
(*iii*) Graphical method
(*iv*) Method of tension co-efficient

15.3. METHOD OF JOINTS

In this method, each and every joint is treated as a free body in equilibrium as shown in below Fig. 15.8. The unknown forces are then determined by equilibrium equations viz., $\Sigma V = 0$ and $\Sigma H = 0$. While selecting the joint, care should be taken that at any instant, the joint should not contain more than two members in which the forces are unknown. At the end a force table is prepared showing the members, magnitude of forces and their nature. Tensile force is represented with (+) sign and the compressive force with a –ive sign.

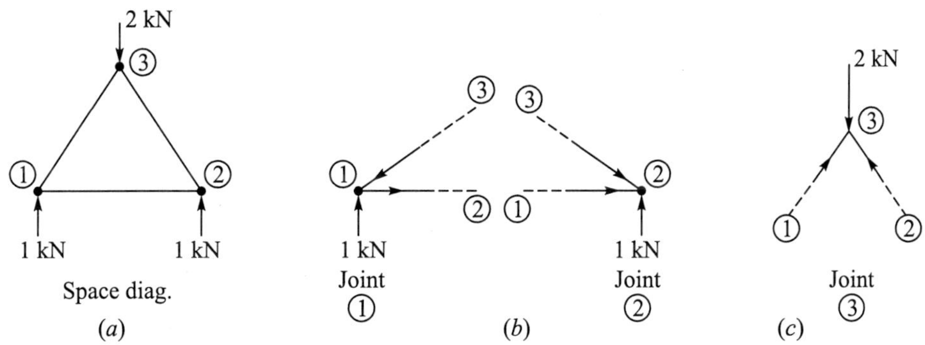

Fig. 15.8

Example 15.1: Find the forces in the members *AB, AC* and *BC*

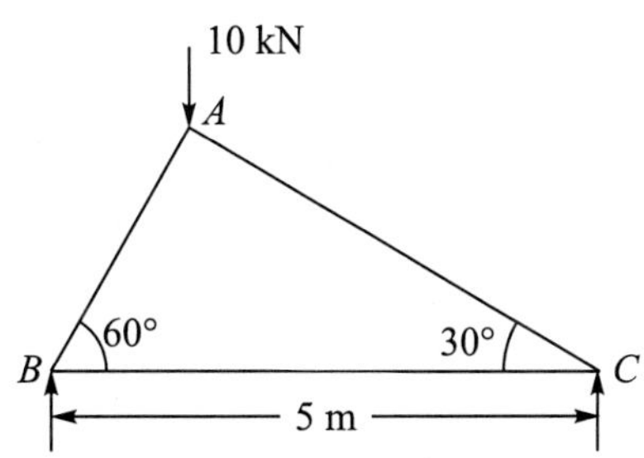

Solution: From the geometry of the truss, we find that the load of 10 kN is acting at a distance 1.25 m from the left hand support *i.e., B* and 3.75 m from *C*. Taking moments about *B* and equating the same,

$$R_C \times 5 = 10 \times 1.25 = 12.5$$

$$\therefore \quad R_C = \frac{12.5}{5} = 2.5 \text{ kN}$$

and $$R_B = 10 - 2.5 = 7.5 \text{ kN}.$$

Next consider the joint *B*. Let the directions of the forces P_{AB} and P_{BC} (or P_{BA} and P_{CB}) be assumed as shown in Fig. 15.9 below,

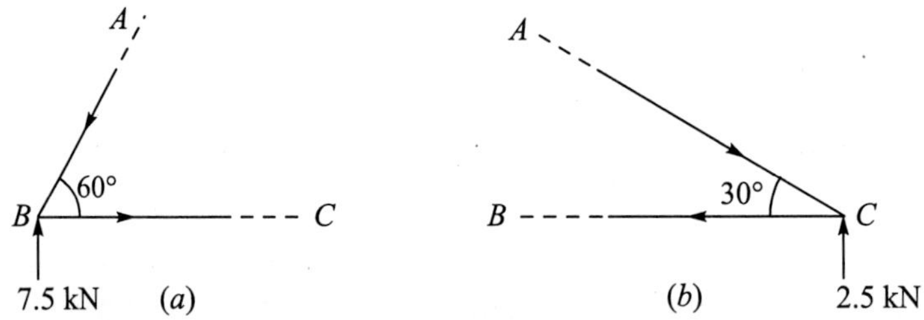

Fig. 15.9

Resolving the forces vertically and equating the same,

$$P_{AB} \sin 60° = 7.5$$

or $$P_{AB} = \frac{7.5}{\sin 60} = \frac{7.5}{0.866} = 8.66 \text{ kN} \quad \text{(Compressive)}$$

Also resolving the forces horizontally and equating the same,

$$P_{BC} = P_{AB} \cos 60° = 8.66 \times 0.5 = 4.33 \text{ kN} \quad \text{(Tension)}$$

Now consider the joint "*C*". Let the direction of forces P_{AC} and P_{BC} (or P_{CA} and P_{CB}) be assumed as shown in figure. Resolving the forces vertically and equating the same,

$$P_{AC} \sin 30 = 2.5$$

$$\therefore \quad P_{AC} = \frac{2.5}{\sin 30} = \frac{2.5}{0.5} = 5.0 \text{ kN} \quad \text{(Compressive)}$$

Resolving the forces horizontally,

$$P_{BC} = P_{AC} \cos 30 = 5 \times 0.866 = 4.33 \text{ kN} \quad \text{(Tensile)}$$

Example 15.2: *Find the forces in all members of the Warren-Girder shown below by method of joints.*

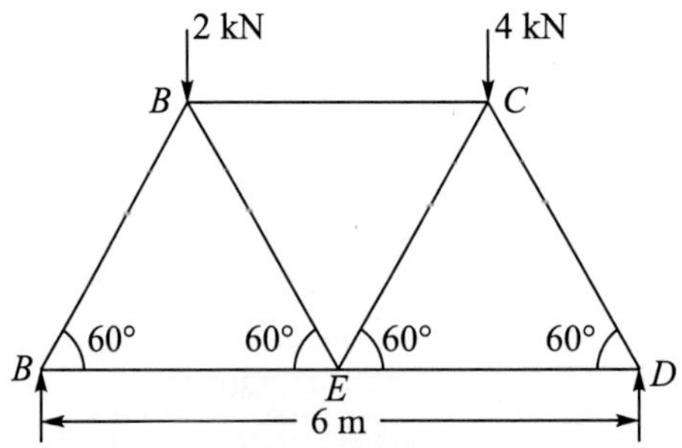

Fig. 15.10 (a)

Solution: Taking moments about *A* and equating the same,

$$R_D \times 6 = (2 \times 1.5) + (4 \times 4.5) = 21$$

$$R_D = \frac{21}{6} = 3.5 \text{ kN}$$

$$\therefore \quad R_A = (2 + 4) - 3.5 = 2.5 \text{ kN}$$

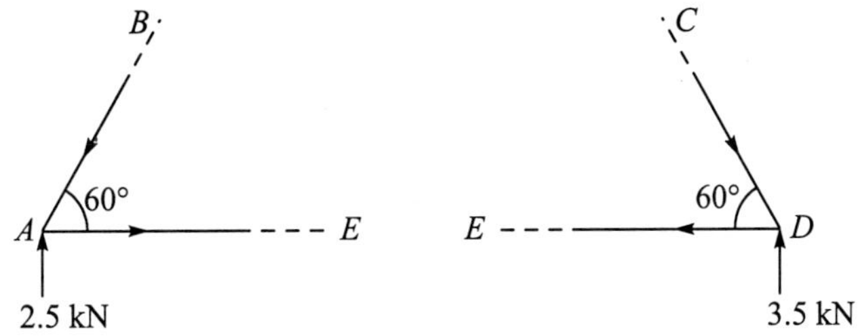

Fig. 15.10 (b)

Consider joint A, let the directions of P_{AB} and P_{AE} be assumed as shown. Resolving the forces vertically

$$P_{AB} \sin 60° = 2.5$$

$$\therefore \quad P_{AB} = \frac{2.5}{0.866} = 2.887 \text{ kN} \quad \text{(Compressive)}$$

Next, resolving the forces horizontally,

$$P_{AE} = P_{AB} \cos 60° = 2.887 \times 0.5 = 1.444 \text{ kN} \quad \text{(Tensile)}$$

Now consider the joint D, Let the direction of forces P_{CD} and P_{ED} be assumed as shown in figure, resolving the forces vertically, we have

$$P_{CD} \times \sin 60 = 3.5$$

$$\therefore \quad P_{CD} = \frac{3.5}{\sin 60} = \frac{3.5}{0.866} = 4.042 \text{ kN} \quad \text{(Compressive)}$$

Now, Resolving the forces horizontally,

$$P_{DE} = P_{CD} \cos 60° = 4.042 \times 0.5 = 2.021 \text{ kN} \quad \text{(Tension)}$$

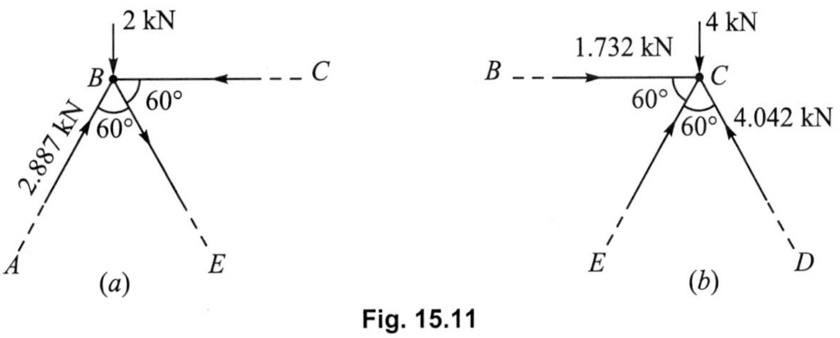

Fig. 15.11

Now consider joint B. We have already found that forces in member AB, *i.e.*, P_{AB} is 2.887 kN (compression). Let the direction of the forces P_{BC} and P_{BE} be assumed as shown in Fig. 15.11.

Resolve the forces vertically,

$$P_{BE} \sin 60 = P_{AB} \sin 60° - 2.0 = 2.887 \times 0.866 - 2.0 = 0.5 \text{ kN}$$

$$\therefore \quad P_{BE} = \frac{0.5}{\sin 60} = \frac{0.5}{0.866} = 0.577 \text{ kN} \quad \text{(Tensile)}$$

Next resolving the forces horizontally,

$$P_{BC} = 2.887 \cos 60° + 0.577 \cos 60° = (2.887 \times 0.5) + (0.577 \times 0.5)$$

$$= 1.732 \text{ kN} \quad \text{(Compressive)}$$

Now consider joint "C", we have already found out that the forces in the members BC and CD are 1.723 kN (Compressive) and 4.042 (Compressive) respectively. Let the direction of P_{CE} be assumed as shown in figure. Resolving the forces vertically,

$$P_{CE} \sin 60° = 4 - P_{CD} \sin 60° = 4 - (4.042 \times 0.866) = 0.5$$

$$\therefore \quad P_{CE} = \frac{0.5}{\sin 60°} = \frac{0.5}{0.866} = 0.577 \text{ kN} \quad \text{(Compressive)}$$

Example 15.3:

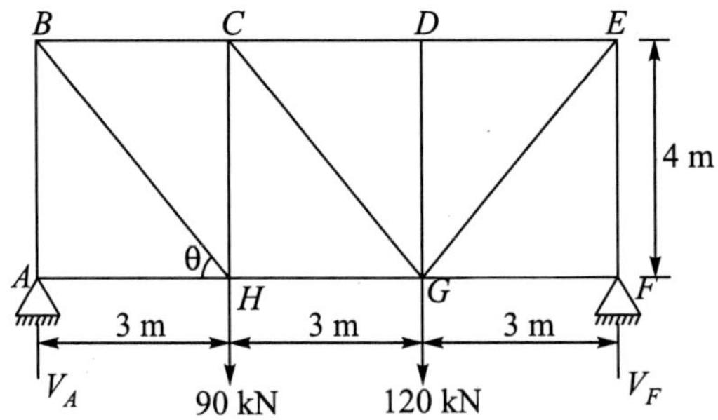

Fig. 15.12

Solution: Let V_A and V_F be the reactions at supports A and F,

M^A, we have,

$$V_F \times 9 = (90 \times 3) + (120 \times 6)$$

$$\therefore \quad V_F = 110 \text{ kN}, \; V_A = 210 - 110 = 100 \text{ kN}$$

Members which do not carry forces,

At the joint A, there are 3 forces, 2 of which are along the same straight line. Hence, the third force $P_{ah} = 0$

Similarly, considering the joint 'F', we have

$$P_{fg} = 0$$

and at joint D, we have $\quad P_{dg} = 0$

Now let us consider the equilibrium of various joints.

Joint A:

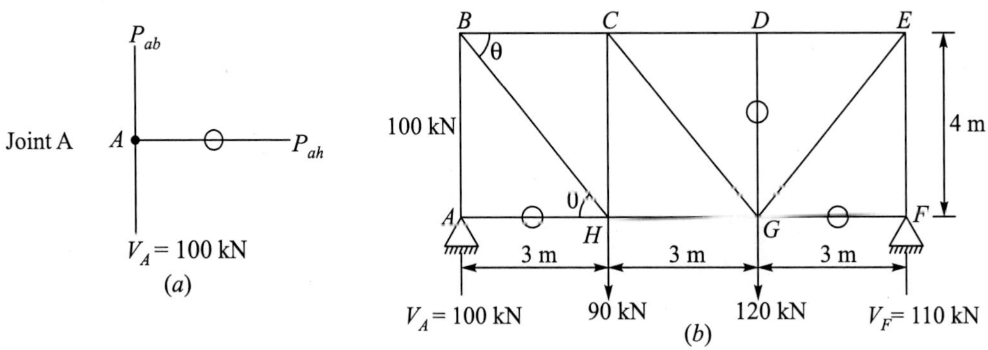

Fig. 15.13

Resolving vertically,

$$P_{ab} = 100 \text{ kN} \quad \text{(Compressive)}$$

Joint B: Resolving vertically,

$$\left\{ \tan\theta = \frac{4}{3} \therefore \sin\theta = \frac{4}{5} \text{ and } \cos\theta = \frac{3}{5} \right\}$$

$$P_{bh} \sin\theta = 100$$

$$P_{bh} = \frac{100}{\sin\theta} = \frac{100}{4/5} = 125 \text{ kN} \quad \text{(Tensile)}$$

Resolving horizontally,

$$P_{bc} = 125 \cos\theta = 75 \text{ kN} \quad \text{(Compressive)}$$

Joint *H*: Resolving vertically, we have the following components,

(*i*) 90 kN ↓

(*ii*) 125 sin θ = 100 kN ↑

Balancing force needed = 10 kN (Compressive)

∴ P_{hc} = 10 kN (Compressive)

Resolving horizontally, P_{hg} = 125 cos θ = 75 kN (Tensile)

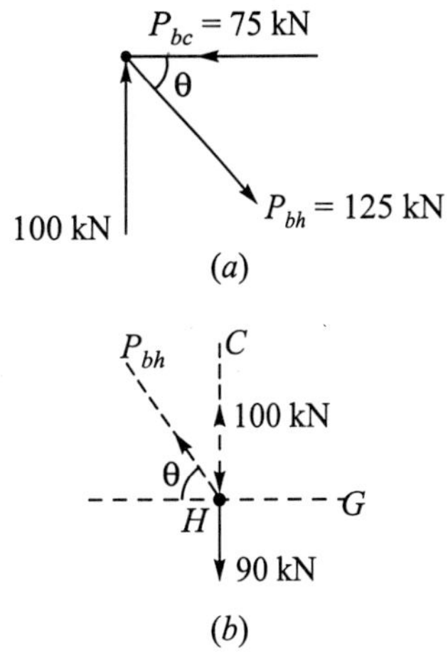

Fig. 15.14

Joint *C*: Resolving vertically

$$P_{eg} \sin\theta = 10$$

$$\therefore \quad P_{eg} = \frac{10}{\sin\theta} = 12.5 \text{ kN (Tenside)}$$

Resolving horizontally,

(*i*) 75 kN → (*ii*) 12.5 cos θ = 7.5 kN →

Balancing horizontally force needed = 82.5 kN ←

∴ P_{cd} = 82.5 kN (Compressive)

Joint *D*: Resolving horizontally,

$$P_{de} = 82.5 \text{ kN} \quad \text{(Compressive)}$$

Joint *G*: Resolving vertically,

(*i*) 120 kN ↓ (*ii*) 12.5 sin θ = 10 kN ↑

∴ Balancing force needed = 110 kN ↑

$$P_{ge} \sin\theta = 110 \text{ kN}$$

$$\therefore \quad P_{ge} = \frac{110}{\sin\theta} = 137.5 \text{ kN} \quad \text{(Tensile)}$$

Joint *E*: Resolving vertically at *E*,

$$P_{ef} = 137.5 \sin\theta$$

$$= 137.5 \times \frac{4}{5} = 110 \text{ kN} \quad \text{(Compressive)}$$

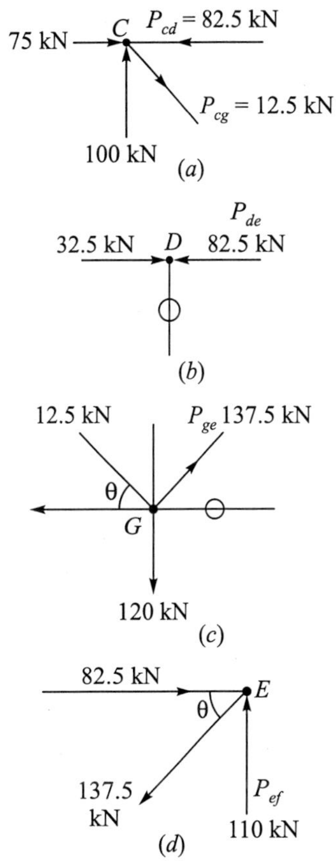

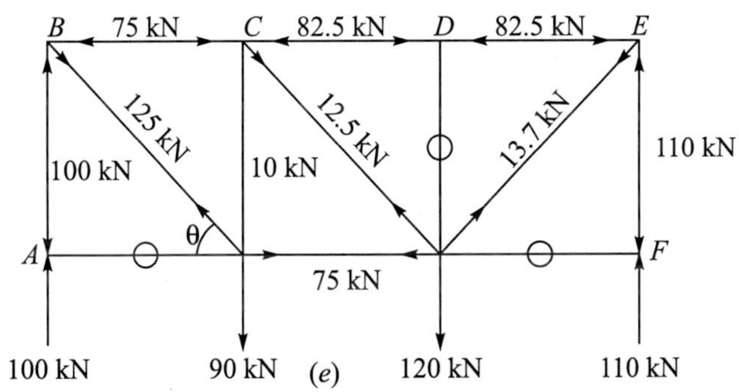

Fig. 15.15

"Abstract of Forces"

Table 15.1

Members	Force in the member	
	Compressive (kN)	Tensile (kN)
AB	100	
BC	75	
CD	82.5	
DE	82.5	
EF	100	
FG	0	
GH		7.5
HA	0	0
BH		125
HC	10	
CG		12.5
DG	0	0
GE		137.5

Example 15.4:

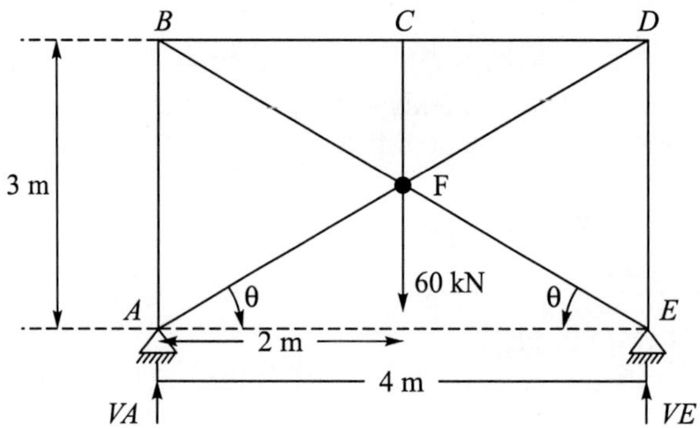

Fig. 15.16

Solution: By symmetry,

$$V_A = V_E = \frac{60}{2} = 30 \text{ kN}$$

Let the inclined members be at θ with the horizontal

$$\tan\theta = \frac{3}{4}, \quad \therefore \quad \sin\theta = \frac{3}{5}, \cos\theta = \frac{4}{5}$$

Joint *A*: Resolving horizontally,

$$P_{af} \cos\theta = 0, P_{af} = 0$$

Resolving vertically, $P_{ab} = 30$ kN (Compressive)

Similarly, considering "**Joint *E***", we conclude

$$P_{ef} = 0 \text{ and } P_{ed} = 30 \text{ kN} \quad \text{(Compressive)}$$

Joint *C*: Resolving vertically,

$$P_{bf} \sin\theta = 30, P_{bf} = 30 \times \frac{5}{3} = 50 \text{ kN} \quad \text{(Tensile)}$$

Resolving horizontally, $P_{bC} = 50 \cos \theta = 50 \times \frac{4}{5} = 40$ kN (Compressive)

Similarly, considering "**Joint *D***",

$P_{df} = 50$ kN (Tensile)

and $P_{de} = 40$ kN (Compressive)

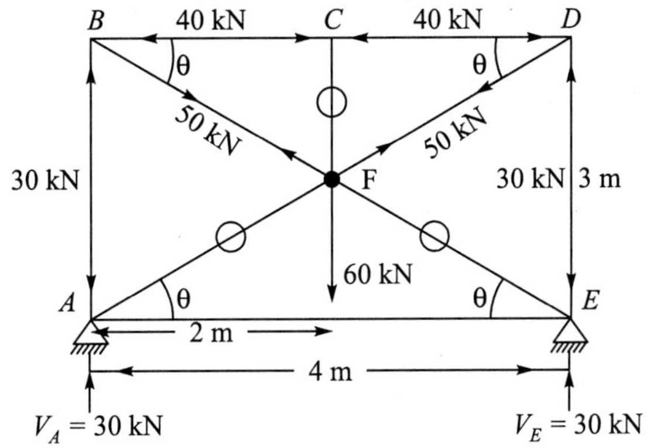

Fig. 15.17

"**Abstract of Forces**"

Table 15.2

Members	Forces in Members	
	Compressive (kN)	Tensile (kN)
AB, DE	30	
BC, CD	40	
BF, DF		50
AF, CF, EF	0	0

Example 15.5:

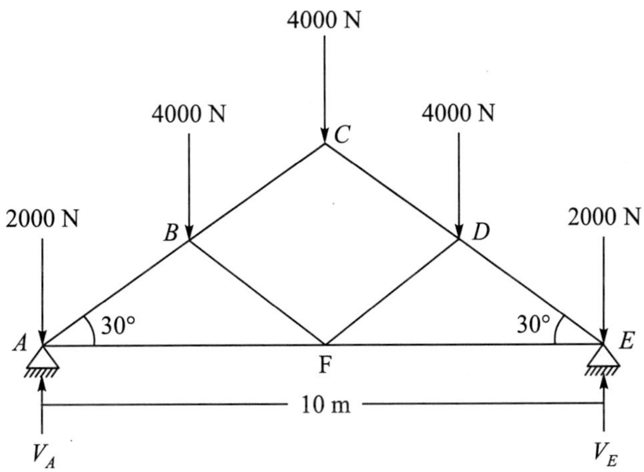

Fig. 15.18

Solution: Each vertical reaction = half the total load

viz. $$V_A = V_E = \frac{1}{2}[2000 + 4000 + 4000 + 4000 + 2000] = 8000 \text{ N}$$

Joint *A*: The forces keeping the joint '*A*' is equilibrium are

(*i*) Downward load = 2000 N (*ii*) V_A = 8000 N

(*iii*) P_{ab} (*iv*) P_{af}

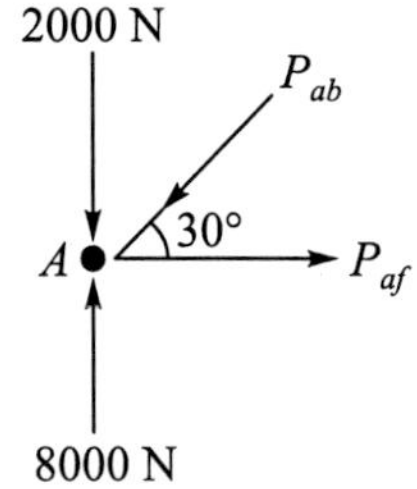

Fig. 15.19

Resolving the forces at *A*, vertically,

(*i*) Downward load = 2000 N

(*ii*) Upward load = 8000 N

Resultant of the two forces is 6000 N ↑ (Hence to keep Joint '*A*' in equilibrium we require a downward force of 6000 N ↓ at '*A*')

If P_{ab} is the force in member *AB*, its vertical component

viz., $$P_{ab} \sin 30° = 6000 \text{ N}$$

or $$P_{ab} = 12000 \text{ N} \quad \text{(Compressive)}$$

Now resolving the forces at *A*, horizontally,

$$P_{ab} \cos 30 = 12000 \times \frac{\sqrt{3}}{2} = 6000\sqrt{3} \text{ N}.$$

This is the force in member *AF*,

or $$P_{af} = 6000\sqrt{3} \text{ N} \quad \text{(Tensile)}$$

Joint *B*: The forces keeping this joint in equilibrium are:

(*i*) 4000 N ↓ (*ii*) P_{ba} = 12000 N

(*iii*) P_{bc} (*iv*) P_{bf}

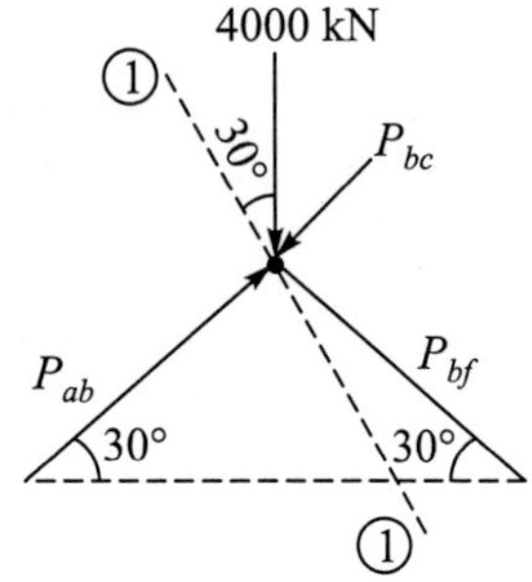

Fig. 15.20

The unkown forces are P_{bc} and P_{bf}. Hence, a direction for resolution will now be choosen at right angles to the line of action of one of the unknowns *e.g.*, consider a line (1) – (1) norlal to P_{bc}.

Hence resolving the forces along (1) – (1) are

$$4000 \cos 30° = P_{bf} \cos 30°$$

$$\therefore \quad P_{bf} = 4000 \text{ N} \quad \text{(Compressive)}$$

Now resolve the forces at '*B*' along the principal Rafter, we have following set of forces along the principal Rafter,

(*i*) P_{ba} of 12000 N ↑

(*ii*) $P_{bf} \cos 60°$ *i.e.*, 4000 × cos 60° = 2000 N ↓

(*iii*) 4000 cos 60° = 2000 N (Component of 4000 N force) ↓

These forces along the principal Rafter have a resultant of 8000 N acting up the line of rafter at *B*. Hence, for equilibrium, we need a force

$$P_{bc} = 8000 \text{ N} \quad \text{(Compressive)}$$

Hence, by symmetry,

Force in *CD* = P_{cd} = 8000 N (Compressive)

Force in *DE* = P_{de} = 12000 N (Compressive)

Force in *FD* = P_{fd} = 4000 N (Compressive)

Force in *FE* = P_{fe} = $6000\sqrt{3}$ (Tensile)

Joint *F*: Resolving the forces at this joint vertically, we have the following components.

(*i*) Vertical component of P_{fb} = 4000 sin 30° = 2000 N ↓

(*ii*) Vertical component of P_{fd} = 4000 sin 30° = 2000 N ↓

The resultant of the two forces is 4000 N ↓. Hence, we require a balance upward force of 4000 N (Tensile) at "*F*".

$$P_{fc} = 4000 \text{ N} \quad \text{(Tensile)}$$

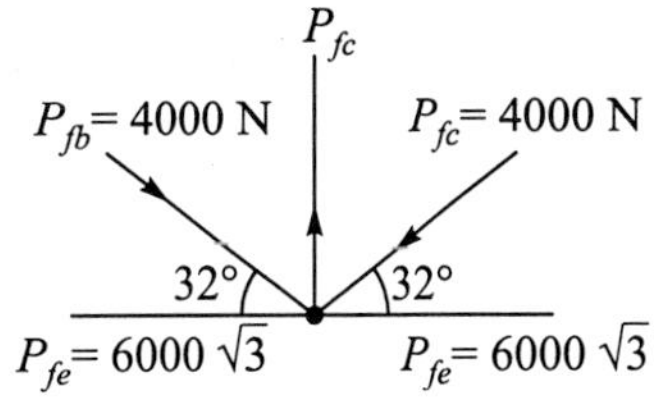

Fig. 15.21

"Abstract of forces"

Table 15.3

Members	Force in members	
	Compressive (N)	Tensile (N)
AB	12000	
BC	8000	
CD	8000	
DE	12000	
EF		$6000\sqrt{3}$
FA		$6000\sqrt{3}$
BF	4000	
DF	4000	
CF		4000

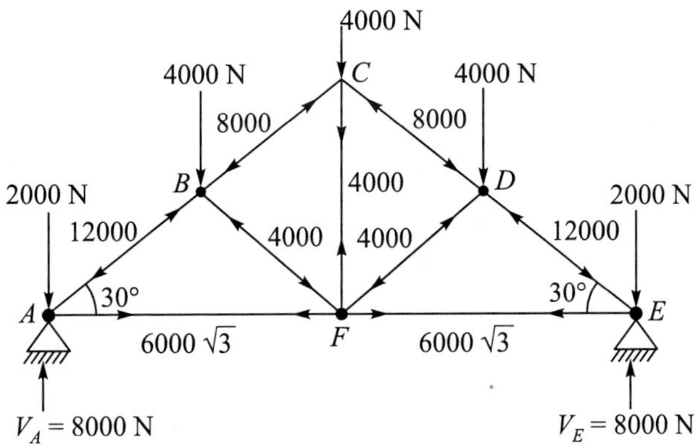

Fig. 15.22

15.4. METHOD OF SECTIONS

When do we use method of sections: This method is particularly convenient when the forces in some selected intermediate members of a frame are required to be found.

In this method, a section line is passed through the member or members in which the forces are required to be found out.

A part of the structure, on any one side of the section line, is then treated as a free body in equilibrium under the action of external forces. The unkown forces are then found out by the application of equilibrium or the principles of statics viz, Σ *M* = 0.

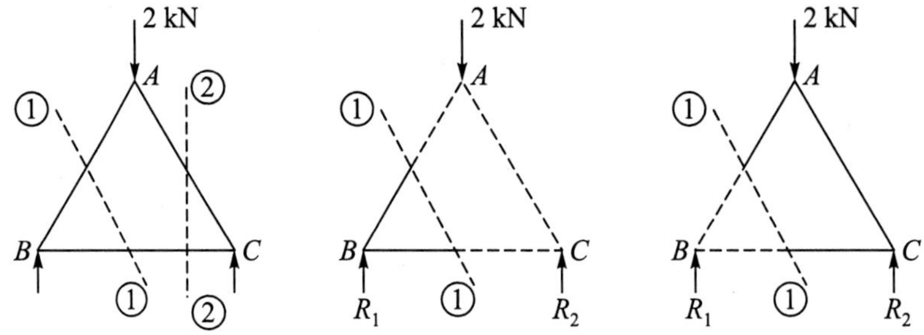

Fig. 15.23

Note I: To start with, we have shown section line (1) – (1) cutting the members *AB* and *BC*. Now in order to find out the forces in the member *AC*, section line (2) – (2) may be drawn.

Note II: While drawing a section line, care should be taken not to cut more than 3 members, in which the forces are unkown.

Example 15.6: *Solve the frame by method of sections*

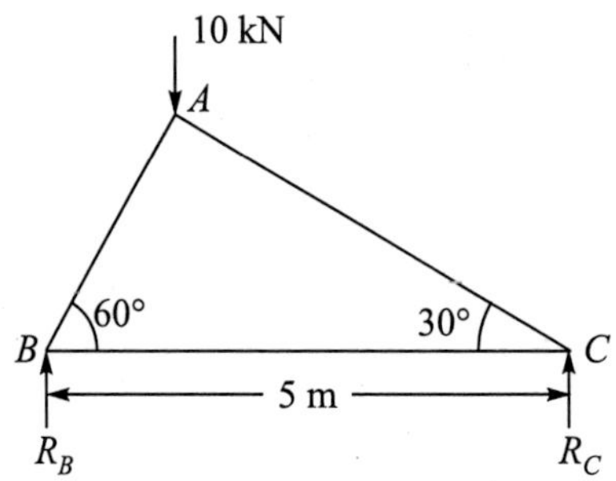

Fig. 15.24

Solution: Calculate Reactions R_B and R_C.

From the geometry of the truss, we find that the load of 10 kN is acting at distance of 1.25 m from the *L.H.* support *B*. ∴ Taking moments about '*B*' and equating the same,

$$R_C \times 5 = 10 \times 1.25 = 1.25 \text{ kN}$$

∴ $$R_C = \frac{12.5}{5} = 2.5 \text{ kN}$$

and $$R_B = 10 - 2.5 = 7.5 \text{ kN.}$$

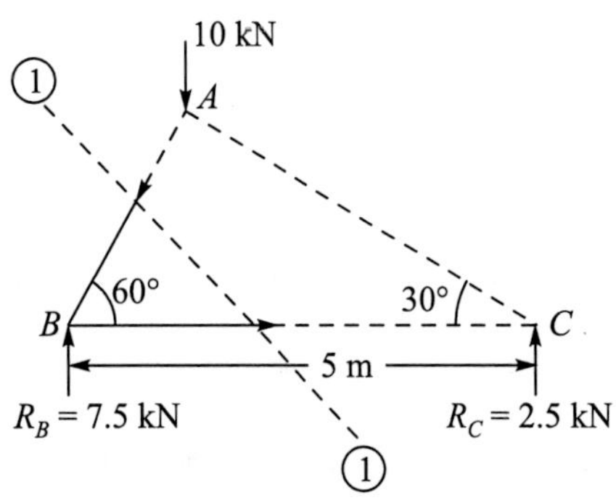

Fig. 15.25 (*a*)

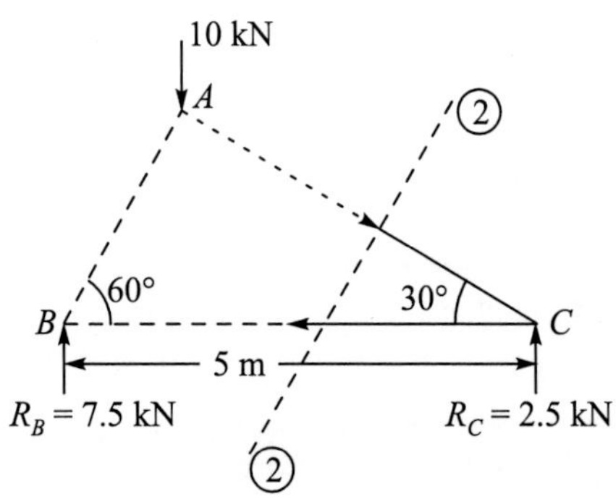

Fig. 15.25 (*b*)

First of all, pass section (1) – (1) cutting the truss into two parts (one part shown by firm lines and the other by dotted lines) through the members *AB* and *BC* of the truss as shown in Fig. 15.25. Now consider the equilibrium of the left part of the truss. Let the directions of the forces P_{AB} and P_{AC} be assumed as shown in figure.

Taking moments of the forces acting in the left part of the truss only about the joint *C* and equating the same.

$$P_{AB} \times 5 \sin 60° = 7.5 \times 5$$

$$\therefore \quad P_{AB} = \frac{7.5 \times 5}{5 \sin 60°} = \frac{7.5}{0.866} = 8.66 \text{ kN} \quad \text{(Compressive)}$$

Next taking moment of the forces acting in the left part of the truss only about the joint *A* and equating the same,

$$P_{BC} \times 1.25 \tan 60° = 7.5 \times 1.25$$

$$\therefore \quad P_{BC} = \frac{7.5 \times 1.25}{1.25 \tan 60°} = \frac{7.5}{1.732} = 4.33 \text{ kN} \quad \text{(Tensile)}$$

Now pass section (2) – (2) cutting the truss into two parts through the members *AC* and *BC*.

Next consider the equilibrium of the right part of the truss. Let the direction of the forces P_{AC} and P_{BC} be assumed as shown in figure.

Taking moments of the force in the right part of the truss only about the joint '*B*',

$$P_{AC} \times 5 \sin 30° = 2.5 \times 5$$

$$\therefore \quad P_{AC} = \frac{2.5}{\sin 30°} = \frac{2.5}{0.5} = 5 \text{ kN} \quad \text{(Compressive)}$$

Next taking moments of the forces acting in the right part of truss about joint '*A*'

$$P_{BC} \times 3.75 \tan 30° = 2.5 \times 3.75$$

$$\therefore \quad P_{BC} = \frac{2.5 \times 3.75}{3.75 \tan 30°} = \frac{2.5}{0.577} = 4.33 \text{ kN} \quad \text{(Tensile)}$$

"Table of Results"

Table 15.4

S.No.	Member	Magnitude of forces (KNS)	Nature of forces
1.	*AB*	8.66	Compression
2.	*BC*	4.33	Tension
3.	*AC*	5.0	Compression

Example 15.7: *Find the forces in all members of the Warren Girder by method of sections.*

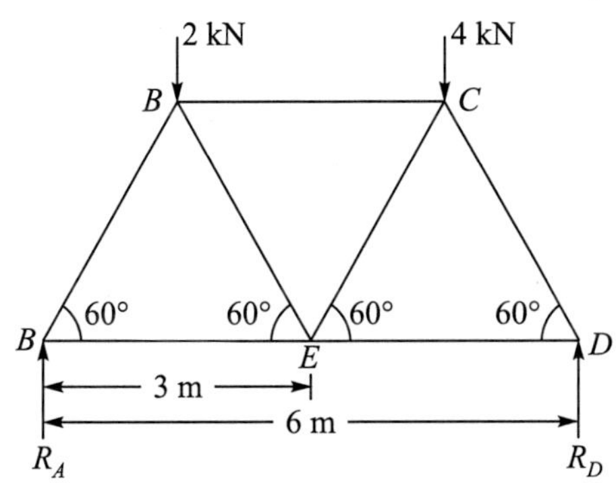

Fig. 15.26

Solution: Calculation of reactions R_A and R_D.

Taking moments about A, and equating the same,

$$R_D \times 6 = (2 \times 1.5) + (4 \times 4.5) = 21$$

$$\therefore \quad R_D = \frac{21}{6} = 3.5 \text{ kN}$$

$$\therefore \quad R_A = (2 + 4) - 3.5 = 2.5 \text{ kN}$$

First of all pass section (1) – (1) through the truss cutting members AB and AE as shown, and next consider equilibrium of the left part of truss.

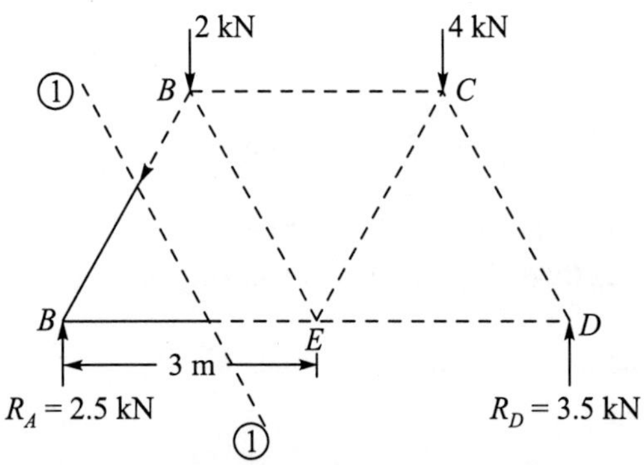

Fig. 15.27 (*a*)

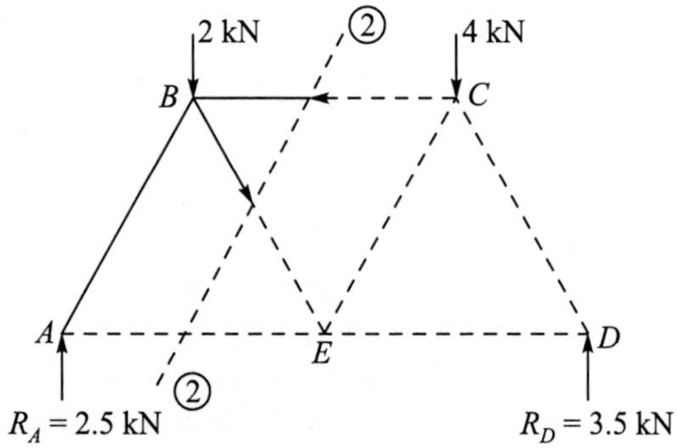

Fig. 15.27 (*b*)

Let the directions of the forces P_{AB} and P_{AE} be assumed as shown in the Fig. 15.27.

Taking moment of the forces acting in the left part of the truss only about joint "E", we have

$$P_{AB} \times 3 \sin 60° = 2.5 \times 3$$

$$\therefore \quad P_{AB} = \frac{2.5}{\sin 60°} = \frac{2.5}{0.866} = 2.887 \text{ kN} \quad \text{(Compression)}$$

Now pass section (2) – (2) cutting the truss through the members BC, BE and AE. Consider the equilibrium of the left of the truss. Let the directions of the force P_{BC} and P_{BE} be assumed as shown in the figure. Taking moments of all the forces acting in left part of the truss about the joint E, we have,

$$P_{BC} \times 3 \sin 60° = (2.5 \times 3) - (2 \times 1.5) = 4.5$$

$$\therefore \quad P_{BC} = \frac{4.5}{3 \sin 60°} = \frac{4.5}{3 \times 0.866} = 1.732 \quad \text{(Compression)}$$

Next take moments of the forces acting in the left part of the truss only about the joint 'A' we have,

$$P_{BE} \times 3 \sin 60° = (P_{BC} \times 3 \sin 60°) - (2 \times 1.5)$$

$$= (1.732 \times 3 \times 0.866) - 3.0 = 1.5$$

$$\therefore \quad P_{BE} = \frac{1.5}{3 \sin 60°} = \frac{1.5}{3 \times 0.866} = 0.577 \text{ kN} \quad \text{(Tensile)}$$

Now pass section (3) – (3) cutting the truss through the members BC, CE and ED. Consider the equilibrium of the right part of the truss. Let the directions of the forces P_{CE} and P_{DE} be assumed as shown in the figure. Taking moments of the forces in the right part of the truss only about the joint D, we have,

$$P_{CE} \times 3 \sin 60° = (4 \times 1.5) - (P_{BC} \times 3 \sin 60°)$$
$$= 6.0 - (1.732 \times 3 \times 0.866) = 1.5$$

$$\therefore \qquad P_{CE} = \frac{1.5}{3 \sin 60°} = \frac{1.5}{3 \times 0.866} = 0.577 \text{ kN} \quad \text{(Compression)}$$

Next taking moment of the forces in the right part of the truss only about joint 'C', we have

$$P_{CE} \times 3 \sin 60° = 3.5 \times 1.5 = 5.25$$

$$\therefore \qquad P_{DE} = \frac{5.25}{3 \sin 60°} = \frac{5.25}{3 \times 0.866} = 2.021 \text{ kN} \quad \text{(Tensile)}$$

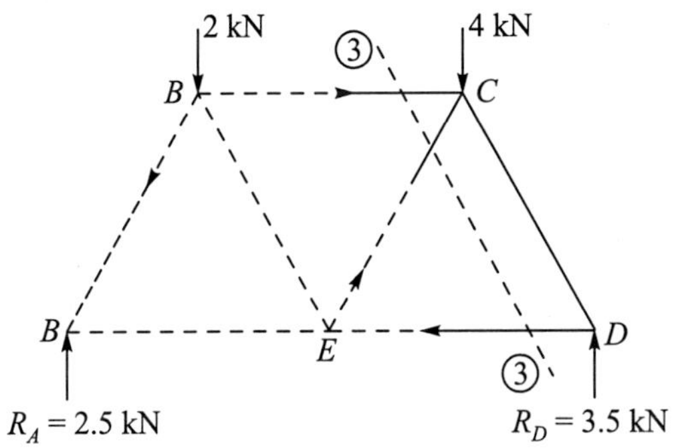

Fig. 15.28 (*a*)

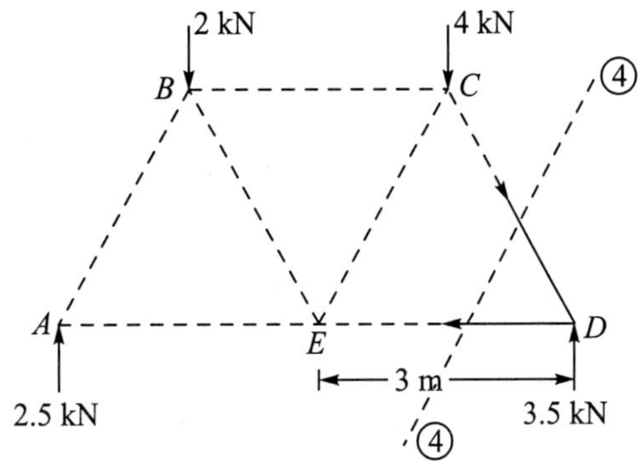

Fig. 15.28 (*b*)

Now pass section (4) – (4) cutting the truss through the member CD and DE. Let the directions of the forces P_{CD} be assumed as shown in the Fig. 15.28 (*a*) (*b*).

Taking moments of the forces acting in the right part of the truss only about the joint 'E', we have

$$P_{CD} \times 3 \sin 60 = 3.5 \times 3$$

$$P_{CD} = \frac{3.5}{\sin 60°} = \frac{3.5}{0.866} = 4.042 \text{ kN} \quad \text{(Compression)}$$

"Table of Results"

Table 15.5

S.No.	Member	Magnitude of forces in kN	Nature of force
1.	*AB*	2.887	Compression
2.	*AE*	1.444	Tension
3.	*CD*	4.042	Compression
4.	*DE*	2.021	Tension
5.	*BE*	0.577	Tension
6.	*BC*	1.732	Compression
7.	*CE*	0.577	Compression

Example 15.8: *Determine the nature and magnitude of the forces in the members 1, 2 and 3 of the plane loaded and supported as shown in the Fig. 15.29.*

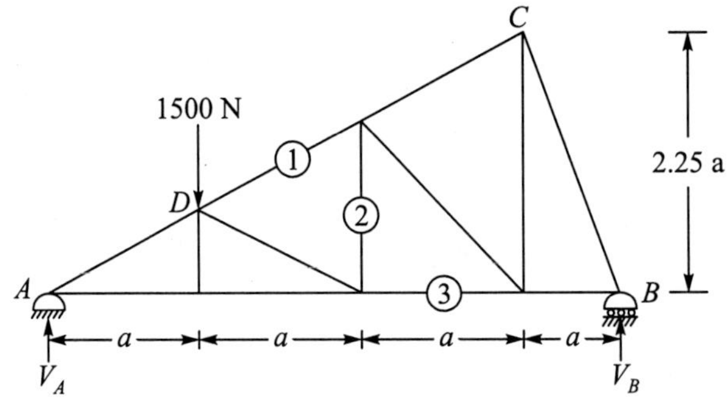

Fig. 15.29

Solution: Taking moments about A, we have

$$V_B \times 4a = 1500 \times a$$

$$\therefore \quad V_B = \frac{1500}{4} = 375 \text{ N} \quad \text{and} \quad V_A = 1500 - 375 = 1125 \text{ N}$$

From the geometry of the Fig. 15.30, we find that,

$$\tan\theta = \frac{2.25a}{3a} = 0.75$$

and

$$\sin\theta = \frac{3}{5} = 0.6 \text{ and } \cos\theta = \frac{4}{5} = 0.8$$

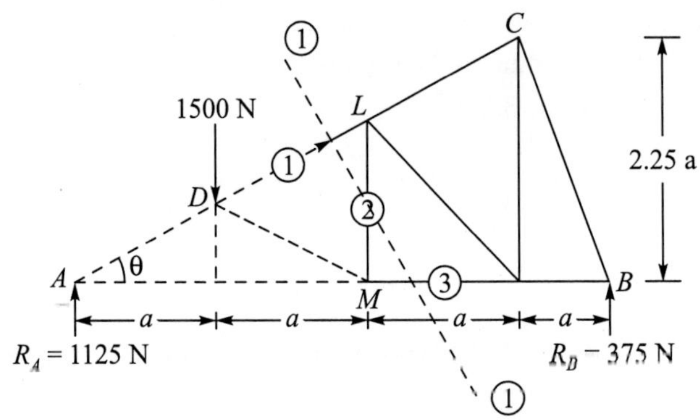

Fig. 15.30

Let one section line (1) – (1) cut the members 1, 2, and 3 in which the forces are required to be found out. Consider the equilibrium of the right part of the truss. Let the directions of P_1, P_2 and P_3 be assumed as shown in the figure.

Taking moments about joint M and equating the same,

$$P_1 \times 2a \sin\theta = 375 \times 2a$$

$$\therefore \quad P_1 = \frac{375}{\sin\theta} = \frac{375}{0.6} = 625 \text{ N} \quad \text{(Compression)}$$

Similarly, taking moments about joint A, we have

$$P_2 \times 2a = 375 \times 4a = 1500a$$

$$\therefore \qquad P_2 = \frac{1500a}{2a} = 750 \text{ N} \quad \text{(Tension)}$$

and next taking moments about the joint L, we have

$$P_3 \times \frac{3a}{2} = 375 \times 2a = 750a$$

$$\therefore \qquad P_3 = \frac{750}{1.5} = 500 \text{ N} \quad \text{(Tension)}$$

15.5. METHOD OF TENSION CO-EFFICIENTS

The method of tension co-efficients first introduced by prof. R.V. Southwell, is infact a neat and systematic presentation of the "method of joints", which is equally applicable to both frames as well as space frames. The method is particularly useful to space frames in which other methods prove to be cumbersome and tedious.

15.6. DEFINITION-TENSION COEFFICIENTS

The tension coefficient for a member of a frame is defined as the pull or tension in that member divided by its length. Thus

$$\text{“}t\text{”} = \frac{T}{L},$$

where t is the tension co-efficient for the member, T is the pull in the member and L is its length.

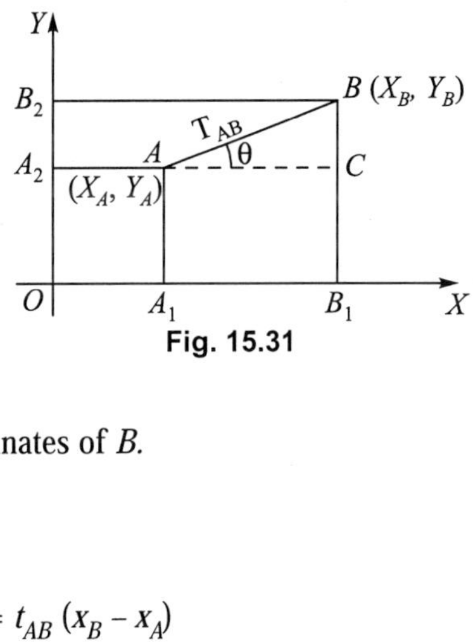

Fig. 15.31

Consider a member AB of a pin-jointed perfect frame in equilibrium, under a given system of external forces acting at the joints. Let T_{AB} be the resulting pull in the member.

Let (x_A, y_A) be the coordinates of A and (x_B, y_B) be the co-ordinates of B.

Referred to suitable axes of reference.

Component of Pull T_{AB} in x-direction is

$$T_{AB} \cdot \cos\theta = T_{AB}\frac{AC}{AB} = T_{AB}\frac{(x_B - x_A)}{L_{AB}} = t_{AB}(x_B - x_A)$$

where t_{AB} = tension co-efficient for $AB = \dfrac{T_{AB}}{L_{AB}}$

Similarly, component of T_{AB} in y-direction is

$$T_{AB} \cdot \sin\theta = T_{AB} \cdot \frac{CB}{AB} = T_{AB}\frac{(y_B - y_A)}{L_{AB}} = t_{AB}(y_B - y_A)$$

Thus we observe that the component of forces in the members along the x and y directions can very easily be expressed in terms of tension. Co-efficients and the co-ordinates of the ends of the members with reference to the chosen reference axes.

The length of the member *AB* is obviously given by

$$L_{AB} = \sqrt{(x_B - x_A)^2 + (y_B - y_A)^2}$$

A member which is in compression will have a negative tension co-efficient.

Example 15.9: *A plane frame consists of two members AB and CB, hinged at A and C to the wall, as shown in Fig. 15.32. Determine the forces in the two members due to vertical force P applied at P.*

Solution: Let us take origin at joint *C*, and *CX* and *CY* be the axes of reference. The co-ordinates of the three joints are *C* (0, 0), *B* (2, 0), *A* (0, 1.5).

There are two members only, hence there will be only two tension-co-efficients.

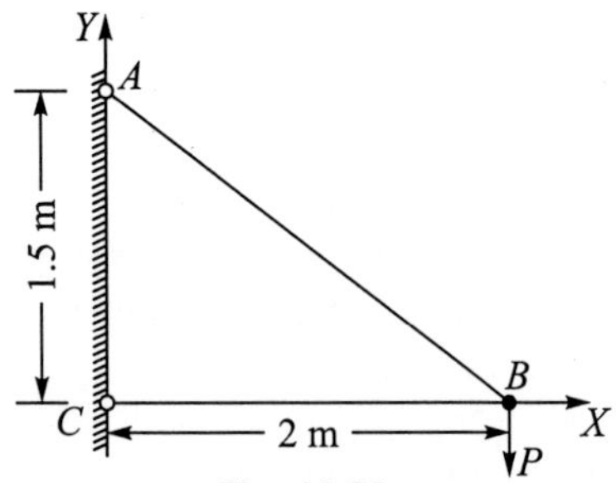

Fig. 15.32

Length $\quad L_{BA} = \sqrt{(0-2)^2 + (1.5-0)^2} = 2.5$ m

$\quad L_{BC} = 2$ m (given)

At the joint *B*, we have the following two = *ns*, in *x* and *y* directions.

$$t_{BA}(x_A - x_B) + t_{BC}(x_C - x_B) + 0 = 0$$

and $$t_{BA}(y_A - y_B) + t_{BC}(y_C - y_B) - P = 0$$

(Negative sign has been placed before *P*, since force *P* acts in the negative *y*-direction)

Substituting the values, we get

$$t_{BA}(0-2) + t_{BC}(0-2) = 0$$

or $$t_{BA} + t_{BC} = 0 \qquad \ldots(15.1)$$

and $$t_{BA}(1.5-0) + t_{BC}(0-0) - P = 0$$

or $$1.5 + t_{BA} = P \qquad \ldots(15.2)$$

Solving (15.1) and (15.2), above,

$$t_{BA} = \frac{P}{1.5}\ \text{kN/m} \quad \text{and} \quad t_{BC} = -\frac{P}{1.5}\ \text{kN/m}$$

Minus sign suggests that member *BC* will be in compression.

$\therefore$ Force in member $\quad BA = T_{BA} = t_{BA} \cdot L_{BA}$

$$= \frac{P}{1.5} \times 2.5 = 1.6667\,P \quad \text{(Tension)}$$

and force in member $\quad BC = T_{BC} = t_{BC} \times L_{BC}$

$$= -\frac{P}{1.5} \times 2 = -\,1.3333\,P \quad \text{(Compression)}$$

15.7. METHOD OF TENSION-COEFFICIENT APPLIED TO "SPACE-FRAMES"

A space frame or space truss is a three dimensional assemblage of line members, each member being jointed at its ends to the foundation or to the other members by frictionless ball and socket joints. The simplest space frame consists of 6 members jointed to from a Tetrahedron.

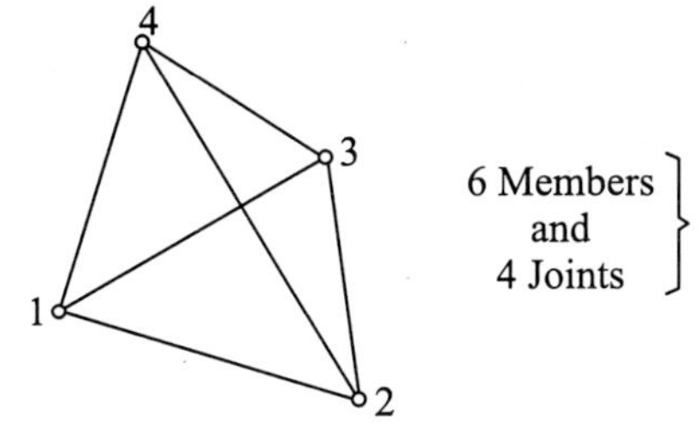

"Basic-Tetrahedron"

Fig. 15.33

15.8. BASIC-TETRAHEDRON

Here the relation between the number of members and number of joints is given by,

$$n - 6 = 3\,(j - 4)$$

or $\boxed{\text{“}n\text{”} = 3j - 6\,,}$ where 'n' stands for number of member and j for number of joints.

This equation gives the minimum number of members required to build a stable space frame.

The method of tension co-efficients can be extended to space frames by writing the equation of equilibrium of forces at each joint, in each direction. However, in space frame there are three axes of reference x–axis, y-axis and z-axis, the first two axes being considered in a horizontal plane and z-axis in a vertical plane. Thus, there will be 3 equations used in space frame in place of 2 equations used for plane frames.

Let us consider a member AB of a space frame, carrying a tensile force T_{AB}. Let (x_A, y_A, z_A) be the co-ordinates of point A, and (x_B, y_B, z_B) be the co-ordinates of point B, referred to the three axes X, Y, Z of reference. Let the line AB be inclined at angles θ_x, θ_y and θ_z respectively to the three reference axes, evidently

$$L_{AB} = [(x_B - x_A)^2 + (y_B - y_A)^2 + (z_B - z_A)^2\,]^{1/2}$$

The direction cosines are

$$l_{AB} = \cos\theta_x,\ m_{AB} = \cos\theta_y,\ n_{AB} = \cos\theta_z,$$

The components of AB along X, Y and Z axes are

$$AL = (x_B - x_A)\ ;\ AM \doteq (y_B - y_A);\ AN = (z_B - z_A)$$

Hence,

$$AL = (x_B - x_A) = AB\cos\theta_x, = l_{AB}\cdot L_{AB}$$

$$AM = (y_B - y_A) = AB\cos\theta_y, = m_{AB}\cdot L_{AB}$$

$$AN = (z_B - y_A) = AB\cos\theta_z, = n_{AB}\cdot L_{AB}$$

The resolved component of force T_{AB} along x-direction is,

$$T_{AB}\cos\theta_x = T_{AB}\frac{(x_B - x_A)}{L_{AB}} = t_{AB}\,(x_B - x_A)$$

where t_{AB} = tension coefficient for AB.

Similarly, the resolved component of force T_{AB} along y and z directions are,

$$T_{AB}\cos\theta_y = T_{AB}\frac{(y_B - y_A)}{L_{AB}} = t_{AB}\,(y_B - y_A)$$

and

$$T_{AB}\cos\theta_z = T_{AB}\frac{(z_B - z_A)}{L_{AB}} = t_{AB}\,(z_B - z_A)$$

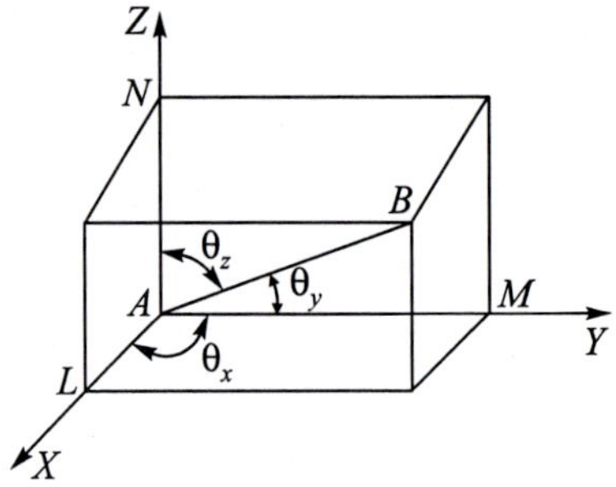

Fig. 15.34

Let P_A be the external force at joint A, and let X_A, Y_A and Z_A be the resolved components of this force in X, Y and Z directions respectively. Let AB, AC, AD, ... be the members meeting at the joint A. Since the joint is in equilibrium under the external force P_A and the pulls in the members meeting at the joint A, the algebraic sum of resolved parts of these forces in any direction must be equal to zero. Hence resolving the forces in X, Y and Z directions and equating the algebraic sum of these forces in each direction to zero, we get the following three equations for joint A,

$$t_{AB}(x_B - x_A) + t_{AC}(x_C - x_A) + t_{AD}(x_D - x_A) + \ldots\ldots + X_A = 0$$
$$t_{AB}(y_B - y_A) + t_{AC}(y_C - y_A) + t_{AD}(y_D - y_A) + \ldots\ldots + Y_A = 0$$
$$t_{AB}(z_B - z_A) + t_{AC}(z_C - z_A) + t_{AD}(z_D - z_A) + \ldots\ldots + Z_A = 0$$

The above equations may also be written as under:

$$\Sigma\, t\,(x_F - x_N) + X_A = 0$$
$$\Sigma\, t\,(y_F - y_N) + Y_A = 0$$
$$\Sigma\, t\,(z_F - z_N) + Z_A = 0$$

where (x_F, y_F, and z_F) are the co-ordinates of the far end of each member and (x_N, y_N, z_N) are the co-ordinates of the near end of each member meeting at the joint. The end of the member at the joint under consideration is known as near end.

Example 15.10: *A pair of shear leg has length of each leg as 5 m, and the distance between their feet is 4 m. The line joining the feet of the legs is 7 m from the foot of the guy rope. If the length of the guy rope is 10 m, find thrust in each leg and pull in the guy rope when a load of 100 kN is suspended from head.*

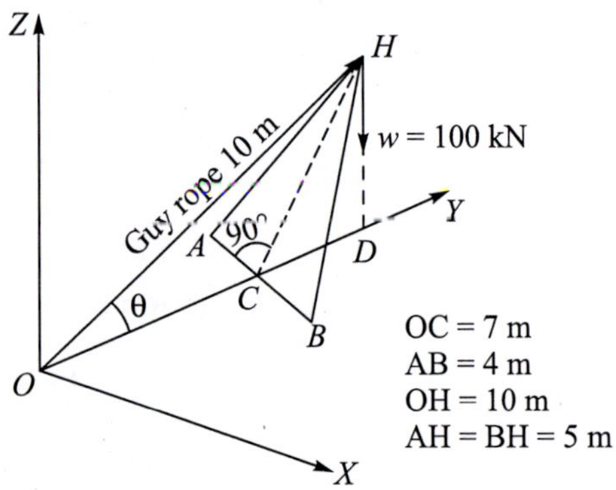

Fig. 15.35

Solution: Let the origin pass through the foot of the guy rope, and let the Y-axis pass through the midpoint C of the line AB joining the feet of the two legs as shown in Fig. 15.35. Along with the direction of X and Z axis. 'H' is the head of the shear leg. Let the guy rope make an angle θ with Y-axis.

Now $\quad CH = \sqrt{AH^2 - AC^2} = \sqrt{5^2 - 2^2}$

$$= \sqrt{21} = 4.58 \text{ m}$$

From $\Delta\, OCH$,

$$\text{Cos } \theta = \frac{OH^2 + OC^2 - CH^2}{2\,OC \cdot OH} = \frac{10^2 + 7^2 - 21}{2 \times 7 \times 10} = 0.9143$$

$$\theta = 23.9°, \sin\theta = 0.4051$$

From H, drop a perpendicular HD on Y-axis.

$$\therefore \quad OD = OH\cos\theta = 10 \times 0.9143 = 9.143 \text{ m}$$
$$HD = OH\sin\theta = 10 \times 0.4051 = 4.051 \text{ m}$$
$$CD = OD - OC = 9.143 - 7 = 2.143 \text{ m}$$

Hence the co-ordinates of various points are as under:

Point	Co-ordinates		
	x	y	z
O	0	0	0
H	0	9.143	4.051
A	–2	7	0
B	+2	7	0

At the head H, the three equations are as under:

$$t_{HA}(x_A - x_H) + t_{HB}(x_B - x_H) + t_{HO}(x_O - x_H) = 0$$
$$t_{HA}(y_A - y_H) + t_{HB}(y_B - y_H) + t_{HO}(y_O - y_H) = 0$$
$$t_{HA}(z_A - z_H) + t_{HB}(z_B - z_H) + t_{HO}(z_O - z_H) + z_H = 0$$

substituting the values, we get

$$-2\, t_{HA} + 2t_{HB} = 0 \qquad \ldots(15.3)$$
$$-2.143\, t_{HA} - 2.143\, t_{HB} - 9.143\, t_{HO} = 0 \qquad \ldots(15.4)$$

and
$$-4.051(t_{HA} + t_{HB} + t_{HO}) - 100 = 0 \qquad \ldots(15.5)$$

From eq. (15.1), $\quad t_{HA} = t_{HB}$

From eq. (15.2), $\quad -(2.143 + 2.143)t_{HA} = 9.143\, t_{HO}$

$\therefore \quad t_{HA} = -2.143\, t_{Ho}$

From eq. (15.3), $\quad -4.051\,(-2.1332 - 2.1332 + 1)\, t_{HO} = 100$

$\therefore \quad t_{HO} = 7.5573$

$\therefore \quad t_{HA} = t_{HB} = -2.1332 \times 7.5573 = -16.1212$

Hence pull in the Guy rope $t_{HO} \times L_{HO}$

$= 7.5573 \times 10 = 75.57$ kN.

Force in each shear leg $\quad = -16.1212 \times 5 = 80.61$ kN

$\therefore \quad$ Thrust in each leg $= 80.61$ kN

SUMMARY

- A frame may be defined as a structure made up of several bars, riveted or welded together. These are made up of angle irons or channel sections and are called the members of the framed structure.
- A member under tension is called tie and under compression is known as strut.
- A perfect frame should satisfy the relation, $m = (2j - 3)$,

  where $\quad m =$ Number of members in the frame

  and $\quad j =$ Number of joints in the frame.
- A frame, in which the number of members < $(2j - 3)$, is known as deficient frame.
- A frame in which the number of members > $(2j - 3)$, is known as redendant frame.

- The forces in various members of a perfect frame may be found out by the following methods :
  (*a*) Graphical method.
  (*b*) Method of joints.
  (*c*) Method of sections.
- Though there are many type of frames, yet from analysis point of view, the frames may be classified as perfect and imperfect frames.
- Following assumptions are made while computing the forces in the members of the perfect frame :
  (*a*) All members are pin jointed.
  (*b*) The frame is loaded only at joints.
  (*c*) The frame is a perfect frame.
  (*d*) The self weight of the members is neglected.

SHORT ANSWER QUESTIONS

1. What is a frame ? How are frames classified ?
2. What is a perfect frame ? How does it differ from an imperfect frame ?
3. Enumirate the assumptions made while finding out the forces in a frame.
4. What is the difference between a cantilever and a simply supported frame ?
5. What is the difference between a defificient and a redundant frame ?

EXERCISE FOR PRACTICE

1. Determine the forces in various members of the truss shown in Fig. 15.36 below. **(MDU 2009)**

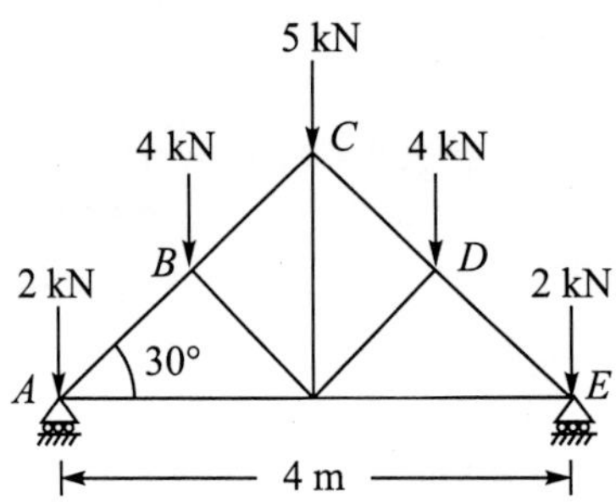

Fig. 15.36.

2. Determine the magnitude and nature of forces in various members of truss shown below : **(MDU 2012)**

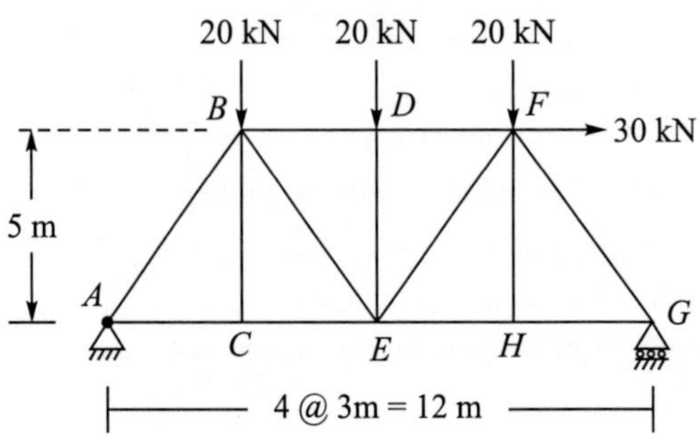

Fig. 15.37.

3. Determine the magnitude and nature of forces in various members of truss shown in Fig. 15.38 below : **(MDU 2008)**

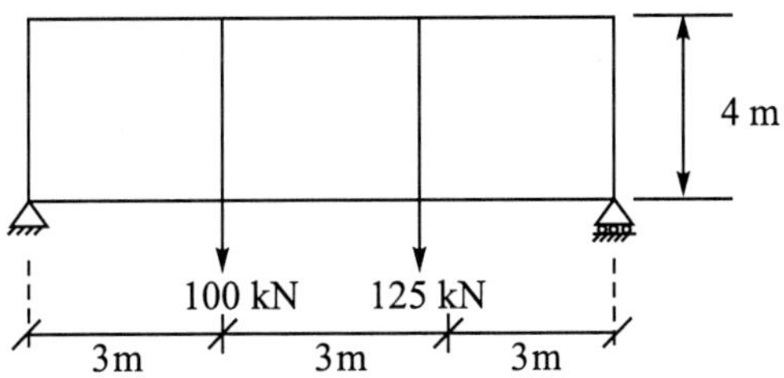

Fig. 15.38

4. Determine the forces in the truss shown below : **(MDU 2011)**

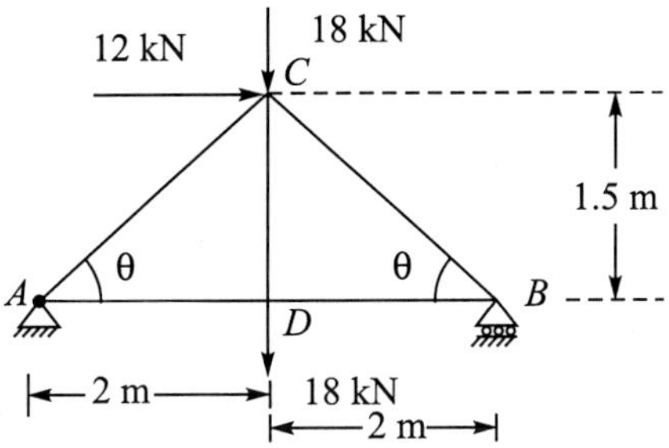

Fig. 15.39

5. Using the method of tension co-efficients, find forces in the members of the crane structure shown in Fig. 15.40 below :

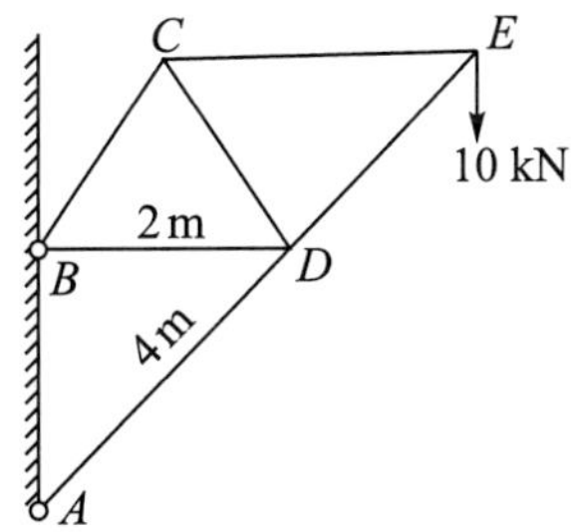

Fig. 15.40

6. Analyse the pin-jointed plane frame shown below by tension co-efficient method. **(KU 2010)**

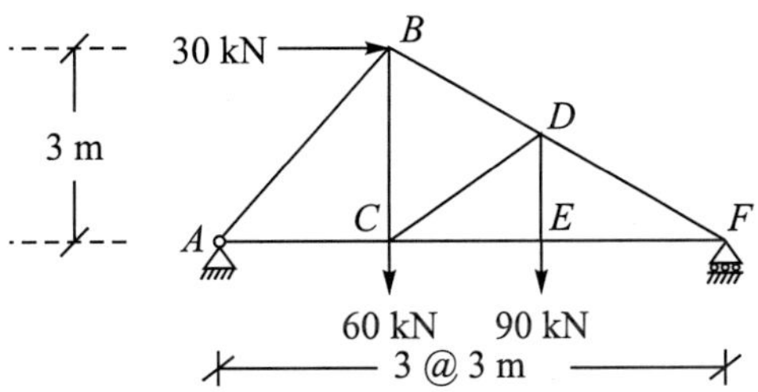

Fig. 15.41

CHAPTER

16 Theorem of Three Moments

16.1. INTRODUCTION

The three moment equations express the relationship between the moments at three successive supports and the loading on the two spans between those three supports, with or without the unequal settlement of the supports. This relation is derived from the consistancy condition that the slope at the middle support, calculated from the left span should be the same as the slope at the middle support calculated from the right span. Using the three moment equations, continuous beams can be analysed.

16.2. CONTINUOUS BEAMS

A beam which is supported on more than two supports, is called a continuous beam. Such a beam when loaded will deflect with convexity upwards, over the intermediate supports and with concavity upwards over the mid of the spans. The intermediate supports of a continuous beams are always subjected to some B.M. The end supports, it simply supported will not be subjected to any B.M. But the end supports, if fixed will be subjected to fixing moments and the slope of the beam, at the fixed end will be zero. The analysis of the continuous beams is similar to the fixed beams. The B.M. diagram for a continuous beam under any system of loading may be drawn in the following two stages :

1. By considering the beam as a series of discontinuous beams, from support to support and drawing the usual μ-diagram due to vertical loads.
2. By superimposing the usual diagram, due to end moments over μ-diagram.

16.3. CLAYPEYRON'S THEOREM OF THREE MOMENTS

It states, "If a beam has n supports, the end ones being fixed, then the same number of equations required to determine the support moments may be obtained from the consecutive pairs of spans, *i.e.*, $AB - BC$, $BC - CD$, $CD - DE$ and so on."

Proof:

Consider a continuous beam ABC, fixed at A and C and supported at B as shown in Fig. 16.1, below:

Let l_1 = span of the beam

I_1 = M.I. of the beam in span AB

l_2 and I_2 = Corresponding values for span BC.

M_A = Support moment of A

M_B = Support moment of B

M_C = Support moment of C

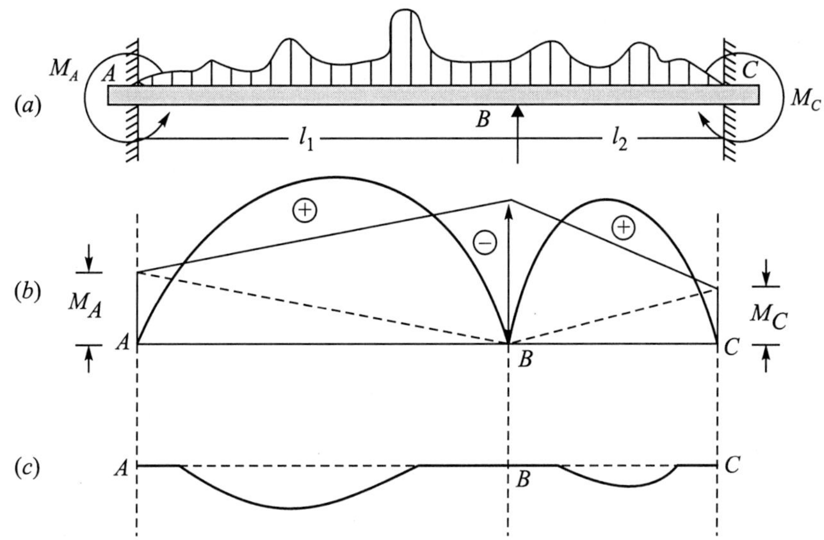

Fig. 16.1

μ_x = B.M. at any section X, considering the beam between two supports as simply supported.

$\mu x'$ = Fixing moment at any section X, of the beam,

We know that in the span AB, the B.M. at any section X at a distance x from A,

$$M_x = \mu_X + \mu_X'$$

$$\therefore \qquad EI_1 \frac{d^2 y}{dx^2} = \mu_X + \mu_X' \qquad \left(\because \frac{M}{EI} = \frac{d^2 y}{dx^2}\right)$$

Multiplying the above equation by x and integrating the same for the whole span AB, *i.e.*, from 0 to l_1.

$$EI_1 \int_0^{l_1} \frac{x \cdot d^2 y}{dx^2} = \int_0^{l_1} x \cdot \mu_x \cdot dx + \int_0^{l_1} x \cdot \mu_x' \cdot dx$$

$$EI_1 \left[x \cdot \frac{dy}{dx} - y \right]_0^{l_1} = a_1 \bar{x}_1 + a_1' \bar{x}_1'$$

or

$$EI_1 \left[l_1 (i_B - y_B) - 0 \, (i_A - y_A) \right] = a_1 \bar{x}_1 + a_1' \bar{x}_1'$$

$$EI_1 \left[l_1 (i_B - y_B) \right] = a_1 \bar{x}_1 + a_1' \bar{x}_1' \qquad \text{...(16.1)}$$

Since y_B is equal to zero, therefore,

$$EI_1 \, l_1 \, i_B = a_1 \bar{x}_1 + a_1' \bar{x}_1' \qquad \text{...(16.2)}$$

where a_1 = Area of the μ-diagram in span AB

$\bar{x}_1$ = Distance of C.G. of μ diagram from A in span AB

$a_1'\overline{x}_1'$ = Corresponding values for the μ′-diagram.

i_B = Slope of the beam AB at B.

We know that the shape of the μ′-diagram is trapezoidal, having end ordinates equal to M_A and M_B as shown in Fig. 16.1. Therefore, splitting up this trapezium into two triangles,

$$a_1' \cdot \overline{x}_1' = \left(M_A \times \frac{l_1}{2} \times \frac{l_1}{3}\right) + \left(M_B \times \frac{l_1}{2} \times \frac{2l_1}{3}\right)$$

$$= (M_A + 2\,M_B)\,\frac{l_1^2}{6}$$

Substituting this value of $a_1' \cdot \overline{x}_1'$ in eq. (2) above,

$$EI_1\, l_1\, i_B = a_1\overline{x}_1 + (M_A + 2M_B)\frac{l_1^2}{6}$$

or $$EI_1\, i_B = \frac{a_1\overline{x}_1}{l_1} + (M_A + 2M_B)\frac{l_1}{6} \qquad ...(16.3)$$

∴ $$E.i_B = \frac{a_1\overline{x}_1}{I_1 l_1} + (M_A + 2M_B)\frac{l_1}{6I_1}$$

Similarly in the span BC, taking C as the origin and x positive to the left,

$$E_{iB}' = \frac{a_2\overline{x}_2}{I_2 l_2} + (M_C + 2M_B)\frac{l_2}{6I_2} \qquad ...(16.4)$$

where,

a_2 = Area of the μ-diagram in span BC

$\overline{x}_2$ = Distance of C.G. of μ-diagram from C in span BC

a_2' = Area of the μ′-diagram in span BC

$\overline{x}_2'$ = Distance of C.G. of the μ′-diagram from C in the span BC

i_B' = Slope of the beam BC at B

Since i_B is equal to $-i_B'$, ∴ $E \cdot i_B = -E \cdot i_B'$

or $$\frac{a_1\overline{x}_1}{I_1 l_1} + (M_A + 2\,M_B)\,\frac{l_1}{6I_1} = -\left[\frac{a_2\overline{x}_2}{I_2 l_2} + (M_C + 2M_B)\frac{l_2}{6I_2}\right]$$

$$(M_A + 2M_B)\,\frac{l_1}{I_1} + (M_C + 2M_B)\,\frac{l_2}{I_2} = -\frac{6a_1\overline{x}_1}{I_1 l_1} - \frac{6a_2\overline{x}_2}{I_2 l_2}$$

$$M_A\frac{l_1}{I_1} + 2M_B\frac{l_1}{I_1} + M_C\frac{l_2}{I_2} + 2M_B\frac{l_2}{I_2} = -\left(\frac{6a_1\overline{x}_1}{I_1 l_1} + \frac{6a_2\overline{x}_2}{I_2 l_2}\right)$$

∴ $$M_A\left(\frac{l_1}{I_2}\right) + 2M_B\left(\frac{l_1}{I_1} + \frac{l_2}{I_2}\right) + M_C\left(\frac{l_2}{I_2}\right) = -\left(\frac{6a_1\overline{x}_1}{I_1 l_1} + \frac{6a_2\overline{x}_2}{I_2 l_2}\right)$$

Notes:

1. For the sake of simplicity, we have considered a continuous beam with two spans only. But this equation can be extended for any number of spans.

2. If moment of inertia of the beam is constant, then

$$M_A l_1 + 2\,M_B\,(l_1 + l_2) + M_C l_2 = -\left(\frac{6a_1\bar{x}_1}{l_1} + \frac{6a_2\bar{x}_2}{l_2}\right).$$

3. The S.F. diagram for the beam may be drawn as usual.

16.4. APPLICATION OF CLAPEYRON'S THEOREM OF THREE-MOMENTS TO VARIOUS TYPES OF CONTINUOUS BEAMS

The types of beams are :

(*a*) Continuous beams with simply supported ends.

(*b*) Continuous beams with fixed end supports.

(*c*) Continuous beams with end span overhanging.

(*d*) Continuous beams with a sinking support.

Clapeyron's theorem application to various cases is illustrated in the following examples :

16.5. CONTINUOUS BEAMS WITH S.S. ENDS

Example 16.1: *A continuous beam ABC 10 m long rests on three supports A, B and C at the same level and is loaded as shown in Fig. 16.2.*

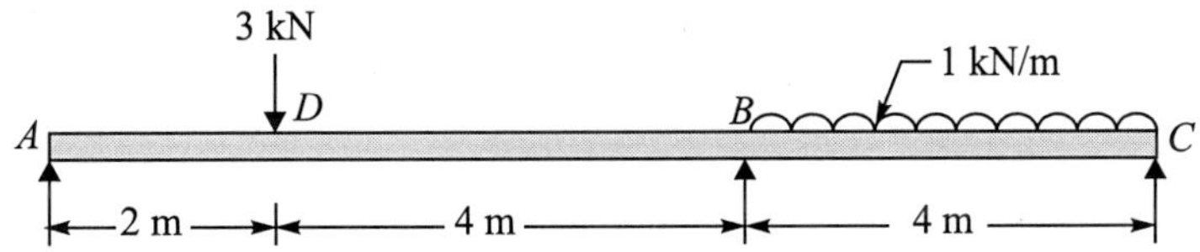

Fig. 16.2.

Determine the moments over the beam and draw the bending moment diagram. Also calculate the reactions at the supports and draw the shear force diagram.

Solution: Given: Length AB (l_1) = 6 m; Length BC (l_2) = 4 m; Point load at D = (W) = 3 kN; Distance AD (a) = 2 m; Distance DB (b) = 4 m and uniformly distributed load in $BC = w = 1$ kN/m.

Moments over the beam

Let $\quad$ **M_A = Fixing moment at A

M_B = Fixing moment at B and

**M_C = Fixing moment at C

First of all, let us consider the beam AB as a simply supported. Therefore, bending moment at D,

$$M_D = \frac{Wab}{l_1} = \frac{3 \times 2 \times 4}{6} = 4 \text{ kN-m}$$

Similarly, bending moment at the mid of the span BC

$$= \frac{wl_2^2}{8} = \frac{1 \times (4)^2}{8} = 2 \text{ kN-m}$$

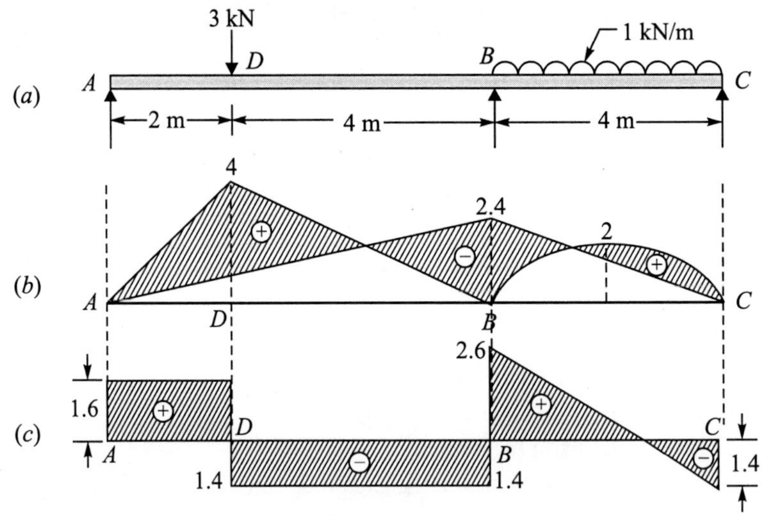

Fig. 16.3.

Now draw the μ-diagrams witht the help of above bending moment as shown in Fig.16.3 (*a*). From the geometry of the above bending moment diagrams, we find that

$$a_1\bar{x}_1 = \left[\left(\frac{1}{2}\times 2\times 4\times\frac{2\times 2}{3}\right)+\left(\frac{1}{2}\times 4\times 4\right)\left(2+\frac{4}{3}\right)\right] = 32$$

and

$$a_2\bar{x}_2 = \left(\frac{2}{3}\times 2\times 4\times 2\right) = \frac{32}{3}$$

Now using three moments equation,

$$M_A l_1 + 2M_B(l_1+l_2) + M_C l_2 = \left(\frac{6a_1\bar{x}_1}{l_1}+\frac{6a_2\bar{x}_2}{l_2}\right)$$

$$0 + 2\,M_B\,(6+4) + 0 = \left(\frac{6\times 32}{6}+\frac{6\times\frac{32}{3}}{4}\right)$$

$\therefore$

$$20\,M_D = -(32+16) = -48$$

or

$$M_B = -\frac{48}{20} = -2.4 \text{ kN-m} \quad \textbf{Ans.}$$

Now complete the bending moment diagram as shown in Fig. 16.3. (*b*)

Shear force diagram

Let R_A = Reaction at *A*,

R_B = Reaction at *B* and

R_C = Reaction at *C*.

Taking moments about *B*,

$$R_A\times 6-(3\times 4) = -2.4 \qquad \ldots(\because M_B = -2.4 \text{ kN-m})$$

$$R_A = \frac{-2.4 + 12.0}{6} = \frac{9.6}{6} = 1.6 \text{ kN} \quad \textbf{Ans.}$$

Similarly, $R_C \times 4 - (4 \times 2) = -2.4$ ...($\because M_B = -2.4$ kN-m)

$$R_C = \frac{-2.4 + 8.0}{4} = \frac{5.6}{4} = 1.4 \text{ kN} \quad \textbf{Ans.}$$

$$R_B = (3 + 1 \times 4) - (1.6 + 1.4) = 4.0 \text{ kN} \quad \textbf{Ans.}$$

Now complete the shear force diagram as shown in Fig. 16.3. (*c*)

Example 16.2: *A continuous beam ABCD, simply supported at A, B, C and D is loaded as shown below:*

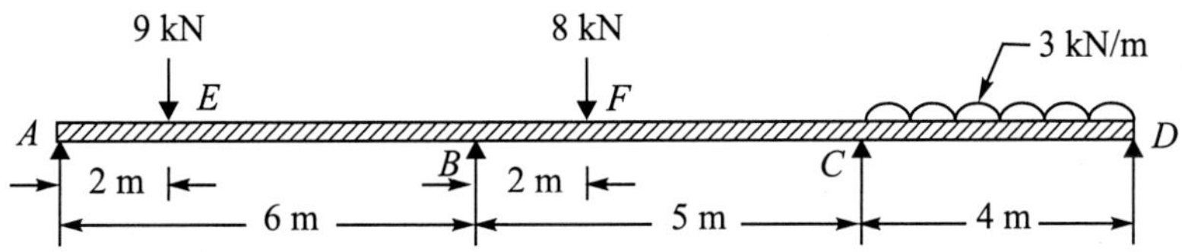

Fig. 16.4.

Find the moments over the beam and draw bending moment and shear force diagrams.

Solution: Given: Length *AB* (l_1) = 6 m; Length *BC* (l_2) = 5 m; Length *CD* (l_3) = 4 m; Load at = 9 kN; distance *AE* (a_1) = 2 m; Distance *EB* (b_1) = 4 m; Load at *F* (W_2) = 8 kN; Distance = 2 m; Distance *FC* (b_2) = 3 m and uniformly distributed load in *CD* (w) = 3 kN/m.

Moments over the beam

Let $^*M_A$ = Fixing moments at *A*,

M_B = Fixing moment at *B*,

M_C = Fixing moment at *C* and

$^*M_D$ = Fixing moment at *D*.

First of all, let us consider the beam *AB* as a simply supported beam. Therefore, bending moment as

$$M_E = \frac{W_1 a_1 b_1}{l_1} = \frac{9 \times 2 \times 4}{6} = 12 \text{ kN-m}$$

Similarly, $$M_F = \frac{W_2\, a_2\, b_2}{l_2} = \frac{8 \times 2 \times 3}{5} = 9.6 \text{ kN-m}$$

and bending moment at the mid of the span *CD*

$$= \frac{w l_3^2}{8} = \frac{3 \times (4)^2}{8} = 6 \text{ kN-m}$$

Now draw the *m*-diagrams with the help of above bending moments as shwon in Fig. 16.5 (*a*). From the geometry of the above bending moment diagrams, we find that for the spans *AB* and *BC*,

$$a_1 \bar{x}_1 = \left[\left(\frac{1}{2} \times 2 \times 12 \times \frac{2 \times 2}{3}\right) + \left(\frac{1}{2} \times 4 \times 12\right)\left(2 + \frac{4}{3}\right)\right] = 96$$

* Since the beam is simply supported at *A* and *D*, therefore fixing moments M_A and M_D will be zero.

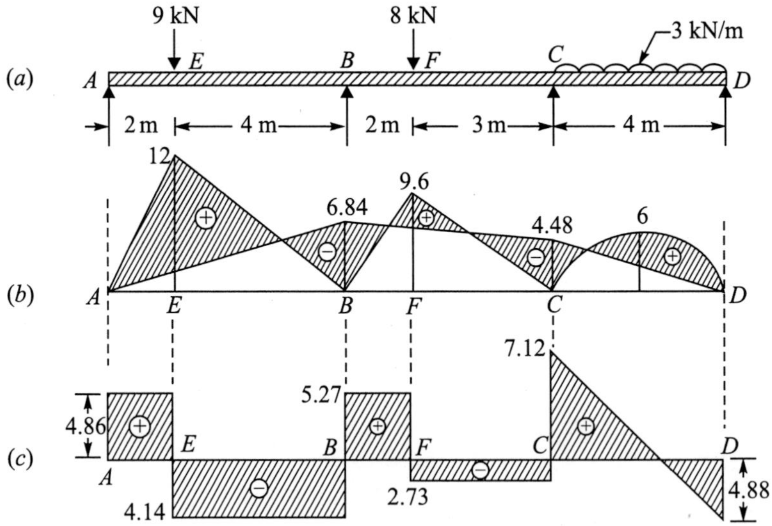

Fig. 16.5

and $$a_2\bar{x}_2 = \left[\left(\frac{1}{2}\times3\times9.6\times\frac{2\times3}{3}\right)+\left(\frac{1}{2}\times2\times9.6\right)\left(3+\frac{2}{3}\right)\right] = 64$$

Similarly, for the spans BC and CD,

$$* \quad a_2\bar{x}_2 = \left[\left(\frac{1}{2}\times3\times9.6\times\frac{2\times3}{3}\right)+\left(\frac{1}{2}\times2\times9.6\right)\left(2+\frac{3}{3}\right)\right] = 56$$

and $$a_3\bar{x}_3 = \left(\frac{2}{3}\times6\times4\times2\right) = 32$$

Now using three moments equation for the spans AB and BC,

$$M_A l_1 + 2\,M_B\,(l_1 + l_2) + M_C\,l_2 = -\left(\frac{6a_1 x_1}{l_1}+\frac{6a_2\bar{x}_2}{l_2}\right)$$

$$0 + 2\,M_B\,(6 + 5) + M_C \times 5 = -\left(\frac{6\times96}{6}+\frac{6\times64}{5}\right)$$

$$22\,M_B + 5\,M_C = -\,172.8$$

Again using three moments equation for the spans BC and CD,

$$M_B\,l_2 + 2\,M_C\,(l_2 + l_3) + M_D\,l_3 = -\left(\frac{6a_2\bar{x}_2}{l_2}+\frac{6a_3\bar{x}_3}{l_3}\right)$$

$$M_B \times 5 + 2\,M_C\,(5 + 4) + 0 = \left(\frac{6\times56}{5}+\frac{6\times32}{4}\right)$$

$$5\,M_B + 18\,M_C = -\,115.2$$

* The previous value of $a_2\bar{x}_2$ is with reference to the support C (being the end support of spans AB and BC). This value of $a_2\bar{x}_2$ is with reference to the support B (being the end support of spans BC and CD).

Solving equations (*i*) and (*ii*),

$$M_B = -6.84 \text{ kN-m} \quad \text{and} \quad M_C = -4.48 \text{ kN-m} \quad \textbf{Ans.}$$

Now complete the bending moment diagram as shown in Fig. 16.5 (*b*).

Sections at the supports

Let R_A = Reaction at *A*,

R_B = Reaction at *B*,

R_C = Reaction at *C* and

R_D = Reaction at *D*.

Taking moments about *B*,

$$R_A \times 6 - (9 \times 4) = -6.84 \qquad ...(\because M_B = -6.84 \text{ kN-m})$$

$$R_A = \frac{-6.84 + 36}{6} = \frac{29.16}{6} = \mathbf{4.86\ kN} \quad \textbf{Ans.}$$

Now taking moments about *C*,

$$R_D \times 4 - (12 \times 2) = -4.48 \qquad ...(\because M_C = -4.48 \text{ kN-m})$$

$$R_D = \frac{-4.48 + 24}{4} = \frac{19.52}{4} = \mathbf{4.88\ kN} \quad \textbf{Ans.}$$

Again taking moments about *C*,

$$R_A \times 11 - (9 \times 9) + R_B \times 5 - (8 \times 3) = -4.48 \qquad ...(\because M_C = -4.48 \text{ kN-m})$$

$$4.86 \times 11 - 81 + 5\,R_B - 24 = -4.48$$

$$R_B = \frac{-4.48 - 53.46 + 81 + 24}{5} = \frac{47.06}{5} = 9.41 \text{ kN} \quad \textbf{Ans.}$$

and $$R_C = (9 + 8 + 12) - (4.86 + 4.88 + 9.41) = 9.85 \text{ kN} \quad \textbf{Ans.}$$

Now draw the shear force diagram as in Fig. 16.5 (*c*).

16.6. CONTINUOUS BEAMS WITH FIXED END SUPPORTS

Sometimes, a continuous beam is fixed at its one or both ends. If the beam is fixed at the left end and then an imaginary zero span is taken to the left of *A* and the three moments theorem is applied as usual. Similarly, if the beam is fixed at the right end, then an imaginary zero span is taken after the right end support and the three moments theorem is applied as usual.

Notes :

1. The fixing moment, at the imaginary support of the zero span *i.e.*, M_0 is always equal to zero.
2. Propped cantilevers and beams may also be analyses by Clapeyron's theorem of three moments.

Example 16.3: *A continuous beam ABC of uniform section, with span AB as 8 m and BC as 8 m, is fixed at A and simply supported at B and C. The beam is carrying a uniformly distributed load of 1 kN/m throughout its length. Find the moments along the beam and reactions at the supports. Also draw the bending moment and shear force diagrams.*

Solution: Given: Length *AB* (l_1) = 8 m; Length *BC* (l_2) = 6 m and uniformly distributed load (*w*) 1 kN/m.

Moments along the beam

Since the beam is fixed at *A*, therefore assume a zero span to the left of *A*.

Let * M_0 = Fixing moment at the left hand support of zero span,

M_A = Fixing moment at *A*,

M_B = Fixing moment at *B* and

* M_C = Fixing moment at *C*.

First of all, consider the beam *AB* as a simply supported beam. Therefore, bending moment at mid of the span *AB*.

$$= \frac{wl_1^2}{8} = \frac{10 + (8)^2}{8} = 8.0 \text{ kN-m}$$

Similarly, bending moment at the mid of the span *BC*

$$= \frac{wl_2^2}{8} = \frac{1 \times (6)^2}{8} = 4.5 \text{ kN-m}$$

Now draw the μ-diagram with the help of above bending moments as shown in Fig. 16.6 (*b*). From the geometry of the above bending moment diagrams, we find that for the span 0*A* and *AB*.

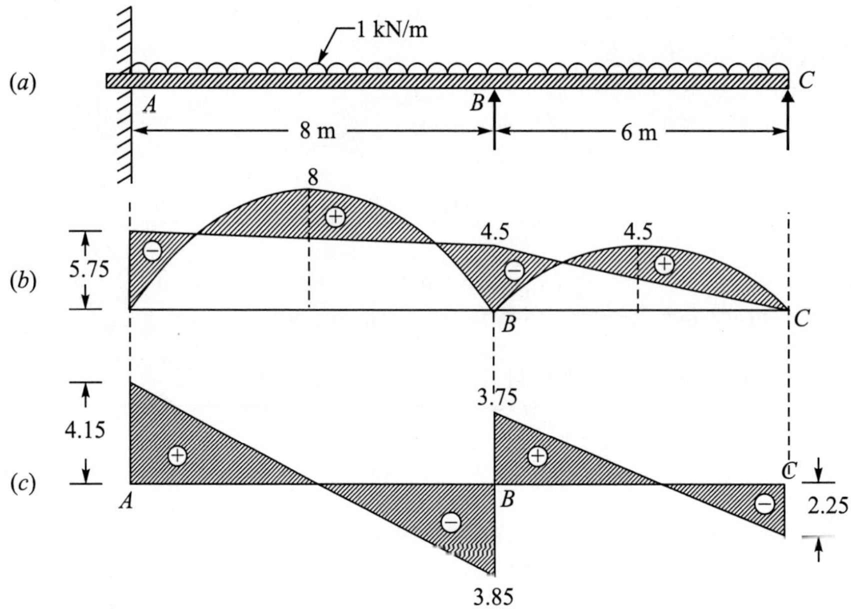

Fig. 16.6.

$$a_0\bar{x}_0 = 0$$

and
$$a_1\bar{x}_1 = \left(\frac{2}{3} \times 8 \times 8 \times 4\right) = \frac{512}{3} = 170.67$$

* Since there is a zero span on the left of *A*, therefore the fixing moment M_0 will be zero. Moreover, as the beam is simply supported at *C*, therefore fixing moment M_C will also be zero.

Similarly, for the span *AB* and *BC*,

$$a_1\bar{x}_1 = \left(\frac{2}{3}\times 8\times 8\times 4\right) = \frac{512}{3} = 170.67$$

and
$$a_2\bar{x}_2 = \left(\frac{2}{3}\times 4.5\times 6\times 3\right) = 54$$

Now using three moments equation for the spans 0*A* and *AB*.

$$M_0\, l_0 + 2\, M_A\,(0 + l_1) + M_B\, l_1 = -\left(\frac{6a_0\bar{x}_0}{l_0} + \frac{6a_1\bar{x}_1}{l_1}\right)$$

$$0 + 2\, M_A\,(0 + 8) + M_B \times 8 = -\left(0 + \frac{6\times 170.67}{8}\right)$$

$$16\, M_A + 8\, M_B = -\,128 \qquad \text{...(i)}$$

Again using three momens equation for the spans *AB* and *BC*,

$$M_A\, l_1 + 2\, M_B\,(l_1 + l_2) + M_C\, l_2 = -\left(\frac{6a_1\bar{x}_1}{l_1} + \frac{6a_2\bar{x}_2}{l_2}\right)$$

$$M_A \times 8 + 2\, M_B\,(8 + 6) + 0 = -\left(\frac{6\times 170.67}{8} + \frac{6\times 54}{6}\right)$$

$$8\, M_A + 28\, M_B = -\,182 \qquad \text{...(ii)}$$

Solving the equations (*i*) and (*ii*)

$M_A = -\,5.75$ kN-m and $M_B = -\,4.5$ kN-m **Ans.**

Now complete the bending moment diagram as shwon in Fig. 16.6 (*b*)

Reactions at supports

Let R_A = Reaction at *A*,
R_B = Reaction at *B*, and
R_C = Reaction at *C*

Taking moments about *B*,

$$R_C \times 6 - (6 \times 3) = -\,4.5 \qquad (\because M_B = -\,4.5 \text{ kN-m})$$

$$\therefore \quad R_C = \frac{-\,4.5 + 18}{6} = \frac{13.5}{6} = 2.2 \text{ kN} \quad \textbf{Ans.}$$

Now taking moments about *A*,

$$R_C \times 14 + R_B \times 8 - (1 \times 14 \times 7) = -5.75 \qquad (\because M_A = -\,5.57 \text{ kN-m})$$

$$(2.25 \times 14) - 8 \cdot R_B - 98 = -5.75$$

$$\therefore \quad R_B = \frac{-5.75 + 98 - 31.5}{8} = 7.6 \text{ kN} \quad \textbf{Ans.}$$

and $\quad {}^*R_A = 14.0 - (2.25 + 7.6) = 4.15$ kN **Ans.**

Now completed the shear force diagram as shown in Fig. 16.6 (*c*).

---

* The reaction at R_A may also be found out by taking mokments about *B*, *i.e.*,

$$R_A \times 8 - (8 \times 4) - M_A = -\,4.5 \qquad \text{...}(\because M_B = -\,4.5 \text{ kN-m})$$

$$8\, R_A - 32 - 5.75 = -\,4.5$$

$$R_A = \frac{-\,4.5 + 32 + 5.75}{8} = \frac{33.25}{8} = 4.15 \text{ kN} \quad \textbf{Ans.}$$

Example 16.4: *Evaluate the bending moment and shear force diagrams of the beam shown in Fig. 16.7 below:*

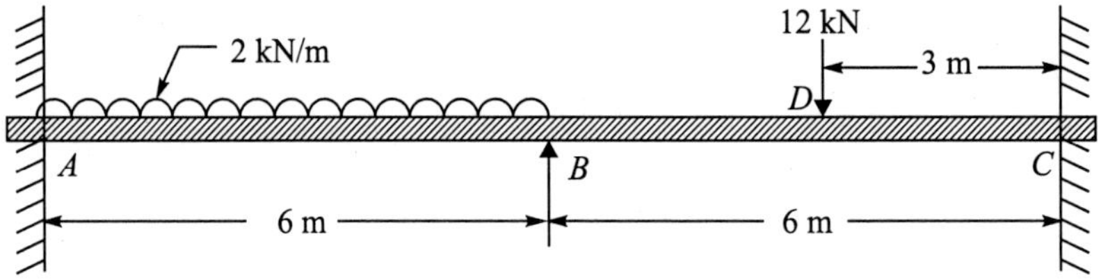

Fig. 16.7.

What are the reactions at the supports ?

Solution: Given: Length AB (l_1) = 6 m; Length BC (l_2) = 6 m; Uniformly distributed load AB (w) = 2 kN/m and point load at D (W) = 12 kN.

Bending moment at *A*, *B* and *C*

Since the beam is fixed at A and C, therefore assume a zero span to the left of A and right of B

Let M_0 = Fixing moments at the imaginary supports of zero span (is left of A and right of B)

M_A = Fixing moment at A,

M_B = Fixing moment at B and

M_C = Fixing moment at C.

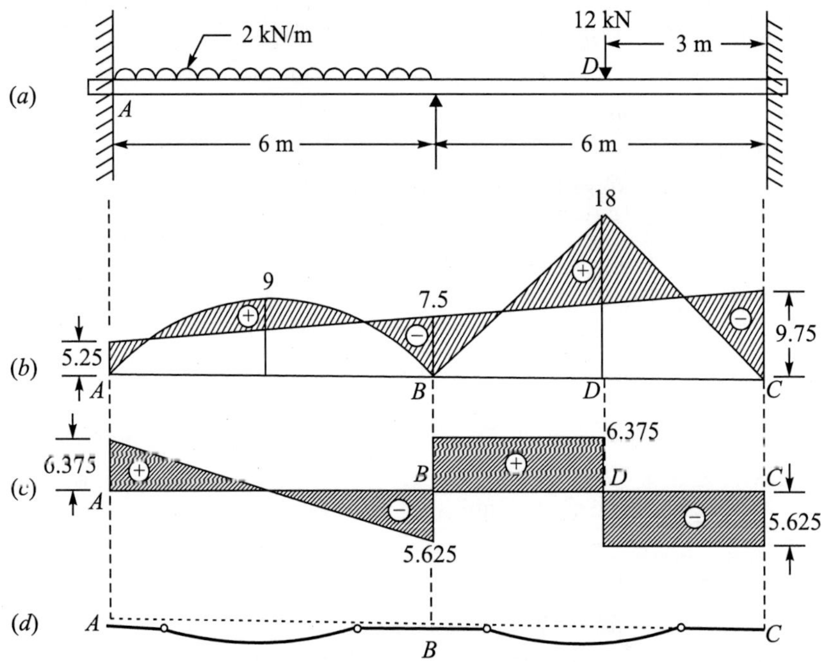

Fig. 16.8.

First of all, consider the beam AB as a simply supported beam. Therefore, bending moment at the mid of span AB

$$= \frac{wl_1^2}{8} = \frac{2 \times (6)^2}{8} = 9 \text{ kN-m}$$

Similarly, bending moment at the mid of the span BC

$$= \frac{Wl}{4} = \frac{12 \times 6}{8} = 18 \text{ kN-m}$$

Now draw the μ-diagram with the help of above bending moments as shown in Fig. 16.8 (*b*). From the geometry of bending moment diagrams, we find that for the spans $0A$ and AB,

$$a_0\bar{x}_0 = 0$$

$$a_1\bar{x}_1 = \left(\frac{1}{2} \times 9.0 \times 6 \times 3\right) = 108$$

Similary, for the span AB and BC,

$$a_1\bar{x}_1 = \left(\frac{2}{3} \times 9.0 \times 6 \times 3\right) = 108$$

$$a_2\bar{x}_2 = \left(\frac{1}{2} \times 18.0 \times 6 \times 3\right) = 162$$

and for span BC and CD, $a_3\bar{x}_3 = \left(\frac{1}{2} \times 18.0 \times 6 \times 3\right) = 162$

$$a_0\bar{x}_0 = 0$$

Now using three moments equation for the spans $0A$ and AB,

$$M_0\, l_0 + 2\, M_A\, (0 + l_1) + M_0\, l_1 = -\left(\frac{6a_0\bar{x}_0}{l_0} + \frac{6a_1\bar{x}_1}{l_1}\right)$$

$$0 + 2\, M_A\, (0 + 6) + M_B \times 6 = \left(0 + \frac{6 \times 108}{6}\right)$$

$$12\, M_A + 6\, M_B = -108$$

or $$2\, M_A + M_B = -18 \qquad ...(i)$$

Again using three moments equation for the spans AB and BC,

$$M_A\, l_1 + 2\, M_B\, (l_1 + l_2) + M_C\, l_2 = \left(\frac{6a_1x_1}{l_1} + \frac{6a_2x_2}{l_2}\right)$$

$$M_A \times 6 + 2\, M_B\, (6 + 6) + M_C \times 6 = -\left(\frac{6 \times 108}{6} + \frac{6 \times 162}{6}\right)$$

$$6\, M_A + 24\, M_B + 6\, M_C = -270$$

or $$M_A + 4\, M_B + M_C = -45 \qquad ...(ii)$$

Again using three moments equation for the spans BC and $C0$.

$$M_B\, l_2 + 2\, M_C\, (l_2 + l_0) + M_C\, l_0 = -\left(\frac{6a_2\bar{x}_2}{l_2} + \frac{6a_0\bar{x}_0}{l_0}\right)$$

$$M_B \times 6 + 2\,M_C\,(6+0) + 0 = -\left(\frac{6 \times 162}{6} + 0\right)$$

$$6\,M_B + 12\,M_C = -162$$

or $$M_B + 2\,M_C = -27 \quad \text{...}(iii)$$

Solving equations (*i*), (*ii*) and (*iii*)

$$M_A = -5.25 \text{ kN-m},$$

$$M_B = -7.5 \text{ kN-m}$$

and $$M_C = -9.75 \text{ kN-m}$$

Now complete the bending moment diagram as shown in Fig. 16.8 (*b*).

Reactions at the supports

Let R_A = Reaction at *A*,

R_B = Reaction at *B* and

R_C = Reaction at *C*

Taking moments about *B* and equating the same,

$$R_A \times 6 + M_B = M_A + (2 \times 6 \times 3)$$

$$R_A \times 6 - 7.5 = -5.25 + 36 = 30.75$$

$$\therefore \quad R_A = \frac{30.75 + 7.5}{6} = 6.375 \text{ kN} \quad \textbf{Ans.}$$

Again taking moments about *B* and equating the same,

$$R_C \times 6 + M_B = M_C + 12 \times 3$$

$$R_C \times 6 - 7.5 = -9.75 + 36 = 26.25$$

$$\therefore \quad R_C = \frac{26.25 + 7.5}{6} = 5.625 \text{ kN} \quad \textbf{Ans.}$$

and $$R_B = (2 \times 6 + 12) - (6.375 + 5.625) = 12 \text{ kN} \quad \textbf{Ans.}$$

Now draw the shear force diagram and elastic curve as shown in Fig. 16.8. (*c*) and (*d*).

16.7. CONTINUOUS BEAMS WITH END SPAN OVERHANGING

Sometimes, a continuous beam is overhanging, at its one or both ends. In such a case, the overhanging part fo the beam behaves like a cantilever. The fixing moments on the end support may be found out by the cantilever action fo the overhanging part of the beam.

Example 16.5: *A beam ABCD 9 m long is simply supported at A, B and C, such that the length AB is 3 m, length BC is 4.5 m and the overhung CD is 1.5 m. It carries a uniformly distributed load of 1.5 kN/m in span AB and a point load of 1 kN at the free end D. The moments of inertia of the beam in span AB and CD is I and that in the span BC is 2I. Draw the bending moment and shear force diagrams for the beam.*

Solution: Given: Length *AB* (l_1) = 3 m; Length *BC* (l_2) = 4.5 m; Length *CD* (l_3) = 1.5 m: Uniformly distributed load in *AB* (w) = 1.5 kN/m; Point load at *D* = 1kN; Moment of inertia of lengths *AB* and *CD* (I_{AB}) = *I* and moment and inertia of length *BC* (I_{BC}) = 2 *I*.

Bending moment diagram

Let $*M_A$ = Fixing moment at A,

M_B = Fixing moment at B and

M_C = Fixing moment at C.

First of all, consider the beam AB as a simply supported beam. Therefore, bending moment at the mid of span AB.

$$= \frac{wl_1^2}{8} = \frac{1.5 \times (3)^2}{8} = 1.69 \text{ kN-m}$$

From the geometry of the figure, we find that the dixing moment at C,

$$M_C = -1.0 \times 1.5 = -1.5 \text{ kN-m}$$

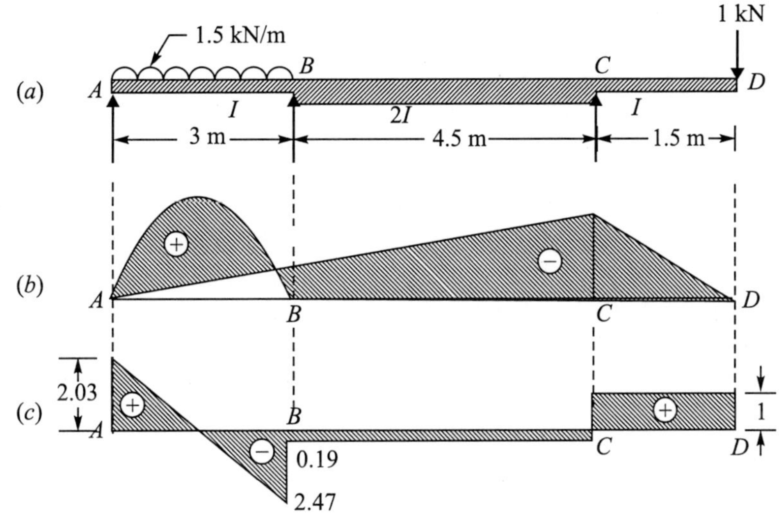

Fig. 16.9.

Now draw the μ-diagram, with the help of above bending moments as shown in Fig. 16.9. (*b*). From the geometry of the above bending moment diagram, we find that for the span AB and BC,

$$a_1\bar{x}_1 = \frac{2}{3} \times 1.69 \times 3 \times 1.5 = 5.07$$

Now using three moments equation for the spans AB and BC,

$$M_A \frac{l_1}{I_1} + 2M_B\left(\frac{l_1}{I_1} + \frac{l_2}{I_2}\right) + M_C \frac{l_2}{I_2} = -\left[\frac{6a_1\bar{x}_1}{I_1 l_1} + \frac{6a_2\bar{x}_2}{I_2 l_2}\right]$$

$$0 + 2M_B\left(\frac{3}{I} + \frac{4.5}{2I}\right) + 1.5 \times \frac{4.5}{2I} = -\frac{6 \times 5.07}{I \times 3}$$

$$\frac{10.5M_B}{I} - \frac{6.75}{2I} = \frac{10.14}{I}$$

or

$$10.5\ M_B - 3.375 = -10.14$$

---

* Since the beam is simply supported at A, therefore fixing moment M_A will be zero.

$$\therefore \qquad M_B = \frac{-10.14 + 3.375}{10.5} = -0.65 \text{ kN-m}$$

Now complete the final bending moment diagram as shown in Fig. 16.9 (*b*).

Shear force diagram

Let R_A = Reaction at A,
R_B = Reaction at B and
R_C = Reaction at C,

Taking moments at B,

$$R_A \times 3 - (1.5 \times 3 \times 1.5) = -0.65 \qquad ...(\because M_B = -0.65 \text{ kN-m})$$

$$\therefore \qquad R_A = \frac{-0.65 + 6.75}{3} = \frac{6.1}{3} = 2.03 \text{ kN}$$

Again taking moments about B,

$$R_C \times 4.5 - (1 \times 6) = -0.65 \qquad ...(\because M_B = -0.65 \text{ kN-m})$$

$$\therefore \qquad R_C = \frac{-0.65 + 6}{4.5} = \frac{5.35}{4.5} = 1.19 \text{ kN}$$

and $\qquad R_B = [(3 \times 1.5) + 1] - (2.03 + 1.19) = 2.28$ kN

Now complete the shear force diagram as shown in Fig. 16.9 (*c*).

Example 16.6: *A beam ABCDE has a built-in support at A and roller supports at B, C and D. DE being an overhung. AB = 7 m, BC = 5 m, CD = 4 m and DE = 1.5 m. The values of moment of inertia of the section over each of these lengths are 3I, 2I, I and I respectively. The beam carries at point load of 10 kN at a point 3 m from A, a uniformly distributed load of 4.5 kN/m over whole of BC and a concentrated load of 9 kN in CD, 1.5 m from C and another point load of 3 kN at E, the tope of overhung as shown in Fig. 16.10.*

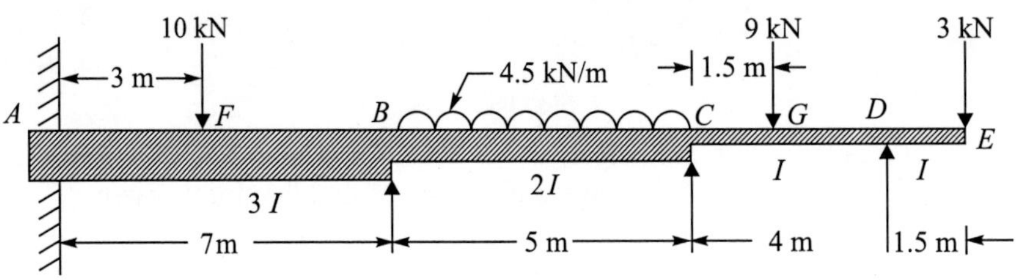

Fig. 16.10.

Determine: (i) moments developed over each support, and (ii) draw diagram for the entire beam, stating values at salient points.

Solution: Given : Length AB (l_1) = 7 m ; length BC (l_2) = 5 m; Length CD (l_3) = 4 m; Length DE (l_4) = 1.5 m; Moment of inertia of length AB (I_{AB}) = 3 I; Moment of inertia for length BC (I_{BC}) = 2I; Moment of inertia of length CD (I_{CD}) = I_{DE} = I ; Point load at F (W_1) = 10 kN; Uniformly distributed load between B and C(w_2) = 4.5 kN/m; Point load at G (W_2) = 9 kN and pont load at E = 3 kN.

(i) Moments developed over each support

Since the beam is fixed at A, therefore, assume a zero span to the left of A.

Let $\qquad$ **M_0 = Fixing moment at left hand support of zero span,
M_A = Fixing moment at A,

M_B = Fixing moment at *B*,
M_C = Fixing moment at *C* and
M_D = Fixing moment at *D*.

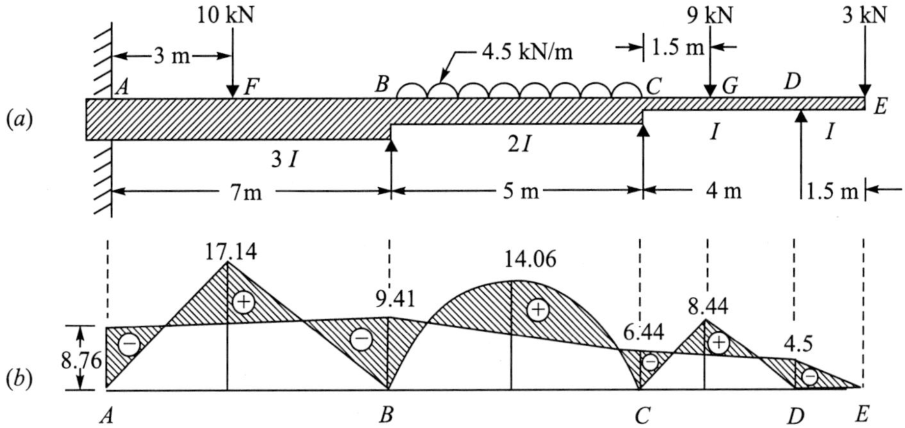

Fig. 16.11.

First of all, let us consider the beam *AB* as a simply supported beam. Therefore, bending moment under the 10 kN load

$$= \frac{W_1 a_1 b_1}{l_1} = \frac{10 \times 3 \times 4}{7} = 17.14 \text{ kN-m}$$

Similarly, bending moment under the 9 kN load in span *CD*

$$= \frac{W_3 a_3 b_3}{l_3} = \frac{9 \times 1.5 \times 2.5}{4} = 8.44 \text{ kN-m}$$

and bending moment at the mid of the span *BC*

$$= -\frac{w_2 l_2^2}{8} = \frac{4.5 \times (5)^2}{8} = 14.06 \text{ kN-m}$$

From the geometry of the figure, we find that the fixing moment at *D*, due to load at *E*,

$$M_D = -3 \times 1.5 = -4.5 \text{ kN-m}$$

Now draw μ-diagram with the help of above bending moments as shown in Fig. 16.11 (*b*). From the geometry of the above bending moment diagram, we find that for the span 0*A* and *AB*,

$$a_0 \bar{x}_0 = 0$$

and $$a_1 \bar{x}_1 = \left[\left(\frac{1}{2} \times 4 \times 17.14 \times \frac{2 \times 4}{3}\right) + \left(\frac{1}{2} \times 3 \times 17.14\right)\left(4 + \frac{3}{3}\right)\right] = 220$$

Similarly, for the spans *AB* and *BC*,

$$a_1 \bar{x}_1 = \left[\left(\frac{1}{2} \times 3 \times 17.14 \times \frac{3 \times 2}{3}\right) + \left(\frac{1}{2} \times 4 \times 17.14\right)\left(3 + \frac{4}{3}\right)\right] = 200$$

and $$a_2 \bar{x}_2 = \frac{2}{3} \times 5 \times 14.06 \times 2.5 = 11.72$$

Similarly, for the spans *BC* and *CD*,

$$a_2\bar{x}_2 = \frac{2}{3} \times 5 \times 14.06 \times 2.5 = 11.72$$

and
$$a_3\bar{x}_3 = \left[\left(\frac{1}{2}\times 2.5\times 8.44\times\frac{2\times 2.25}{3}\right)+\left(\frac{1}{2}\times 1.5\times 8.44\right)\left(2.5+\frac{1.5}{3}\right)\right]$$

Now using three moments equation for the spans 0*A* and *AB*,

$$M_0\frac{l_0}{I_0}+2M_A\left(\frac{l_0}{I_0}+\frac{l_1}{I_1}\right)+M_B\frac{l_1}{I_1} = -\left(\frac{6a_0\bar{x}_0}{I_0\,l_0}+\frac{6a_1\bar{x}_1}{I_1\,l_1}\right)$$

$$0+2M_A\times\frac{7}{3I}+M_B\frac{7}{3I} = -\left(0+\frac{6\times 220}{3I\times 7}\right)$$

$$\frac{14\,M_A}{3I}+\frac{7M_B}{3I} = -\frac{1320}{3I\times 7}$$

or
$$2\,M_A + M_B = -26.94$$

Again using three moments equation for the span *AB* and *BC*,

$$M_A\frac{l_1}{I_1}+2M_B\left(\frac{l_1}{I_1}+\frac{l_2}{I_2}\right)+M_C\frac{l_2}{I_2} = -\left(\frac{6a_1\bar{x}_1}{I_1\,l_1}+\frac{6a_2\bar{x}_2}{I_2\,l_2}\right)$$

$$M_A\frac{7}{3I}+2M_B\left(\frac{7}{3I}+\frac{5}{2I}\right)+M_C\frac{5}{2I} = -\left(\frac{6\times 200}{3I\times 7}+\frac{6\times 117.2}{2I\times 5}\right)$$

$$\frac{7M_A}{3I}+\frac{29M_B}{3I}+\frac{5M_C}{2I} = -\left(\frac{171.4}{3I}+\frac{140.6}{2I}\right)$$

$$14\,M_A + 58\,M_B + 15\,M_C = -\,764.6$$

Again using three moments equation for the spans *BC* and *CD*,

$$M_B\frac{l_2}{I_2}+2M_C\left(\frac{l_2}{I_2}+\frac{l_3}{I_3}\right)+M_D\frac{l_3}{I_3} = \left(\frac{6a_2\bar{x}_2}{I_2\,l_2}\right)+\frac{6a_3\bar{x}_3}{I_3\,l_3}$$

$$M_B\frac{5}{2I}+2M_C\left(\frac{5}{2I}+\frac{4}{I}\right)-4.5\frac{4}{I} = -\left(\frac{6\times 117.2}{2I\times 5}+\frac{6\times 36.6}{I\times 4}\right)$$

$$\frac{5M_B}{2I}+\frac{26M_C}{2I}-\frac{18}{I} = -\left(\frac{140.6}{2I}+\frac{54.9}{I}\right)$$

$$5\,M_B + 26\,M_C = -\,214.4 \qquad \ldots(iii)$$

Now solving equations (*i*), (*ii*) and (*iii*), we get

$M_A = -\,8.76$ kN-m **Ans.**

$M_B = -\,9.41$ kN-m **Ans.**

$M_C = -\,6.44$ kN-m **Ans.**

Now complete the bending moment diagram as shown in Fig. 16.11 (*b*).

16.8. CONTINUOUS BEAMS WITH A SINKING SUPPORT

Sometimes, one of the supports of a continuous beam sinks down due to loading with respect to the other supports which remain at the same level. The sinking of a support effects the moments at the supports.

Now consider a continuous beam *ABC* fixed at *A* and *C* and supported at *B* as shown in Fig. 16.12. Let the support *B** sink down through a more distance than the support *C* sinks down from its original position (or in other words from the support *A*),

Let δ_A = Height of support *A* from *B*

and δ_C = Height of support *C* from *B*.

We have already discussed that in the span *AB*

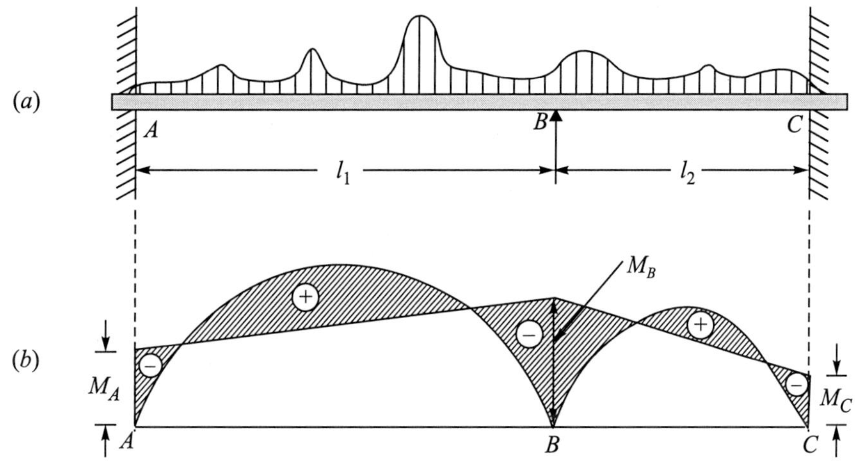

Fig. 16.12.

$$dEI_1\,[l_1\,i_B - y_B] = a_1\bar{x}_1 + a_1' \cdot \bar{x}_1'$$

or
$$EI_1\,[l_1\,i_B - d_A] = a_1\bar{x}_1 + a_1' \cdot \bar{x}_1'$$

We have also discussed that

$$a_1' + \bar{x}_1' = (M_A + 2\,M_B)\,\frac{l_1^2}{6}$$

$\therefore$
$$EI_1\,[l_1\,i_B + \delta_A] = a_1\bar{x}_1 + (M_A + 2M_B)\frac{l_1^2}{6}$$

$$EI_1\,l_1\,i_B + EI_1\,\delta_A = a_1\bar{x}_1 + (M_A + 2M_B)\frac{l_1^2}{6}$$

or
$$Ei_B + \frac{E\delta_A}{l_1} = \frac{a_1\bar{x}_1}{I_1 l_1} + (M_A + 2M_B)\frac{l_1}{6I_1}$$

or
$$Ei_B = \frac{a_1\bar{x}_1}{I_1 l_1} + (M_A + 2M_B)\frac{l_1}{6I_1} - \frac{E\delta_A}{l_1}$$

* It is also possible if all three supports sink down. But the support *B* sinks down more than *A* and *C*.

Similary, in the span *BC* taking *C* as the origin and *x* positive to the left,

$$Ei'_E = \frac{a_2 x_2}{I_2 l_2} + (M_C + 2M_B)\frac{l_2}{6I_2} - \frac{E\delta_C}{l_2}$$

Since i_B is equal to $-i'_B$, therefore Ei_B is equal to $-Ei'_B$.

or
$$\frac{a_1\bar{x}_1}{I_1 l_1} + (M_A + 2M_B)\frac{l_1}{6I_1} - \frac{E\delta_A}{l_1} = -\left[\frac{a_2\bar{x}_2}{I_2 l_2} + (M_C + 2M_B)\frac{l_2}{6I_2} - \frac{E\delta_C}{l_2}\right]$$

$$M_A\frac{l_1}{6I_1} + 2M_B\frac{l_2}{6I_1} + M_C\frac{l_2}{6I_2} + 2M_B\frac{l_2}{6I_2} = -\left(\frac{a_1\bar{x}_1}{I_1 l_1} + \frac{a_2\bar{x}_2}{I_2 l_2}\right) + \frac{E\delta_A}{l_2} + \frac{E\delta_C}{l_2}$$

$$M_A\frac{l_1}{I_1} + 2M_B\left(\frac{l_1}{I_1} + \frac{l_2}{I_2}\right) + M_C\frac{l_2}{I_2} = -\left(\frac{6a_1\bar{x}_1}{I_1 l_1} + \frac{6a_2\bar{x}_2}{I_1 l_2}\right) + \frac{6E\delta_A}{l_1} + \frac{6E\delta_C}{l_2}$$

Notes :

1. If the moment of inertia of the beam is constant, then

$$M_A l_1 + 2M_B(l_1 + l_2) + M_C l_2 = -\left(\frac{6a_1\bar{x}_1}{l_1} + \frac{6a_2\bar{x}_2}{l_2}\right) + \frac{6EI\delta_A}{l_1} + \frac{6EI\delta_C}{l_2}$$

2. The above formula has been derived by taking δ_A and δ_C as positive. But while solving the numericals in sinking supports, care should always be taken to use the proper sign. The following guide rules should be kept in mind for the purpose :
   (*i*) The three moments equation always refer to two adjacent spans. The reference level should always be taken as that of the common support.
   (*ii*) The sign of δ for the left and right support, should then be used by comparing its height with the central support (positive for higher and negative for lower), *e.g.*, Consider a continuous beam *ABC* in which let the support *B* sink down by an amount equal to δ. The three supports, namely *A*, *C* and *B* will remain at the same level. Now, while using three moments equation for the spans *AB* and *BC*, the values of δ_A and δ_C will be positive (because both the supports *A* and *C* are at higher level than that of *B*). But while using three moments equation for the spans *BC* and *CD*, the value of δ_B will be negative (because the support *B* is at a lower level than that of *C*) and the value of δ_D will be zero (because the support *D* is at the same level as that of *C*).

Example 16.7: *A continuous beam ABC, shown in Fig. 16.13 carries a uniformly distributed load of 50 kN on AB and BC. The support B sinks by 5 mm below A and C and the values of EI is constant throughout the beam.*

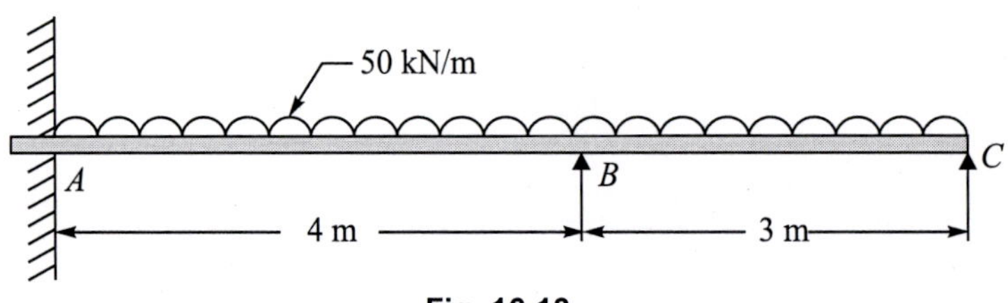

Fig. 16.13.

Taking E = 200 GP$_a$ and I = 332 × 10^6 mm^4, find the bending moment at supports A and B draw the bending moment diagram.

Solution: Given : Length AB (l_1) = 4 m ; Length BC (l_2) = 3 m; Uniformly distributed load over $AC(w)$ = 50 kN-m; Sinking of support of $(\delta_B) = -5$ mm $= -0.005$ m or $\delta_A = \delta_C = +0.005$ m; Modulus of elasticity (E) 200 GPa $= 200 \times 10^6$ kN/m^2 and moment of inertia $(I) = 332 \times 10^6$ mm$^4 = 332 \times 10^{-6}$ m^4.

M_A = Fixing moment at A,

M_B = Fixing moment at B, and

M_C = Fixing moment at C,

First of all, consider the beam AB as a simply supported. Therefore, bending moment at the mid of the span AB

$$= \frac{wl_1^2}{8} = \frac{50 \times (4)^2}{8} = 100 \text{ kN-m}$$

Similarly, bending moment at the mid of the span BC

$$= \frac{wl_2^2}{8} = \frac{50 \times (3)^2}{8} = 56.25 \text{ kN-m}$$

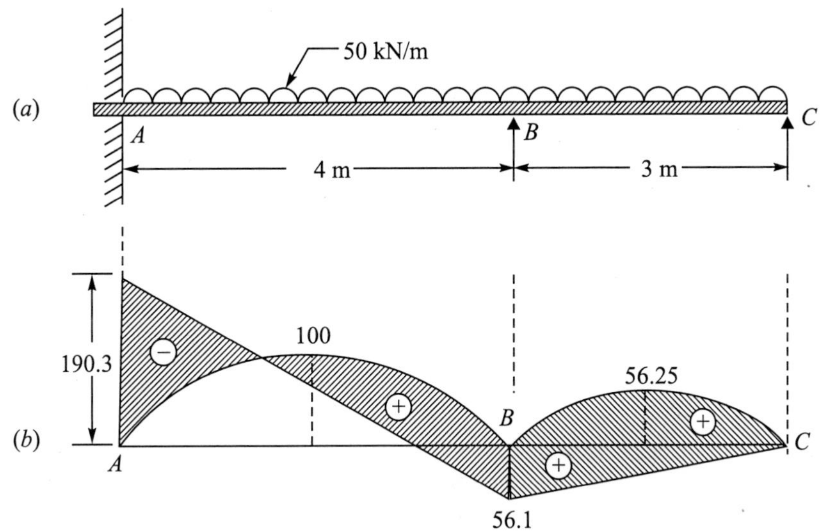

Fig. 16.14.

Since the beam is fixed at A, therefore let us assume a zero span to the left of A,

Now draw μ-diagram, with the help of above bending moments as shown in Fig. 16.14 (*b*). From the geometry of the above bending moment diagram, we find that for the spans $0A$ and AB,

$$a_0\bar{x}_0 = 0$$

and

$$a_1\bar{x}_1 = \frac{2}{3} \times 100 \times 4 \times 2 = \frac{1600}{3} = 533.3$$

Similarly, for the spans AB and BC,

$$a_1\bar{x}_1 = \frac{2}{3} \times 100 \times 4 \times 2 = \frac{1600}{3} = 533.3$$

and $$a_2\bar{x}_2 = \frac{2}{3} \times 56.25 \times 3 \times 1.5 = 168.75$$

Now using three moments equation for the spans $0A$ and AB,

$$M_0\, l_0 + 2\, M_A\, (0 + l_1) + M_B\, l_1 = -\left[\frac{6a_0\bar{x}_0}{l_0} + \frac{6a_1\bar{x}_1}{l_1}\right] + \frac{6EI\delta_0}{l_1} + \frac{6EI\delta_B}{l_2}$$

$$0 + 2\, M_A\, (0 + 4) + (M_B \times 4) = -\left[0 + \frac{6 \times 533.3}{4}\right]$$

$$+ 0 + \frac{6 \times (200 \times 10^6) \times (332 \times 10^{-6})(-0.005)}{4}$$

$$8\, M_A + 4\, M_B = -\, 800 - 498 = -\, 1298$$

$\therefore$ $$2\, M_A + M_B = -\, 324.5 \qquad \ldots(i)$$

Now using three moments equation for the spans AB and BC,

$$M_A\, l_1 + 2\, M_B\, (l_1 + l_2) + M_C\, l_2 = -\left[\frac{6a_1\bar{x}_1}{l_1} + \frac{6a_2\bar{x}_2}{l_2}\right] + \frac{6EI\delta A}{l_1} + \frac{6EI\delta C}{l_2}$$

$$(M_A \times 4) + 2\, M_B\, (4 + 3) + 0 = -\left[\frac{6 \times 533.3}{4} + \frac{6 \times 168.75}{3}\right]$$

$$+ \frac{6 \times (200 \times 10^6) \times (332 \times 10^{-6})\, (+\, 0.005)}{4}$$

$$+ \frac{6 \times 200 \times 10^6 \times 332 \times 10^{-6}\, (+\, 0.005)}{3}$$

$$4\, M_A + 14\, M_B = -\, (800 + 337.5) + 498 + 664 = 24.5$$

$\therefore$ $$2\, M_A + 7\, M_B = 12.25$$

Solving equations (*i*) and (*ii*),

$$M_A = -\, 190.3 \text{ kN-m} \quad \textbf{Ans.}$$

and $$M_B = 56.1 \text{ kN-m} \quad \textbf{Ans.}$$

Now complete the bending moment diagram as shown in Fig. 16.14 (*b*).

Notes :

1. While considering the spans 0A and AB, the assumed support 0 and fixed support A are at the same level. But the support B has sunk down. Therefore, the values of δ_0 is taken as zero and that for δ_B as negative.
2. While considering the spans AB and BC, the supports A and C are at the same level. But the support B has sunk down. Therefore the values δ_A and δ_C are taken as positive.

16.9. CONTINUOUS BEAMS SUBJECTED TO A COUPLE

Sometimes, a continuous beam is subjected to a couple in one (or more) of the spans. Such a sign is also analyses in the similar manner. The couple will cause negative moments in one part and negative moments in the other part of the span. While taking moment of the bending moment diagram, care should be taken for the positive and negative bending moments.

Example 16.8: *A continuous beam ABC of constant moment of inertia is simply supported at A, and C. the beam carries a central point load of 4 kN in span AB and a central clockwise couple of statement 30 kN-m in span BC as shown in Fig. 16.15.*

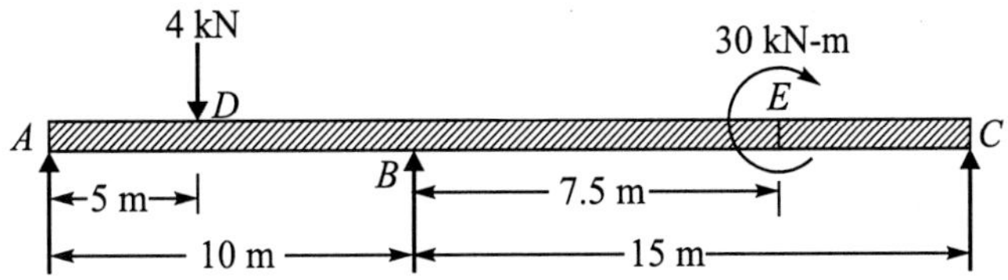

Fig. 16.15.

Find the support moments and plot the shear force and bending moment diagrams.

Solution: Given : Length AB (l_1) = 10 m ; Length BC (l_2) = 15 m; Load at D (w) = 4 kN and sample at $E(\mu)$ = 30 kN-m.

Support Moments

Let $*M_A$ = Support moment at A,

M_B = Support moment at B and

$*M_C$ = Support moment at C.

First of all, consider the beam AB and BC as a simply supported beam. Therefore, bending module at D,

$$M_D = \frac{Wl_1}{4} = \frac{4 \times 10}{4} = 10 \text{ kN-m}$$

We know that the **moment just on the right side of E

$$= \frac{\mu}{2} = \frac{30}{2} = 15 \text{ kN-m}$$

and **moment just on the left side of E

$$= -\frac{\mu}{2} = -\frac{30}{2} = 15 \text{ kN-m}$$

Now draw the μ-diagram with the help of above bending moments as shown in Fig. 16.16 (*b*).

From the geometry of the bending moment diagrams, we find that for spans AB and BC,

$$a_1\bar{x}_1 = \frac{1}{2} \times 10 \times 10 \times 5 = 250$$

$$a_2\bar{x}_2 = 0 \qquad (\because + \text{B.M.} = - \text{B.M.})$$

Now using three moments equation for the spans AB and BC,

$$M_A l_1 + 2 M_B (l_1 + l_2) + M_C l_2 = -\left(\frac{6a_1\bar{x}_1}{l_1} + \frac{6a_2\bar{x}_2}{l_2}\right)$$

$$0 + 2 M_B (10 + 15) + 0 = -\left(\frac{6 \times 250}{10} + 0\right)$$

$\therefore$ $$50 M_B = -150$$

or $$M_B = -3 \text{ kN-m}$$

Now complete the bending moment diagram as shown in Fig. 16.16 (*b*).

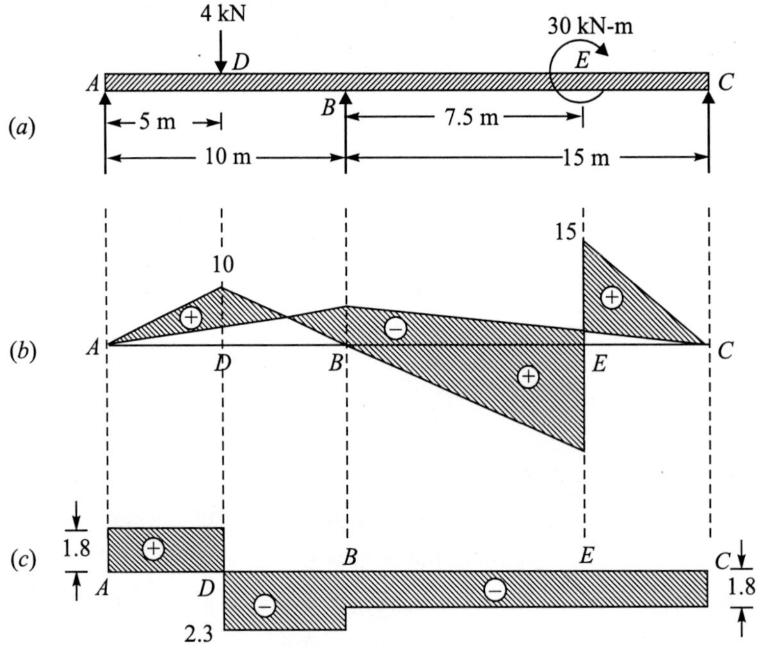

Fig. 16.16.

Shear force diagram

Let R_A = Reaction at A,

R_B = Reaction at B, and

R_C = Reaction at C

Taking moments about B,

$$R_A \times 10 - 4 \times 5 = -3 \qquad ...(\because M_B = -3 \text{ kN-m})$$

$$\therefore \quad R_A = \frac{-3+20}{10} = 1.7 \text{ kN}$$

Similarly, $$R_C \times 15 - 30 = -3 \qquad ...(\because M_B = -3 \text{ kN-m})$$

$$\therefore \quad R_C = \frac{-3+30}{15} = 1.8 \text{ kN}$$

and $$R_B = 4 - (1.7 + 1.8) = 0.5 \text{ kN}$$

Now draw the shear force diagram as shown in Fig. 16.16 (*c*).

SUMMARY

- Clapeyron's theorem of three moments states : "If a beam has '*n*' supports, the end ones being fixed, then the same number of equations required to determine the support moments, may be obtained from the consecutive pairs of spans *i.e., AB – BC, BC – CD, CD – DE* and so on."
- A beam, which is supported on more than two supports, is called a continuous beam. The B.M. diagram for a continuous beam is drawn in the following two stages :

(*a*) By considering the beam as a series of discontinuous beams, from support to support and drawing the μ-diagram due to vertical loads.

(*b*) By superimposing the usual μ′-diagram due to end moments over μ-diagram.

- When a continuous beam is simply supported on its one, or both the end supports, the fixing end moments on the simply supported end is zero.
- When a beam is fixed at its one or both ends, then an imaginary zero span is taken and the three moment theorem is applied as usual.
- When a beam is overhanging at its one or both ends, the overhanging beam behaves like cantilever. The fixing moment on the end support is found out by the cantilever action of the overhanging part of the beam.
- When one of the supports of a continuous beam sinks down, it affects the fixing moments on the supports.

SHORT ANSWER QUESTIONS

1. Explain the theorem of three moments.
2. Prove the clapeyron's theorem of three moments.
3. How will you apply theorem of three moments ot a fixed beam?
4. Explain the effect on a continuous beam, when one of the intermediate supports sinks down.
5. Describe the effect of a couple acting in one of the spans of continuous beam.

CHAPTER

17

Gravity Dams

17.1. INTRODUCTION

A dam is constructed to store large quantity of water, which is used for the purpose of irrigation and power generation. A dam may be of any cross-section, but the dams of trapezoidal cross-section are very popular these days. Though there are many types of dams, yet the following are important from the subject point of view:

1. Rectangular dams
2. Trapezoidal dams having water face vertical
3. Trapezoidal dams having water face inclined

A dam constructed with earth is called an **earthen dam**, whereas a dam constructed with cement concrete is called a **concrete dam** or **gravity dam**.

Unless mentioned, otherwise, the specific weight of water is taken as 9.8 kN/m^3.

17.2. TRAPEZOIDAL DAMS WITH WATER FACE VERTICAL

Consider a unit length of a trapezoidal dam having its water face vertical as shown in the Fig. 17.1.

Let, a = Top width of the dam

b = Bottom width of the dam

H = Height of the dam

p = Specific weight of the dam masonry

h = Height of water retained by the dam

w = Specific weight of water

We know that the weight of the dam per unit length

$$W = p \times \frac{(a+b)}{2} \times H \qquad ...(17.1)$$

Like a rectangular dam, the total pressure on a unit length of the trapezoidal dam,

$$P = \frac{wh^2}{2} \qquad ...(17.2)$$

and the horizontal distance between the centre of gravity (C.G.) of the dam and the point, where the resultant R cuts the base,

$$x = \frac{P}{W} \times \frac{h}{3} \qquad ...(17.3)$$

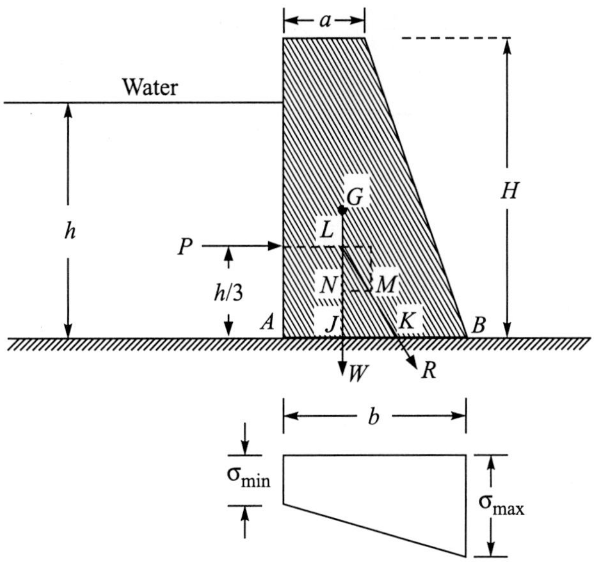

Fig. 17.1.

The distance between the tore of the dam A and the point where the resultant R cuts the base (*i.e.*, distance AK)

$$\text{“}d\text{”} = AJ + JK$$

$$= AJ + \left(\frac{P}{W} \times \frac{h}{3}\right) \qquad \text{...(17.4)}$$

Now the distance AJ may be found out by spliting the dam section into a rectangle and a triangle. Now taking their moments about A and equating the same with the moment of the dam section about A.

$\therefore$ Eccentricity “e” $= d - AJ$

[The distance AJ may also be found out from the relation, $AJ = \dfrac{a^2 + ab + b^2}{3(a+b)}$]

The stress across the base at B will be maximum, whereas the stress across the base at A will be minimum such that,

$$\sigma_{max} = \frac{W}{b}\left(1 + \frac{6e}{b}\right)$$

and $$\sigma_{min} = \frac{W}{b}\left(1 - \frac{6e}{b}\right)$$

When the reservoir is empty, there will be no water pressure on the dam. In this case, the eccentricity of the weight of dam

$$\text{“}e\text{”} = \frac{b}{2} - AJ$$

Since the eccentricity in this case will be minus therefore, the total stress across the base at B will be minimum, whereas the stress across the base at A, will be maximum, such that

$$\sigma_{min} = \frac{W}{b}\left(1 - \frac{6e}{b}\right)$$

and
$$\sigma_{max} = \frac{W}{b}\left(1 + \frac{6e}{b}\right)$$

17.3. SOLVED EXAMPLES

Example 17.1: *A concrete dam of trapezoidal section having water on vertical face is 16 m high. The base of the dam is 8 m wide and top 3 m wide. Find:*

(a) resultant thrust on the base per metre length of the dam,

(b) point, where the resultant thrust cuts the base, and

(c) intensities of maximum and minimum stresses across the base.

Take weight of the concrete as 25 kN/m³ and the water level coinciding with the top of the dam.

Solution: Given; Height of the dam (H) = 16 m

Height of water retained by the dam (h) = 16 m

Bottom width of the dam (b) = 8 m

Top width of the dam (a) = 3 m

and Weight of concrete (ρ) = 25 kN/m³

(*a*) Resultant Thrust on the Base Per Metre Length: We know that total water pressure per metre length of the dam,

$$P = \frac{wh^2}{2} = \frac{9.81 \times (16)^2}{2} \text{ kN}$$

$$= 1255.7 \text{ kN} \qquad \ldots(i)$$

and weight of concrete per metre, length,

$$W = \rho \times \left(\frac{a+b}{2}\right) \times H$$

$$= 25 \times \left(\frac{3+8}{2}\right) \times 16 \text{ kN}$$

$$= 2200 \text{ kN} \qquad \ldots(ii)$$

∴ Resultant thrust per metre length,

$$R = \sqrt{P^2 + W^2} = \sqrt{(1255.7)^2 + (2200)^2}$$

$$= 2533 \text{ kN} \quad \textbf{Ans.}$$

(*b*) Point, where the Resultant Cuts the Base: Let the resultant (R) cut the base at K as shown in Fig. 17.2 (*a*). First of all, let us find out the position* of the centre of gravity of the dam section. Now taking moment of the area about A and equating the same,

$$\left[(16 \times 3) + \left(\frac{16 \times 5}{2}\right)\right] AJ = \left(16 \times 3 \times \frac{3}{2}\right) + \left[16 \times \frac{5}{2}\left(3 + \frac{5}{3}\right)\right]$$

$$88 \times AJ = 72 + 186.7 = 258.7$$

or $$*AJ = \frac{258.7}{88} = 2.94 \text{ m}$$

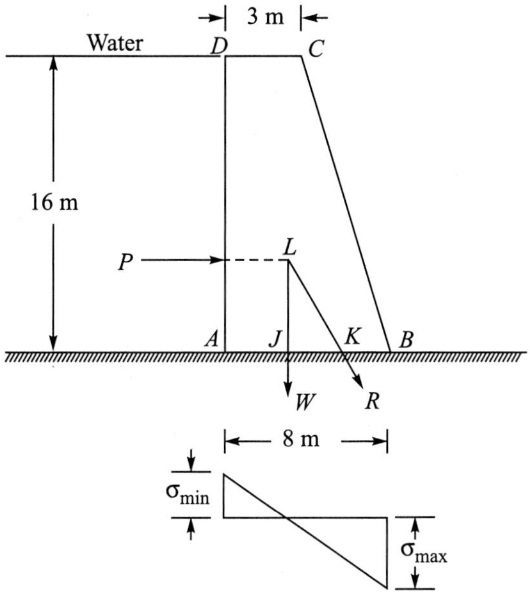

Fig. 17.2.

We know that horizontal distance between the centre of gravity of dam section and the point, where the resultant cuts the base (*i.e.*, distance *JK*),

$$x = \frac{P}{W} \times \frac{h}{3} = \frac{1255.7}{2200} \times \frac{16}{3} = 3.04 \text{ m}$$

∴ Horizontal distance AK,

$$**d = AJ + x = 2.94 + 3.04 = \mathbf{5.98 \text{ m}} \quad \textbf{Ans.}$$

(*c*) **Intensities of Maximum and Minimum Stresses Across the Base:** We know that eccentricity of the resultant,

$$e = d - \frac{b}{2} = 5.98 - \frac{8}{2} = 1.98 \text{ m}$$

---

*The horizontal ditance may also be found out from the following relation.

$$AJ = \frac{a^2 + ab + b^2}{3(a+b)} = \frac{(3)^2 + (3 \times 8) + (8)^2}{3(3+8)}$$

$$= \frac{97}{33} = 2.94 \text{ m}$$

**The horizontal distance *d* may also be found out by taking moment about *A* and equating the same, *i.e.*,

$$W.\,d = \left(P \times \frac{h}{3}\right) + (W \times AJ)$$

or $$d = AJ + \frac{P}{W} \times \frac{h}{3} = 2.94 + \left(\frac{1255}{2200} \times \frac{16}{3}\right) = 5.98 \text{ m}$$

∴ Intensity of maximum stress across the base,

$$\sigma_{max} = \frac{W}{b}\left(1+\frac{6e}{b}\right) = \frac{2200}{8}\left(1+\frac{6\times1.98}{8}\right) \text{ kN/m}^2$$

$$= 683.3 \text{ kN/m}^2 = \mathbf{683.3 \text{ kPa}} \quad \textbf{Ans.}$$

and intensity of minimum stress across the base,

$$\sigma_{min} = \frac{W}{b}\left(1+\frac{6e}{b}\right) = \frac{2200}{8}\left(1+\frac{6\times1.98}{8}\right) \text{ kN/m}^2$$

$$= -\ 133.4 \text{ kN/m}^2 = \mathbf{133.4 \text{ kPa (tension)}} \quad \textbf{Ans.}$$

Example 17.2: *A masonry trapezoidal dam 4 m high, 1m wide as its top and 3 m wide at its bottom retains water on its vertical face. Determine the maximum and minimum stresses at the base: (i) when the reservoir is full and (ii) when the reservoir is empty. Take weight of water as 10 kN/m³ and that of masonry as 24 kN/m³.*

Solution: Given; Height of the dam (H) = 4 m

Top width of the dam (a) = 1 m

Bottom width of the dam (b) = 3 m

Weight of water (w) = 10 kN/m³

and Weight of masonry (ρ) = 24 kN/m³

(*i*) Maximum and Minimum Stresses at the Base when the Reservoir is Full: Let the resultant (R) cut the base at K as shown in Fig. 17.3. (*a*).

We know that the total pressure of water per metre length of the dam,

$$P = \frac{wh^2}{2} = \frac{10\times(4)^2}{2} = 80 \text{ kN} \qquad ...(i)$$

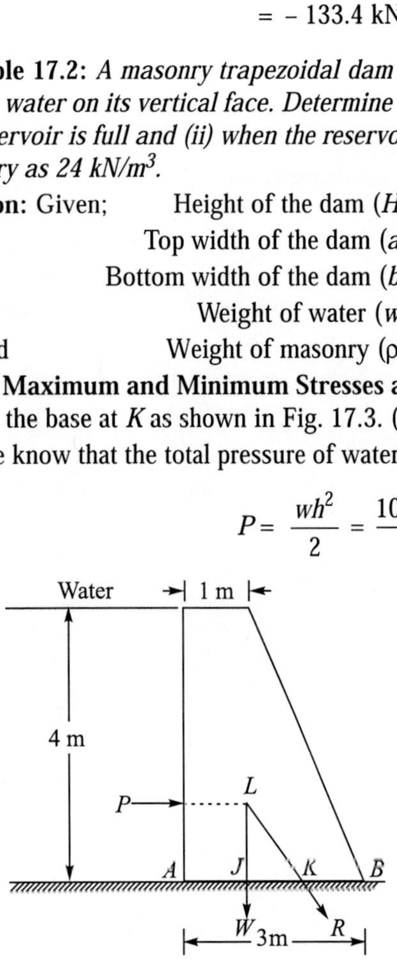

Fig. 17.3. (a) When the reservoir is full

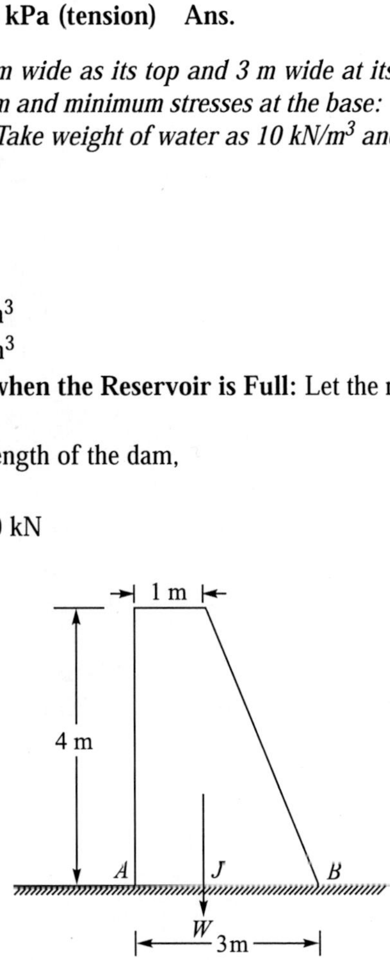

Fig. 17.3. (b) When the reservoir is empty

and weight of concrete per metre length,

$$W = \rho\left(\frac{a+b}{2}\right) \times H = 24 \times \left(\frac{1+3}{2}\right) \times 4 \text{ kN} = 192 \text{ kN} \qquad ...(ii)$$

Now let us find out the position* of the centre of gravity of the dam section. Taking moments of the area about A and equating the same,

$$\left(4\times1+\frac{4\times2}{2}\right)AJ = \left(4\times1+\frac{1}{2}\right)+\left[\frac{4\times2}{2}\left(1+\frac{2}{3}\right)\right]AJ$$

$$8AJ = 2 + 6.67 = 8.67$$

or $$*AJ = \frac{8.67}{8} = 1.08 \text{ m}$$

We know that horizontal distance between the centre of gravity of the dam section and the point, where the resultant cuts the base (*i.e.,* distance JK),

$$x = \frac{P}{W}\times\frac{h}{3} = \frac{80}{192}\times\frac{4}{3} = 0.56 \text{ m}$$

∴ Horizontal distance AK,

$$d = AJ + x = 1.08 + 0.56 = 1.64 \text{ m}$$

and eccentricity, $$e = d - \frac{b}{2} = 1.64 - \frac{3}{2} = 0.14 \text{ m}$$

We also know that maximum stress at the base

$$\sigma_{max} = \frac{W}{b}\left(1+\frac{6e}{b}\right) = \frac{192}{3}\left(1+\frac{6\times0.14}{3}\right) \text{ kN/m}^2$$

$$= 81.92 \text{ kN/m}^2 = \mathbf{81.92\ kPa} \quad \textbf{Ans.}$$

and minimum stress at the base,

$$\sigma_{min} = \frac{W}{b}\left(1-\frac{6e}{b}\right) = \frac{192}{3}\left(1-\frac{6\times0.14}{3}\right) \text{ kN/m}^2$$

$$= 46.08 \text{ kN/m}^2 = \mathbf{46.08\ kPa} \quad \textbf{Ans.}$$

(*ii*) Maximum and Minimum Stresses at the Base when the Reservoir is Empty: In this case, there is no water presure on the dam as shown in Fig. 17.3. (*b*). Therefore eccentricity,

$$e = d - \frac{b}{2} = 1.08 - \frac{3}{2} = -0.42 \text{ m}$$

(Minus sign indicates that the stress at *A* will be more than that at *B*).

We know that maximum stress at the base (*A*),

$$\sigma_{max} = \frac{W}{b}\left(1+\frac{6e}{b}\right) = \frac{192}{3}\left(1+\frac{6\times0.42}{3}\right) \text{ kN/m}^2$$

$$= 117.76 \text{ kN/m}^2 = \mathbf{117.76\ kPa} \quad \textbf{Ans.}$$

and minimum stress at the base, (*B*)

$$\sigma_{min} = \frac{W}{b}\left(1-\frac{6e}{b}\right) = \frac{192}{3}\left(1-\frac{6\times0.42}{3}\right) \text{ kN/m}^2$$

$$= 10.24 \text{ kN/m}^2 = \mathbf{10.24\ kPa} \quad \textbf{Ans.}$$

---

*The distance AJ may also be found out from the following relation:

$$AJ = \frac{a^2+ab+b^2}{3(a+b)} = \frac{(1)^2+(1\times3)+(3)^2}{3(1+3)} = \frac{13}{12} = 1.08 \text{ m}$$

Example 17.3: *A masonry dam as shown in Fig. 17.4, has a total height of 20 m with a top width of 5 m and a free board of 2 m. Its upstream face is vertical while the downstream face has a batter of 0.66 horizontal to 1.0 vertical. The specific gravity of masonry may be taken as 2.4.*

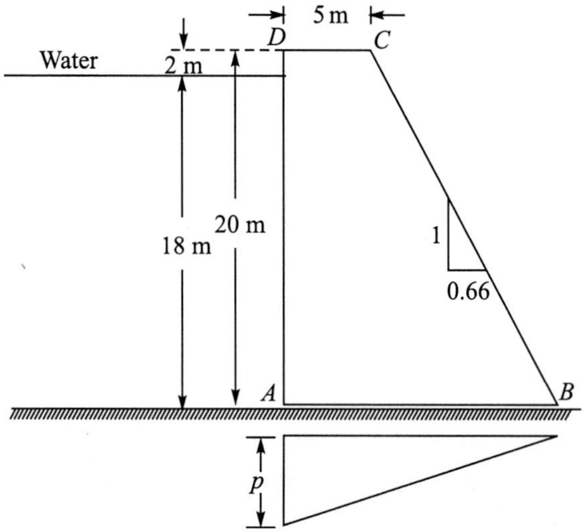

Fig. 17.4.

In addition to the hydrostatic pressure on the upstream face, there is an uplift pressure at the foundation, which may be taken to vary linearly from a value equal to the hydrostatic pressure at the upstream end, to zero at the downstream end.

Calculate the extree values of the normal stresses on the foundation, when the reservoir is full. Take specific weight of water as 10 kN/m³.

Solution: Given: Height of the dam $(H) = 20$ m

Top width of the dam $(a) = 5$ m

Free board = 2 m

Slope of downward face = 0.66 horizontal to 1.0 vertical

Specific gravity of mnasonry = 2.4

Uplift pressure at the downstream point = 0

and Specific weight of water $(w) = 10$ kN/m^3

We know that height of water, $(h) = 20 - 2 = 18$ m

and bottom width $(b) = 5 + 20 \times \dfrac{0.66}{1.0} = 18.2$ m

From the geometry of the uplift pressure, we know that pressure at the upstream (A),

$$p = wh = 10 \times 18 = 180 \text{ kN/m}^2$$

Let the resultant (R) cut the base at K as shown in Fig. 17.5. Let d be the horizontal distance AK.

We know that total water pressure per metre length of the dam,

$$P = \frac{wh^2}{2} = \frac{10 \times (18)^2}{2} = 1620 \text{ kN}$$

and net weight of the dam per metre length,

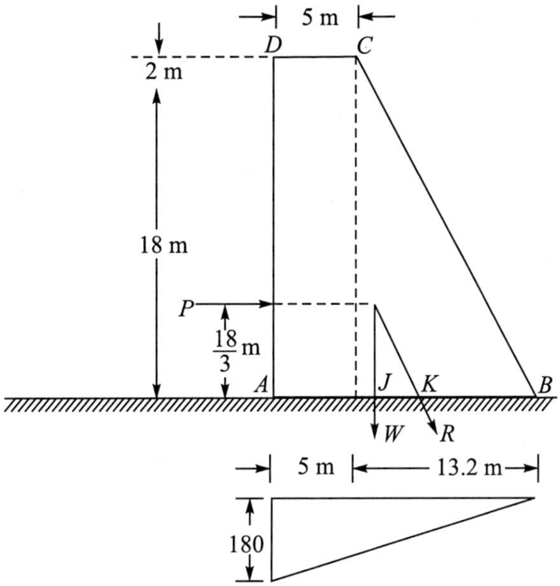

Fig. 17.5.

$$W = \text{Weight of the dam section} - \text{Uplift pressure}$$

$$= \left[2.4 \times 10 \times \left(\frac{10 \times 18.2}{2}\right) \times 20\right] - \left(\frac{180 \times 18.2}{2}\right) \text{ kN}$$

$$= 5568 - 1638 = 3930 \text{ kN}$$

Now let is find out the point K, where the resultant cuts the base. Taking moments of the dam section about A and equating the same,

$$W \times d = [(1620 \times 6)] + [(2.4 \times 10) \times (20 \times 5 \times 2.5)]$$

$$+ \left[(2.4 \times 10) \times \frac{20 \times 13.2}{2}\left(5 + \frac{13.2}{3}\right)\right] - \left[\frac{180 \times 18.2}{2} \times \frac{18.2}{3}\right]$$

$$3930\, d = 9720 + 6000 + 29780 - 9937 = 35563$$

$$d = \frac{35563}{3930} = 9.05 \text{ m}$$

and eccentricity, $$e = d - \frac{b}{2} = 9.05 - \frac{18.2}{2} = -\,0.05 \text{ m}$$

(Minus sign indicates that the stress at A will be more than that at B).

$\therefore$ Maximum stress at the base point (A),

$$\sigma_{max} = \frac{W}{b}\left(1 + \frac{6e}{b}\right) = \frac{3930}{18.2}\left(1 + \frac{6 \times 0.05}{18.2}\right) \text{ kN/m}^2$$

$$= 219.5 \text{ kN/m}^2 = \mathbf{219.5\ kPa} \quad \textbf{Ans.}$$

and minimum stress at the base point (B),

$$\sigma_{min} = \frac{W}{b}\left(1 - \frac{6e}{b}\right) = \frac{3930}{18.2}\left(1 - \frac{6 \times 0.05}{18.2}\right) \text{ kN/m}^2$$

$$= 212.4 \text{ kN/m}^2 = \mathbf{212.4\ kPa} \quad \textbf{Ans.}$$

17.4. TRAPEZOIDAL DAMS WITH WATER FACE INCLINED

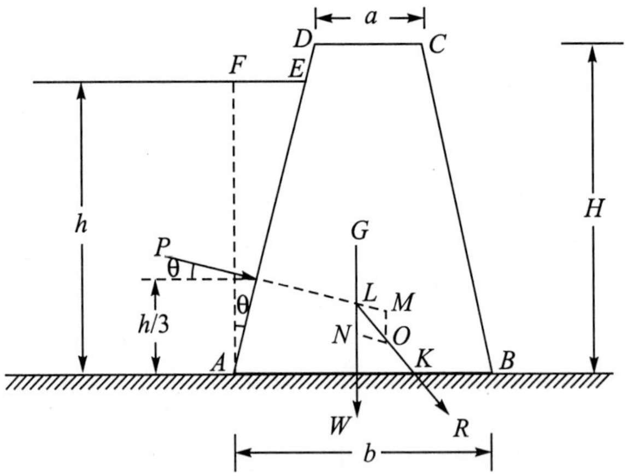

Fig. 17.6.

Consider a unit length of a trapezoidal dam, having its water surface inclined as shown in Fig. 17.6.

Let
a = Top width of the dam
b = Bottom width of the dam
H = Height of the dam
p = Specific weight of dam masonry
h = Height of water retained by dam
w = Specific weight of water
θ = Inclination of the water face with the vertical

∴ Length of the sloping side AE, which is subjected to water pressure,

$$l = \frac{h}{\cos\theta}$$

Now we see that the weight of the dam per unit length

$$W = p \times \frac{(a+b)}{2} \times H \quad ...(17.5)$$

The intensity of the water pressure will be zero at the water surface and will increase by a straight line law to weight at the bottom.

∴ The total pressure on a unit length of the dam

$$P = \frac{wh}{2} \times l = \frac{whl}{2} \quad ...(17.6)$$

This water pressure P will act at a height of $\frac{h}{3}$ from the bottom of the dam as shown in Fig. 17.6.

∴ Horizontal component of this water pressure,

$$P_H = P_{\cos}\theta = \frac{whl}{2} \times \frac{h}{l} = \frac{wh^2}{2} \quad ...(17.7)$$

and vertical component of this water pressure

$$P_V = P_{\sin}\,\theta = \frac{whl}{2} \times \frac{EF}{l} = \frac{w}{2} \times EF \times h$$

$$= \text{weight of the wedge } AEF \text{ of water}$$

It is thus obvious that such a dam may be taken to have a horizontal water pressure = $\frac{wh^2}{2}$ on the imaginary vertical face *AF*. The weight of wedge *AEF* of water may be considered as a part of the weight of dam, while finding out the C.G. of the dam section.

Now the distance between the C.G. of the dam section and the point, where the resultant *R* cuts the base will be given by the relation,

$$x = \frac{P}{W} \times \frac{h}{3}$$

∴ Total stress across the base at *B*,

$$\sigma_{max} = \frac{W}{b}\left(1 + \frac{6e}{b}\right)$$

and the total stress across the base at *A*,

$$\sigma_{min} = \frac{W}{b}\left(1 - \frac{6e}{b}\right)$$

When the reservoir is empty, there will be neither water pressure on the dam nor there will be the weight of wedge *AEF* of water. In this case the eccentricity of the weight of the dam,

$$e = \frac{b}{2} - AJ$$

Since the eccentricity will be minus, therefore total stress across the base at *B*,

$$\sigma_{min} = \frac{W}{b}\left(1 - \frac{6e}{b}\right)$$

and the total stress across the base at *A*,

$$\sigma_{max} = \frac{W}{b}\left(1 + \frac{6e}{b}\right)$$

17.5. SOLVED EXAMPLES

Example 17.4: *An earthen dam of trapezoidal section is 10 m high. It has top width of 1 m and bottom width 7 m. The face exposed to water has a slope of 1 horizontal to 10 vertical as shown in Fig. 17.7.*

Calculate the maximum and minimum stresses on the base, when the water level coincides with the top of the dam. Take weight of the masonry as 20 kN/m³ and that of water as 10 kN/m³.

Solution: Given: Height of the dam (H) = 10 m

Top width of the dam (a) = 1 m

Bottom width of the dam (b) = 7 m

Height of water retained by the dam (h) = 10 m

Weight of masonry (ρ) = 20 kN/m^3

and Weight if ater (w) = 10 kN/m^3

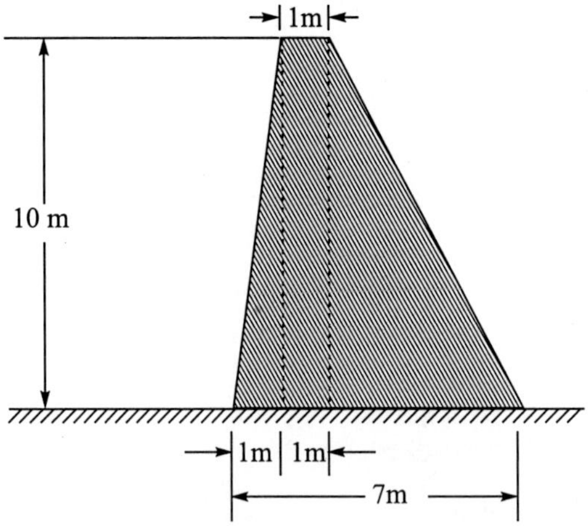

Fig. 17.7.

Let the resultant (R) cut the base at K as shown in Fig. 17.8. We know that total water pressure per metre length of the dam,

$$*P = \frac{wh^2}{2} = \frac{10 \times (10)^2}{2} = 500 \text{ kN}$$

and weight of the dam per metre length (including wedge AED of water)

$$W = \left(w \times \frac{h}{2} \right) + \left(\rho \times \frac{a+b}{2} \times H \right)$$

$$= \left(10 \times \frac{10}{2} \right) \times \left(20 \times \frac{1+7}{2} \times 10 \right) \text{ kN}$$

$$= 50 + 800 = 850 \text{ kN}$$

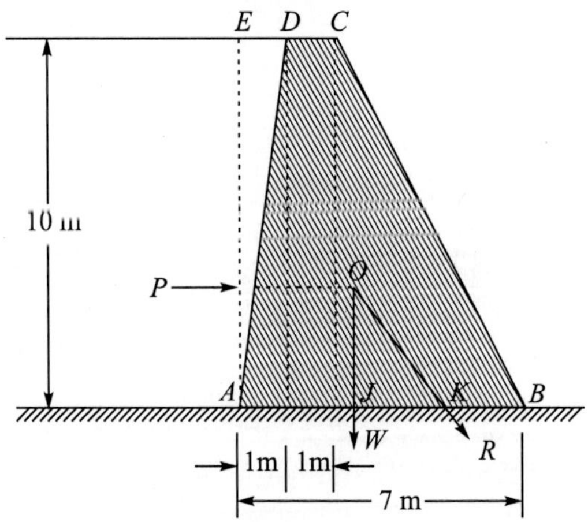

Fig. 17.8.

*Strictly speaking, the total pressure is acting normally to the face AD of the dam. But here we shall assume the pressure to act normally to the imaginary vertical plane AE.

Now let us find out the centre of gravity of the dam section (including wedge AED of water). Taking moments about A and equating the same,

$$W \times AJ = \left(10 \times \frac{10}{2} \times \frac{1}{3}\right) + \left(20 \times \frac{10}{2} \times \frac{2}{3}\right) + (20 \times 10 \times 1.5) + \left(20 \times \frac{10 \times 5}{2} \times \frac{11}{3}\right)$$

$$850\ AJ = \frac{50}{3} + \frac{200}{3} + 300 + \frac{5500}{3} = \frac{6650}{3}$$

$$\therefore \qquad AJ = \frac{6650}{3} \times \frac{1}{850} = 2.61 \text{ m}$$

We know that horizontal distance between the centre of gravity of the dam section and the points where the resultant cuts the base (*i.e.*, distance JK).

$$x = \frac{P}{W} \times \frac{h}{3} = \frac{500}{850} \times \frac{10}{3} = 1.96 \text{ m}$$

$\therefore$ Horizontal distance, $AK = d = AJ + x = 2.61 + 1.96 = 4.57$ m

and eccentricity, $\quad e = d - \dfrac{b}{2} = 4.57 - \dfrac{7}{2} = 1.07$ m

We also know that maximum stress at the base,

$$\sigma_{max} = \frac{W}{b}\left(1 + \frac{6e}{b}\right) = \frac{850}{7}\left(1 + \frac{6 \times 1.05}{7}\right) \text{ kN/m}^2$$

$$= 232.8 \text{ kN/m}^2 = \mathbf{232.8\ kPa} \quad \textbf{Ans.}$$

and minimum stress at the base point (B),

$$\sigma_{min} = \frac{W}{b}\left(1 - \frac{6e}{b}\right) = \frac{850}{7}\left(1 - \frac{6 \times 1.07}{7}\right) \text{ kN/m}^2$$

$$= 10.06 \text{ kN/m}^2 = \mathbf{10.06\ kPa} \quad \textbf{Ans.}$$

17.6. CONDITIONS FOR THE STABILITY OF A DAM

In the previous articles, we used to derive a realtion for the position of a point through which the resultant R (of the water pressure P and the weight of dam W) cuts the base. The position of this point helps us in finding out the total stresses across the base, at toe and heel of the dam. Apart from finding out the total stresses, this point helps us in checking the stability of the dam. In general, a dam is checked for the following conditions of stability:

1. To avoid tension in the masonry at the base of the dam,
2. To safeguard the dam from overturning,
3. To prevent the sliding of dam, and
4. To prevent the crushing of masonry at the base of the dam.

Now we shall discuss all the above conditions of stability one by one.

17.7. CONDITIONS OF AVOID TENSION IN THE MASONRY OF THE DAM AT ITS BASE

We know that the water pressure acting on one side of the dam, produces bending stress; whereas the weight of the dam produces direct stress at the bottom of the dam. We have also seen that on one side of the dam, there is a maximum stress (equal to sum of the direct and bending stress); whereas on other side of the dam, there is a minimum stress (equal to direct stress minus bending stress). A little consideration will show that so long as the bending stress remains less than the direct stress, the resultant stress is compressive. But when the bending stress is equal to the direct stress, there will be zero stress on one side. But when the bending stress exceeds the direct stress, there will be a tensile stress on one side. Though cement concrete can take up a small amount of tensile stress, yet it is desirable to avoid tension in the masonry of the dam at its base.

It is thus obvious, that in order to avoid the tension in the masonry of the dam at its base, the bending stress should be less than the direct stress or it may be equal to the direct stress, *i.e.*,

$$\sigma_b \leq \sigma_0$$

$$\frac{6\,W \cdot e}{b^2} \leq \frac{W}{b}$$

or $$e \leq \frac{b}{6}$$

It means that the eccentricity of the resultant can be equal to $\frac{b}{6}$ on either side of geometrical axis of base section. Thus the resultant must lie within the middle third of the base width in order to avoid tension.

17.8. CONDITION TO PREVENT OVERTURNING OF THE DAM

We have already discussed that when a dam is retainign water, it is subjected to some water pressure. We can easily find out the resultant R of the water presure P and the weight of dam W. Since the dam is in equilibrium, therefore the resultant R must be balanced by equal and opposite reaction actiong at K. This reaction may be split up into two components *viz.,* horizontal and vertical. The horizontal component must be equal to the water pressure P, whereas the vertical component must be equal to the weight W. Thus the following four forces acting on the dam, keep it in equilibrium:

1. Water pressure P
2. Horizontal component of the reaction
3. Weight of the dam W and
4. Vertical component of the reaction.

These four forces may be grouped into two sets or couples. The moment of a couple consisting of the first two forces.

$$M_1 = \text{Force} \times \text{Arm} = P \times \frac{h}{3} \quad \text{...(17.8)}$$

Similarly, moment of a couple consisting of the last two forces,

$$M_2 = W \times JK$$

A little consideration will show that the moment of the first two forces will tend to overturn the dam about B; whereas the moment of the last two forces will tend to restore the dam. Since the dam is in equilibrium and a couple can only be balanced by a couple, therefore overturning moment must be equal to the restoring moment, *i.e.*,

$$P \times \frac{h}{3} = W \times JK$$

or $$JK = \frac{P}{W} \times \frac{h}{3}$$

Since the dam will tend to overturn about B, therefore balancing moment about B,

$$M_3 = W \times JB$$

Now, we see that the dam is safe against overturning, so long as the balancing moment is more than the overturning moment (or restoring moment, which is equal to overturning moment), *i.e.*,

$$W \times JB > W \times JK$$

or $$JB > JK$$

It is thus obvious that the condition to prevent the dam from overturning, is that the point K should be between J and B or more precisely between A and B.

As a matter of fact, this is superfluous condition. We know that to avoid tension in the masonry of a dam at its bottom, the resultant must lie within the middle third of the base width. Since we have to check the stability of a dam for tension in the base masonry, therefore the stability of the dam for overturning is automatically checked.

17.9. CONDITION TO PREVENT THE SLIDING OF DAM

We have already discussed that there are four forces which act on a dam and keep it in equilibrium. Out of these four forces, two are vertical and the following two are horizontal:

(*a*) Water pressure P and

(*b*) Horizontal component of the reaction.

A little consideration will show that the horizontal component of the reaction will be given by the frictional force at the base of the dam.

Let μ = Coefficient of friction between the base of dam and the soil.

We know that the maximum available force of firction,

$$F_{max} = \mu W$$

It is thus obvious that so long as F_{max} is more* than the water pressure P, the dam is safe against sliding.

17.10. CONDITION TO PREVENT CRUSHING OF MASONRY AT THE BASE OF THE DAM

We know that whenever a dam is retaining water, the masonry of dam at its bottom is subjected to some stress. This stress varied from σ_{max} to σ_{min} by a straight line law. A little consideration will show that the condition to prevent the crushing of masonry at the base of the dam, is that the maximum stress σ_{max} should be less than the permissible stress in the masonry.

*Some authorities feel that the dam will be safe, when the force of friction is at least 1.5 times the total water pressure per metre length, *i.e.*, $\frac{\mu W}{P} = 1.5$.

Example 5: *A masonry wall 5 metres high and 1.8 metre wide is containing water up to a height of 4 metres. If the coefficient of friction between the wall and the soil is 0.6, check the stability of the wall. Take weight of the masonry and water as 22 kN/m³ and 9.81 kN/m³.*

Solution: Given: Height of the wall (H) = 5 m

Width of the wall (b) = 1.8 m

Height of water (h) = 4 m

Coefficient of friction between the wall and the soil (μ) = 0.6

and Weight of masonry (ρ) = 22 kN/m³

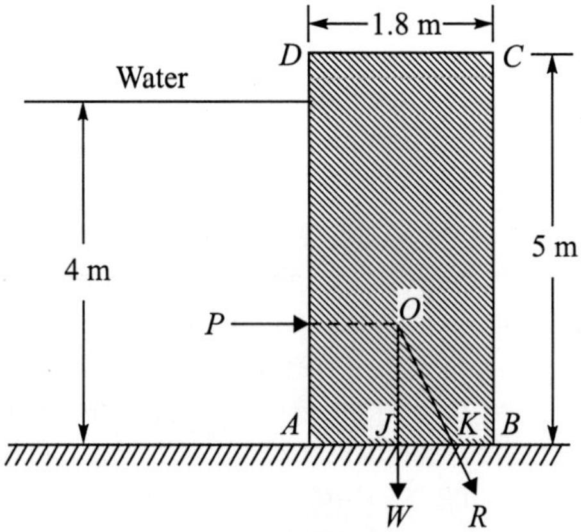

Fig. 17.9.

Let the resultant (R) cut the base at K as shown in Fig. 17.9. We know that total water pressure per metre length of the wall,

$$P = \frac{wh^2}{2}$$

$$= \frac{9.81 \times (4)^2}{2} = 78.48 \text{ kN}$$

and weight of the wall per metre length.

$$W = 22 \times 5 \times 1.8 = 198 \text{ kN}$$

1. Check for Tension in the Masonary at the Base: We know that horizontal distance between the centre of gravity of the wall and point, where the resultant thrust (R) cuts the base,

$$x = \frac{P}{W} \times \frac{h}{3} = \frac{78.48}{198} \times \frac{4}{3} = 0.53 \text{ m}$$

$$\therefore \qquad AK = AJ + x = 0.9 + 0.53 = 1.43 \text{ m}$$

Since the resultant thrust lies beyond the middle third of the base width (*i.e.*, from 0.6 to 1.2 m), therefore the wall shall fail due to tension in its base. **Ans.**

2. Check for Overturning: Since the resultant thrust is passing within the base as obtained above, therefore the wall is safe against overturning. **Ans.**

3. Check for Sliding the Wall: We know that horizontal pressure due to water, (P) = 7.48 kN.

And the frictional force = $\mu W = 0.6 \times 198 = 118.8$ kN

Since the frictional force (118.8 kN) is * more than the horizontal pressure (78.48 kN), therefore the wall is safe against sliding. **Ans.**

Example 17.6: *A trapezoidal masonry dam having 3 m top width, 8 m bottom width and 12 m high is rtaining water as shown in Fig. 17.10.*

Check the stability of the dam, when it is retaining water to a height of 11 m. The masonry weights 20 kN/m³ and coefficient of friction between the dam masonry and soil is 0.6. Take the allowable compressive stress as 400 kN/m².

*Certain authorities on the subejct of the opinion that magnitude of the weight should preferably be 1.5 times the horizontal pressure due to water.

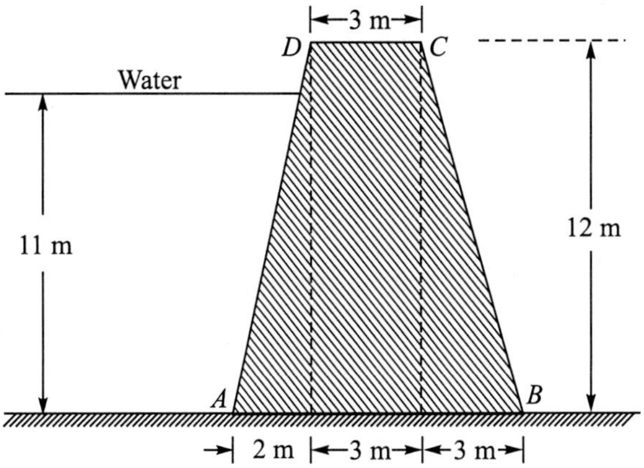

Fig. 17.10.

Solution: Given: Top width of the dam (a) = 3 m

Bottom width of the dam (b) = 8 m

Height of the dam (H) = 12 m

Height of water retained by the dam (h) = 11 m

Weight of masonry (ρ) = 20 kN/m^3

Coefficient of friction between the dam masonry and soil (μ) = 0.6

and Allowable compressive stress (σ_{max}) = 400 kN/m^3.

1. Check for Tension in the Masonry at its Base: Let the resultant (R) cut the base at K as shown in Fig. 17.11.

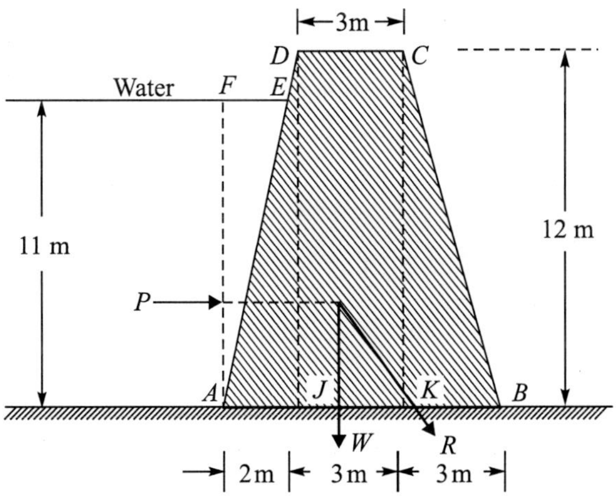

Fig. 17.11.

We know that water pressure per metre length of the dam,

$$P = \frac{wh^2}{2} = 10 \times \frac{(11)^2}{2} = 605 \text{ kN}$$

and weight of the dam per metre length (including the wedge AED of water),

$$W = \left(10\times\frac{1}{2}\times11\times\frac{11}{6}\right) + \left(20\times\frac{3+8}{2}\times12\right)\text{ kN}$$

$$= 100.8 + 1320 = 1420.8 \text{ kN}$$

Now let us find out the position of the centre of gravity of the dam section. Taking moments of the weight of the dam section about A and equating the same,

$$W\times AJ = \left(10\times\frac{11}{2}\times\frac{11}{6}\times\frac{2}{3}\right) + \left(20\times\frac{12+2}{2}\times\frac{4}{3}\right) + \left(20\times12\times3\times\frac{7}{2}\right) + \left(20\times\frac{12\times3}{2}\times6\right)$$

$$1420.8\times AJ = \frac{1210}{18} + 320 + 2520 + 2160 = 5067$$

$$\therefore \qquad AJ = \frac{5067}{1420.8} = 3.57 \text{ m}$$

We know that horizontal distance between the centre of gravity of the dam section and the point, where the resultant cuts the base (*i.e.*, distance JK),

$$x = \frac{P}{W}\times\frac{h}{3} = \frac{605}{1420.8}\times\frac{11}{3} = 1.56 \text{ m}$$

Horizontal distance AK, $d = AJ + x = 3.57 + 1.56 = 5.13$ m

Since the resultant force lies within the middle third of the base width (*i.e.*, from 2.67 m to 5.33 m), therefore the dam is safe against the tension in its masonry at the base. **Ans.**

2. Check for Overturning: Since the resultant force lies within the base AB and obtained above, therefore the dam is safe against overturning. **ans.**

3. Check for Sliding of the Dam: We know that the frictional force at the base

$$= \mu W = 0.6\times1420.8 = 852.5 \text{ kN}$$

Since the frictional force (852.5 kN) is more than the horizontal pressure (605 kN), therefore the dam is safe against sliding. **ans.**

4. Check for Crushing of the Masonry at the Base of the Dam: We know that eccentricity,

$$e = d - \frac{b}{2} = 5.13 - \frac{8}{2} = 1.13 \text{ m}$$

and maximum stress,
$$\sigma_{max} = \frac{W}{b}\left(1+\frac{6e}{b}\right) = \frac{1420.8}{8}\left(1+\frac{6\times1.13}{8}\right)\text{ kN/m}^2$$

$$= 328.1 \text{ kN/m}^2$$

Since the maximum stress (328.1 kN/m^2) is less than the allowable stress (400 kN/m^2), therefore masonry of the dam is safe against crushing. **Ans.**

17.11. MINIMUM BASE WIDTH OF A DAM

We have already discussed the general conditions for the stability of a dam, when the section is given. But sometimes, while designing a dam, we have to calculate its necessary base width. This can be easily found out by studying the conditions for the stability of a dam. Thus the base width (b) of a dam may be obtained from the following three conditions:

1. To avoid tension in the masonry at the base of the dam, the eccentricity $(e) = \frac{b}{6}$. In this case, the maximum stress $\sigma_{max} = \frac{2W}{b} = 5$ and the minimum stress $\sigma_{min} = 0$.
   The stress diagram at the base will be a triangle.
2. To avoid the sliding of dam, the force of friction between the dam and soil, is at least 1.5 times the total water pressure per metre length, *i.e.*,
   $$\frac{\mu W}{P} = 1.5$$
3. To prevent the crushing of masonry at the base of the dam, the maximum stress should be less than the permissible stress of the soil.

Note: If complete data of a dam is given, then the base width for all the above three conditions should be found out separately. The maximum value of the base width from the above three conditions will give the necessary base width of the dam. But sometimes, sufficient data is not given to find out the values of base width for all the above mentioned three conditions. In such a case, the value of minimum base width may be found out, for any one of the above three conditions.

Example 17.7: *A mass concrete dam shown in Fig. 17.12 below has a trapezoidal cross-section. The height above the foundation is 64 m and its water face is vertical. The width at the top is 4.5 m.*

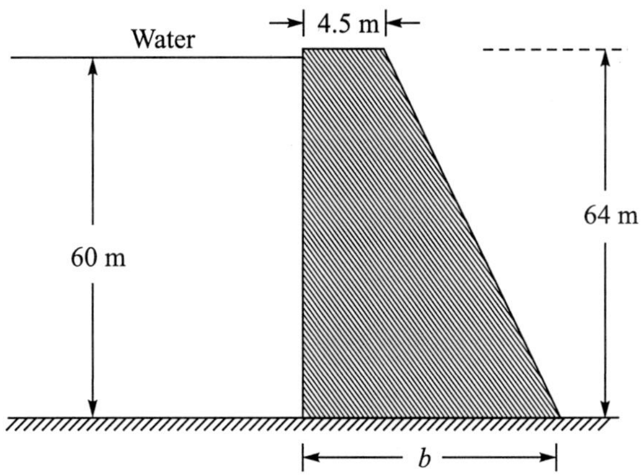

Fig. 17.12.

Calculate the necessary minimum width of the dam at its bottom, to ensure that no tension should be developed when water is stored up to 60 metres. Draw the pressure diagram at the base of the dam, for this condition and indicate the maximum pressure developed.

Take density of concrete as 22.6 kN/m³ and density of water as 9.81 kN/m³.

Solution: Given: Height of dam $(H) = 64$ m

Top width of dam $(a) = 4.5$ m

Height of water restored by the dam $(h) = 60$ m

Density of concrete $(\rho) = 22.6$ kN/m^3

and Density of water $(w) = 9.81$ kN/m^3

Minimum Width of the Dam at its Bottom:

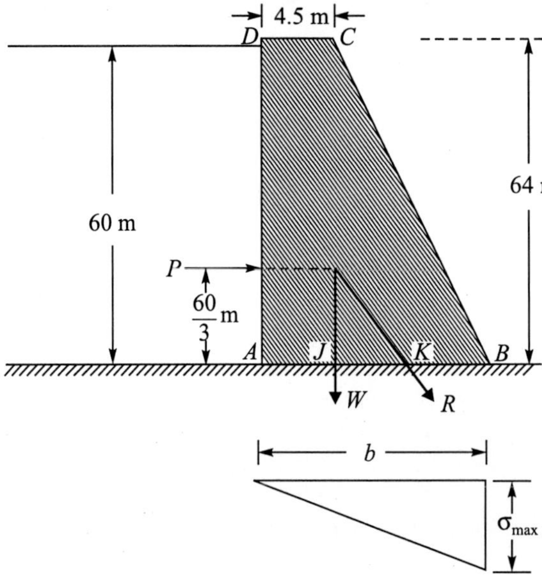

Fig. 17.13.

Let b = Minimum width of the dam at its bottom in metres

We see that the minimum width of the dam at its bottom is to be found out only for one condition *i.e.,* no tension shall be developed at the base.

Let the resultant (R) cut the base at K as shown in Fig. 17.13.

We know that water pressure per metre length of the dam,

$$P = \frac{wh^2}{2} = \frac{9.81 \times (60)^2}{2} = 17660 \text{ kN} \quad ...(i)$$

and weight of the dam per metre length,

$$W = \rho \times \frac{(a+b)}{2} \times H = 22.6 \times \frac{(4.5+b)}{2} \times 64 \text{ kN}$$

$$= 723.2\ (4.5\ b) \text{ kN} \quad ...(ii)$$

Now let us find out the position of the centre of gravity of the dam section. We know that the distance AJ,

$$AJ = \frac{a^2 + ab + b^2}{3(a+b)} = \frac{(4.5)^2 + 4.5b + b^2}{3(4.5+b)} = \frac{20.25 + 4.5b + b^2}{3(4.5+b)}$$

We know that horizontal distance between the centre of gravity of the dam section and the point, where the resultant cuts the base (*i.e.,* distance JK),

$$x = \frac{P}{W} \times \frac{h}{3} = \frac{17660}{723.2(4.5+b)} \times \frac{60}{3} = \frac{488}{(4.5+b)}$$

∴ Horizontal distance AK,

$$d = AJ + x = \frac{20.25 + 4.5b + b^2}{3(4.5+b)} + \frac{488}{4.5+b}$$

$$= \frac{20.25 + 4.5b + b^2 + 1464}{3(4.5+b)} = \frac{1484.25 + 4.5b + b^2}{3(4.5+b)}$$

∴ Eccentricity of the resultant,

$$e = d - \frac{b}{2} = \frac{1484.25 + 4.5b + b^2}{3(4.5+b)} - \frac{b}{2}$$

We know that in order to avoid tension in the masonry at the base of the dam, the eccentricity,

$$e = \frac{b}{6}$$

or $$\frac{1484.25 + 4.5b + b^2}{3(4.5+b)} - \frac{b}{2} = \frac{b}{6}$$

$$\frac{1484.25 + 4.5b + b^2}{3(4.5+b)} = \frac{b}{6} + \frac{b}{2} = \frac{2b}{3}$$

$$1484.25 + 4.5b + b^2 = 2b\,(4.5 + b) = 9b + 2b^2$$

or $$b^2 + 4.5b - 1484.25 = 0$$

Solving this equation as a quadratic equation for b, we get

$$b = \frac{-4.5 \pm \sqrt{(4.5)^2 + (4 \times 1484.25)}}{2} = \mathbf{36.35\ m} \quad \textbf{Ans.}$$

Pressure Diagram: Substituting the value of b in equation (*ii*)

$$W = 723.2\,(4.5 + 36.35) = 29.540 \text{ kN}$$

Since no tension should be developed at the base, therefore pressure diagram will be a triangle with zero pressure stress at B. Therefore pressure at the base

$$\sigma_{max} = \frac{2W}{b} = \frac{2 \times 29.540}{36.35} = \mathbf{1625\ kPa} \quad \textbf{Ans.}$$

$$= \mathbf{1625\ kPa} \quad \textbf{Ans.}$$

$$\sigma_{min} = 0$$

The pressure diagram at the base of the dam is shown in Fig. 17.13. **Ans.**

Example 17.8: *A concrete dam has its upstream face vertical and a top width of 3 m. Its downstream face has a uniform batter. It stores water to a depth of 15 m with a free board of 2 m as shown in Fig. 17.14.*

The weights of water and concrete may be taken as 10 kN/m^3 and 25 kN/m^3. Calculate:

(i) The minimum dam width at the bottom for no tension in concrete. Neglect uplift.

(ii) The extreme intensities of pressure on the foundation, when the reservoir is empty.

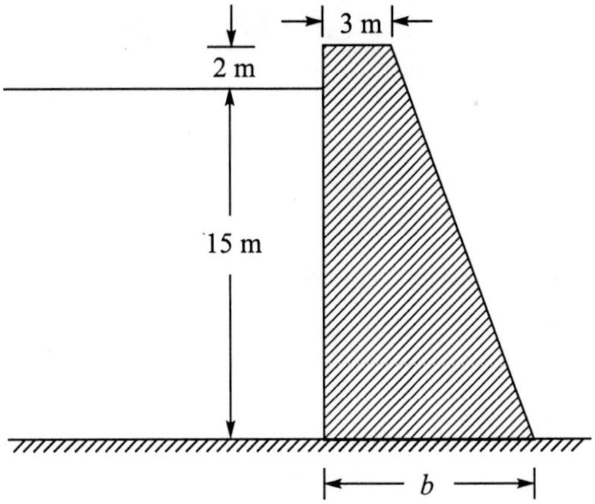

Fig. 17.14.

Solution: Given: Top width of the dam $(a) = 3$ m

Height opf water retained by the dam $(h) = 15$ m

Height of the dam $(H) = 15 + 2 = 17$ m

Weight of water $(w) = 10$ kN/m^3

and Weight of concrete $(\rho) = 25$ kN/m^3

Minimum Dam Width at the Bottom:

Let b = Minimum dam width at bottom

We see that the minimum dam width at the bottom is to be found out only for one condition *i.e.*, no tension should be developed at the base.

We also know that total pressure on the dam per metre length,

$$P = \frac{wh^2}{2} = \frac{10 \times (15)^2}{2}$$

$$= 1125 \text{ kN} \quad ...(i)$$

and weight of concrete of the dam per metre length

$$W = \rho \times \left(\frac{a+b}{2}\right) \times H$$

$$= 25 \times \left(\frac{3+b}{2}\right) \times 17$$

$$= 212.5\,(3 + b) \text{ kN}$$

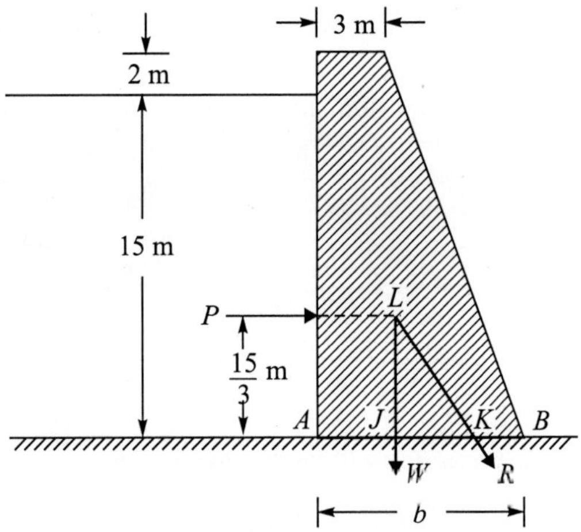

Fig. 17.15.

Now, let us find out the position of the centre of gravity of the dam section. We know that the distance AJ

$$= \frac{a^2 + ab + b^2}{3\,(a+b)} = \frac{(3)^2 + 3b + b^2}{3\,(3+b)} = \frac{9 + 3b + b^2}{3\,(3+b)}$$

We know that horizontal distance between the centre of gravity of the dam section and the point, where the resultant cuts the base (*i.e.*, distance JK),

$$x = \frac{P}{W} \times \frac{h}{3} = \frac{1125}{212.5\,(3+b)} \times \frac{15}{3} = \frac{26.5}{(3+b)}$$

∴ Horizontal distance *AK*,

$$d = AJ + x = \frac{9+3b+b^2}{3\,(3+b)} + \frac{26.5}{(3+b)}$$

$$= \frac{9+3b+b^2+79.5}{3\,(3+b)} = \frac{88.5+3b+b^2}{3\,(3+b)}$$

and eccentricity of the resultant,

$$e = d - \frac{b}{2} = \frac{88.5+3b+b^2}{3\,(3+b)} - \frac{b}{2}$$

We know that in order to avoid tension in the concrete at the dam base, the eccentricity,

$$e = \frac{b}{6}$$

or
$$\frac{88.5+3b+b^2}{3\,(3+b)} - \frac{b}{2} = \frac{b}{6}$$

∴
$$\frac{88.5+3b+b^2}{3\,(3+b)} = \frac{b}{6} + \frac{b}{2} = \frac{2b}{3}$$

$$88.5 + 3b + b^2 = 2b\,(3+b) = 6b + 2b^2$$

$$b^2 + 3b - 88.5 = 0$$

Solving the equation, as a quadratic equation for *b*, we get

$$b = \frac{-3 \pm \sqrt{(3)^2 + (4 \times 88.5)}}{2} = \frac{-3 \pm 19}{2}$$

$$= \mathbf{8\ m} \quad \textbf{Ans.}$$

Extreme Intensities of Pressure on the Foundation when the Reservoir is empty: We know that the weight of dam per metre length,

$$W = 25 \times \frac{(3+8)}{2} \times 17 \text{ kN}$$

$$= 2337.5 \text{ kN}$$

We also know that distance *AJ*,

$$d = \frac{a^2 + ab + b^2}{3\,(a+b)}$$

$$= \frac{(3)^2 + (3 \times 8) + (8)^2}{3\,(3+8)}$$

$$= 2.94 \text{ m}$$

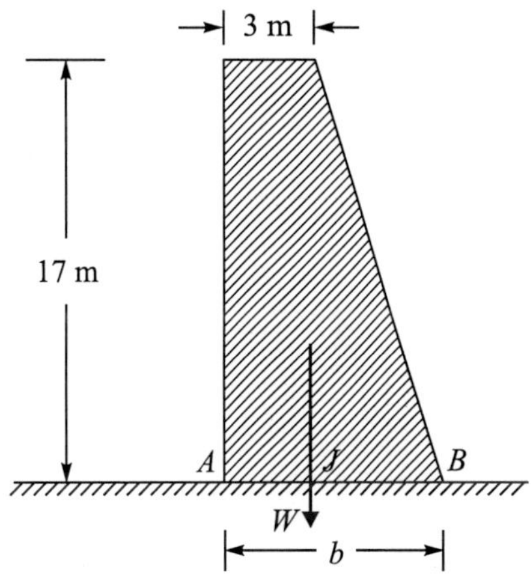

Fig. 17.16.

and eccentricity,
$$e = d - \frac{b}{2}$$

$$= 2.94 - \frac{8}{2} = -1.06 \text{ m}$$

(Minus sign indicates that the pressure at point will be more than that at point *B*).

We also know that maximum intensity of pressure at point *A*,

$$\sigma_{max} = \frac{W}{b}\left(1+\frac{6e}{b}\right) = \frac{2337.5}{8}\left(1+\frac{6\times1.06}{8}\right) \text{ kN/m}^2$$

$$= 524.5 \text{ kN/m}^2 = \textbf{524.5 kPa} \quad \textbf{Ans.}$$

and minimum intensity of pressure at point *B*,

$$\sigma_{min} = \frac{W}{b}\left(1-\frac{6e}{b}\right) = \frac{2337.5}{8}\left(1-\frac{6\times1.06}{8}\right) \text{ kN/m}^2$$

$$= 59.9 \text{ kN/m}^2 = \textbf{59.9 kPa} \quad \textbf{Ans.}$$

17.12. MAXIMUM HEIGHT OF A DAM

We have already discussed the various conditions for the minimum base width of a dam. The same conditions also hold good for the maximum height of a dam.

Example 17.9: *Assuming uniformly varying stress across the base, find the limit of height of a triangular masonry dam, with water upto the top of the vertical face, in order that the vertical compressive stress across the base shall not exceed 1 MPa, the masonry weights = 20 kN/m³.*

Solution: Given: Maixmum compressive stress (σ_{max}) = 1 MPa = 1000 kPa = 1000 kN/m²

and Weight of masonry (ρ) = 20 kN/m³

Let e = Eccentricity of the resultant,

H = Height of the dam in metres, and

b = Bottom width of the dam in metres

and weight of dam per metre length,

$$W = 20 \times \frac{bH}{2} = 10\, bH$$

We know that eccentricity of the resultant for maximum stress,

$$e = \frac{b}{6}$$

We also know that maximum stress across the base (σ_{max}),

$$1000 = \frac{W}{b}\left(1+\frac{6e}{b}\right) = \frac{10\, bH}{b}\left(1+\frac{6\times\frac{b}{6}}{b}\right)$$

$$= 10\, H\,(1 + 1) = 20\, H$$

$$H = \frac{1000}{20} = \textbf{50 m} \quad \textbf{Ans.}$$

SUMMARY

- A dam is constructed to store large quantity of water, which is used for the purpose of irrigation and power generation.
- There are many types of dams viz. Rectangular and trapezoidal, but trapezoidal dams are common. A dam constructed with earth is called an earthen dam, whereas a dam constructed with cement concrete is called a concrete dam or gravity dam.
- The water pressure on a dam per unit length,

$$P = \frac{wh^2}{2},$$

where w = specific weight of water, generally taken as 1000 kg/m^3

and h = height of water

- The horizontal distance between the C.G. of the dam section and the point, where the resultant cut the base,

$$x = \frac{P}{W} \times \frac{h}{3}$$

where P = Water pressure on the dam per unit length

W = Weight of dam per unit length

h = Height of water

- The maximum and minimum intensities of stress at the base of dam,

$$\sigma_{max} = \frac{W}{b}\left(1 + \frac{6e}{b}\right)$$

and

$$\sigma_{min} = \frac{W}{b}\left(1 - \frac{6e}{b}\right),$$ where

W = Weight of the dam per unit length

b = Bottom width of dam

e = Electricity of dam

- The minimum base width of a dam is found out by studying the conditions for the stability of a dam.

SHORT ANSWER QUESTIONS

1. What do you understand by the term dam? Name the various types of dams commonly used these days.
2. Name the various conditions for the stability of a dam. Describe any two of them.
3. How will you find out:
   (*i*) Minimum base width,
   (*ii*) Maximum height of a dam
4. Derive an equation for the maximum and minimum intensities of stress at the base of a trapezoidal dam.

❑❑❑

CHAPTER

18 Two Hinged Arch

These are structural members, curved in elevation used to support heavy loads on large spans. Generally an arch receives its load mainly in compression. A two hinged arch is commonly made from metal or timber. It has four unknown reaction components, two at each support. Three equations of static equilibrium viz., $\Sigma H = 0$, $\Sigma V = 0$ & $\Sigma M = 0$ are thus insufficient, hence forth equations is obtained from deflection conditions (Strain Energy Criterion).

Fig. 18.1

18.1. CALCULATION OF VERTICAL REACTIONS AT HINGED SUPPORTS

In case the supports of the two-hinged arch are at the same level, the vertical reactions are statically determinate and can be found by taking moments about one support.

Thus, for a two-hinged arch as shown below, the vertical reactions at "A" and "B" will be $\frac{P(l - a)}{l}$ and $\frac{Pa}{l}$ respectively.

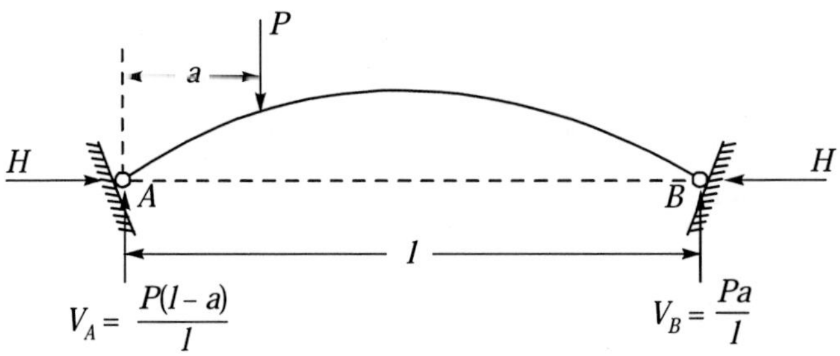

Fig. 18.2

18.2. TO FIND THE HORIZONTAL THRUST "*H*"

The horizontal thrust at each support, may be determined from the condition that the horizontal displacement of either hinge with respect to the other is zero.

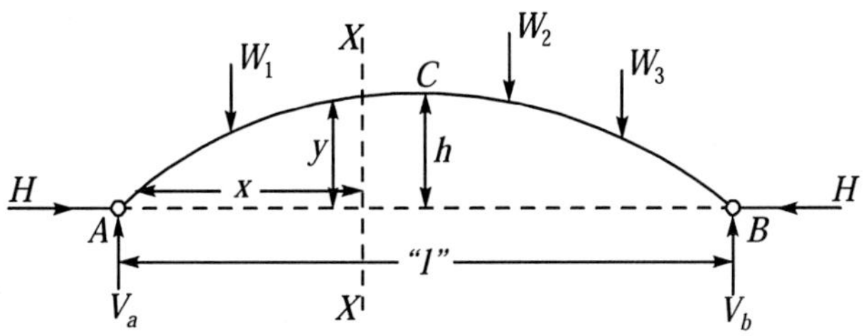

Fig. 18.3

Let "*M*" be the beam moment at any section *XX* (*i.e.*, *B.M.* at any section ignoring the horizontal thrust at the supports). Let the co-ordinates of the "*X*" be (x, y) with end "*A*" as origin.

The actual *B.M.* at the section is given by,

$$M_X = (M - Hy)$$

∴ Total strain energy stored by the whole arch

$$w_i = \int M^2 \times \frac{ds}{2\,EI} = \int (M - Hy)^2 \times \frac{ds}{2\,EI}$$

By the first theorem of castigliano, the horizontal movement of either end relative to the other is given by $\frac{\partial\, w_i}{\partial\, H}$.

Since such a relative horizontal displacement of one end with respect to the other end is not possible in the two hinged arch, we have,

$$\frac{\partial\, w_{i,}}{\partial\, H} = 0$$

$$\therefore \quad \frac{\partial\, w_i}{\partial\, H} = \int 2\,(M - Hy)\,(-y)\,\frac{ds}{2\,EI} = 0$$

$$= \int \frac{My\,ds}{EI} - H \int \frac{y^2\,ds}{2\,EI} = 0$$

$$\therefore \quad H = \frac{\int \frac{My\,ds}{EI}}{\int \frac{y^2\,ds}{2\,EI}},$$

$$\therefore \quad \boxed{\text{“}H\text{”} = \frac{\int My \cdot ds}{\int y^2 \cdot ds}} \quad \text{when } EI \text{ is constant}$$

18.3. CALCULATION OF SHEAR FORCE AND NORMAL THRUST

Consider a portion of two-hinged arch.

Let 'θ' be the slope at the section where *B.M.* and *S.F.* are required.

Let 'V' be the vertical force and H the horizontal thrust resisted at the section as shown in the section.

18.4. SHEAR FORCE AT THE SECTION

S.F., $$\text{“}F\text{”} = V\cos\theta - H\sin\theta$$

Normal thrust at the section

$$\text{“}N\text{”} = V\sin\theta + H\cos\theta$$

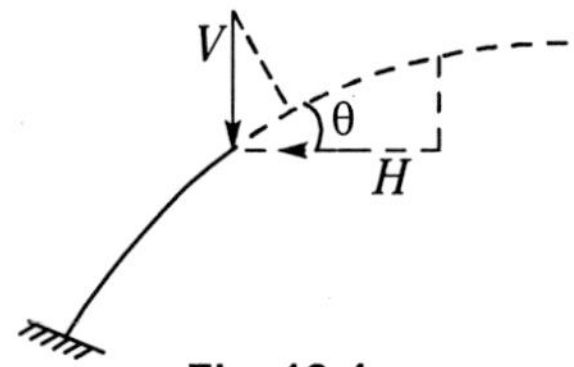

Fig. 18.4

Case of Parabolic Arch

The beam moment at any section "X" (between A and C) distant "x" from "A"

i.e., $$M = \frac{W}{2} \times x$$

Also, $$y = \frac{4h}{l^2}\, x\,(l - x) \qquad \text{(for a parabola)}$$

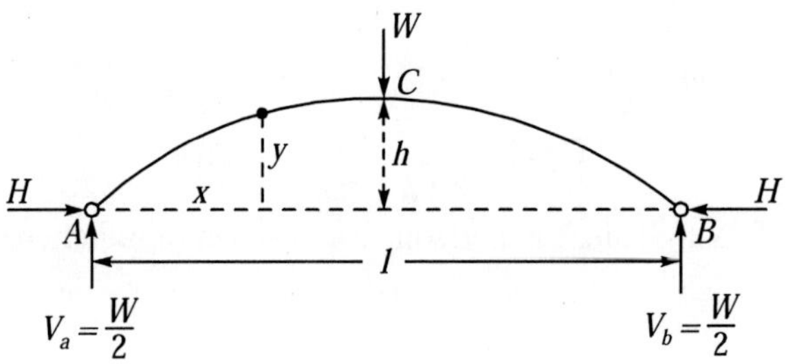

Fig. 18.5

$$H = \frac{\int My \cdot dx}{\int y^2 \cdot dx}$$

$$H = \frac{\int_0^{l/2} \frac{W}{2} x \times \frac{4h}{l^2} x\,(l - x)\, dx}{\int_0^{l/2} \frac{16h^2}{l^4} x^2\,(l - x)^2\, dx}$$

$$\text{“}H\text{”} = \frac{\frac{2Wh}{l^2}\int_0^{l/2} x^2\,(l - x)\, dx}{\frac{16h^2}{l^4}\int_0^{l/2} x^2\,(l - x)^2\, dx} \qquad \ldots(18.1)$$

Numerator of eq. (18.1),

$$= \frac{2Wh}{l^2}\int_0^{l/2} x^2\,(l - x)\, dx$$

$$= \frac{2Wh}{l^2}\left[l \cdot \frac{1}{3} \cdot \frac{l^3}{8} - \frac{1}{4} \cdot \frac{l^4}{16} \right]$$

$$= \frac{2\,Wh}{l^2} \times l^4 \left[\frac{1}{24} - \frac{1}{64}\right] = \frac{2\,Whl^2}{192}\ (8-3)$$

$$= \frac{5}{96}\ whl^2$$

Denominator of eq. (1),

$$= \frac{16\,h^2}{l^4} \int_0^{l/2} x^2\,(l-x)^2\,dx$$

$$= \frac{16\,h^2}{l^4} \int_0^{l/2} [l^2\,x^2 - 2lx^3 + x^4]\,dx$$

$$= \frac{16\,h^2}{l^4} \left[l^2 \cdot \frac{1}{3} \cdot \frac{l^3}{8} - 2l \cdot \frac{1}{4} \cdot \frac{l^4}{16} + \frac{1}{5} \cdot \frac{l^5}{32} \right]$$

$$= \frac{16\,h^2}{l^4} \cdot \frac{l^5}{480}\ [20 - 15 + 3] = \frac{4}{15}\ h^2\,l$$

$\therefore$ $$\text{“}H\text{”} = \frac{\text{Numerator}}{\text{Dinominator}} = \frac{5}{96}\ Whl^2 \times \frac{5}{4\,h^2 l}$$

$\therefore$ $$\boxed{\text{“}H\text{”} = \frac{25}{128}\ \frac{Wl}{h}}$$

18.5. PARABOLIC ARCH WITH UDL ON ENTIRE SPAN

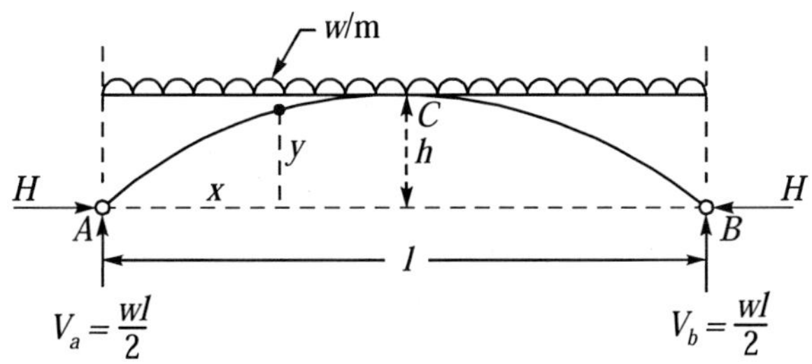

Fig. 18.6

The *B.M.* at any section “*X*” ignoring the horizontal thrust *i.e.,* the beam moment at any section “*X*”

or $$\text{“}M\text{”} = \frac{wl}{2} \times x - \frac{wx^2}{2} = \frac{w}{2} x\,(l-x)$$

$$y = \frac{4\,h}{l^2}\ x\,(l-x)$$

$\therefore$ Horizontal thrust

$$H = \frac{\int My\,ds}{\int y^2\,ds} = \frac{\int_{x=0}^{x=l} \frac{wx}{2}\,(l-x) \cdot \frac{4\,h}{l^2} \times x\,(l-x)\,ds}{\int_{x=0}^{x=l} \frac{16\,h^2}{l^4}\,x^2\,(l-x)^2\,ds}$$

$$= \frac{wl^2 \int_{x=0}^{x=l} x^2 (l - x^2)\, ds}{8h \int_{x=0}^{x=l} x^2 (l - x^2)\, ds}$$

$$\boxed{\text{"}H\text{"} = \frac{wl^2}{8h}}$$

[Since the expressions for integration are the same in numerator and denominator for with same limits]

18.6. PARABOLIC ARCH WITH CONCENTRATED LOAD (NOT AT CENTRE)

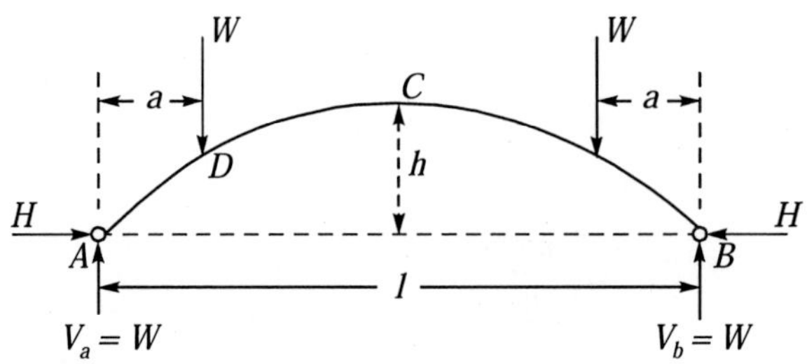

Fig. 18.7

The beam moment at any section "X" distant x from either support ($x > a$ and $< l/2$),

i.e., $$M = Wx - W(x - a) = wa \quad \text{(Constant)}$$

and $$y = \frac{4h}{l^2} x(l - x)$$

∴ Horizontal thrust

$$2H = \frac{\int My\, dx}{\int y^2\, dx}$$

$$2H = 2\frac{\left[\int_0^a Wx \times \frac{4h}{l^2} x(l-x)\, dx + \int_0^{l/2} Wa \cdot \frac{4h}{l^2} x(l-x)\, dx\right]}{\int_0^{l/2} \frac{16h^2}{l^4} x^2 (l-x)^2\, dx}$$

$$2H = \frac{\frac{4Wh}{l^2}\int_0^a x^2 (l-x)\, dx + 4\frac{Wha}{l^2}\int_a^{l/2} x(l-x)\, dx}{\frac{16h^2}{l^4}\int_0^{l/2} x^2 (l-x)^2 \cdot dx} \quad \ldots(18.2)$$

Numerator of eq. (18.2),

$$= \frac{4Wh}{l^2}\left[\frac{la^3}{3} - \frac{a^4}{4}\right] + \frac{4Wha}{l^2}\left[\frac{l}{2}\left(\frac{l^2}{4} - a^2\right) - \frac{1}{3}\left(\frac{l^3}{8} - 3\right)\right]$$

$$= \frac{4\,Wha}{l^2}\left[\frac{la^3}{3} - \frac{a^3}{4} + \frac{l^3}{8} - \frac{la^2}{2} - \frac{l^3}{24} + \frac{a^3}{3}\right]$$

$$= \frac{4\,Wha}{l^2} \cdot \frac{1}{24}\,[8la^2 - 6a^3 + 3l^3 - 12la^2 - l^3 + 8a^3]$$

$$= \frac{Wha}{6\,l^2}\,[2l^3 - 4la^2 + 2a^3] = \frac{Wha}{6\,l^2} \times 2\,(l^3 + 2la^3 + a^3)$$

$$= \frac{Wha}{3\,l^2}\,[l^3 - 2la^2 + a^3] = \frac{Wha}{3\,l^2}\,[l^3 - la^2 + la^2 + a^3]$$

$$= \frac{Wha}{3\,l^2}\,[l\,(l^2 - a^2) - a^2\,(l + a)] = \frac{Wha}{3\,l^2}\,(l - a)\,(l^2 + la - a^2)$$

11[ly] Denominator of eq. (18.1),

$$= \frac{16\,h^2}{l^4}\int_0^{l/2}(l^2\,x^2 - 2lx^3 + x^4)\,dx$$

$$= \frac{16\,h^2}{l^4}\left[l^2 \cdot \frac{1}{3} \cdot \frac{l^3}{8} - 2l \cdot \frac{1}{4} \cdot \frac{l^4}{16} + \frac{1}{5} \cdot \frac{l^5}{32}\right] = \frac{4}{15}\,h^2 l$$

$$\therefore \qquad 2\text{ “}H\text{”} = \frac{N}{D} = \frac{Wha}{3\,l^2}\,(l - a)\,(l^2 + la - a^2) \times \frac{15}{4\,h^2 l}$$

$$\boxed{\text{“}2H\text{”} = \frac{5}{4} \times \frac{W}{hl^3} \times a\,(l - a)\,(l^2 + la - a^2)}$$

Horizontal thrust when one of the two point loads is acting,

$$\boxed{\text{“}H\text{”} = \frac{5}{8} \times \frac{W}{hl^3} \times a\,(l - a)\,(l^2 + la - a^2)}$$

Example 18.1:

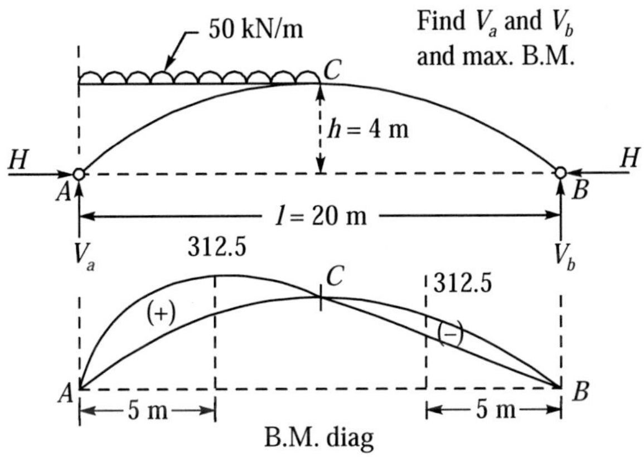

Fig. 18.8

Solution: M^A, $V_b \times 20 = 50 \times 10 \times 5$

$$V_b = 125 \text{ kN}$$

$$\therefore \quad V_a = (50 \times 10) - 125 = 375 \text{ kN}$$

Since one half of the span is loaded,

$$H = \frac{1}{2} \times \frac{wl^2}{8h} = \frac{wl^2}{16h} = \frac{50 \times 20^2}{16 \times 4} = 312.5 \text{ kN}$$

max (+) *B.M.* this will occurs in section *AC* of arch. At any section "*X*" distant *x* from *A*,

$$M_X = \text{Beam moment} - H\text{-moment}$$

$$= \text{Beam moment} - H \times \frac{4h}{l^2} x(l - x)$$

$$\therefore \quad M_X = 375x - 25x^2 - 312.5 \times \frac{4 \times 4}{20 \times 20} \times x(20 - x)$$

$$\therefore \quad M_X = 125x - 12.5x^2$$

For max. *B.M.*, $\frac{dM_X}{dx} = 0$

$$\therefore \quad \frac{dM_x}{dx} = 125 - 25x = 0.$$

$$\therefore \quad x = 5 \text{ m.}$$

$$\therefore \quad \text{Max} = 125 \times 5 - 12.5 \times 5^2 = 625 - 312.5$$

$$= 312.5 \text{ kN-m}$$

Example 18.2:

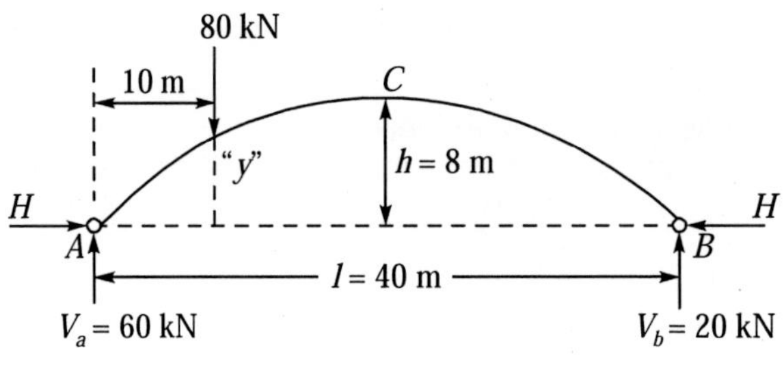

Fig. 18.9

Solution: $V_a = \frac{80 \times 30}{40} = 60 \text{ kN}, \; V_b = 20 \text{ kN}$

$$H = \frac{5}{8} \frac{w}{hl^3} a(l - a)(l^2 + la - a^2)$$

$$= \frac{5}{8} \times \frac{80}{8 \times 40^3} \times 10(40 - 10)(40^2 + 40 \times 10 - 10^2)$$

$$= 55.664 \text{ kN}$$

Max. (+) *B.M.* this will occur under the load, ordinate of the arch under the load

$$y = \frac{4 \times 8}{40 \times 40} \times 10 \times 30 = 6 \text{ m.}$$

$\therefore$ max. (+) $B.M. = 60 \times 10 - 55.664 \times 6$
$= (+)\ 266.016$ kN-m

max. (–) *B.M.*
In section *DB*, distant *x* from *B*

$$M_X = V_b x - Hy = 20x - 55.664 \times \frac{4 \times 8}{40^2} x\,(40 - x)$$

$$M_X = 1.11328x^2 - 24.5312x$$

Fa max *B.M.*

$$\frac{dm}{dx} = 0 = 1.11328 \times 2x - 24.5312 = 0.$$

$$x = 11.0175 \text{ m.}$$

$$\therefore \quad M_{max} = 1.11328\,(11.0175)^2 - 24.5312\,(11.0175)$$
$$= (-)\ 135.1356 \text{ kN-m}$$

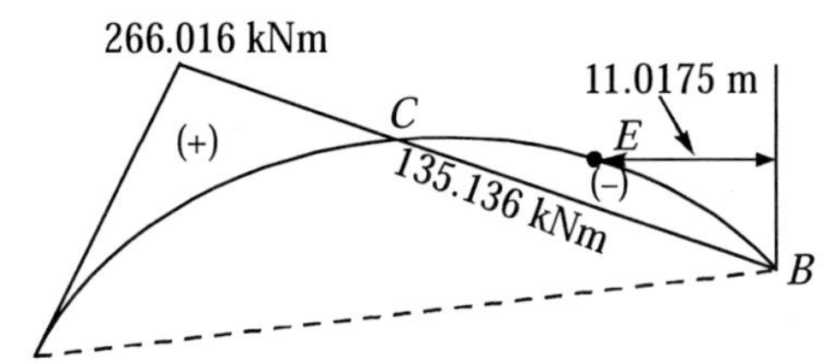

Fig. 18.10

Example 18.3:

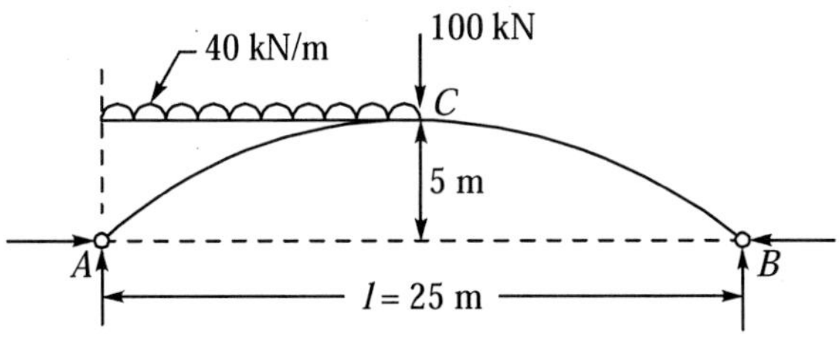

Fig. 18.11

Solution: $$H_1 = \frac{Wl^2}{16\,h} = \frac{40 \times 25^2}{16 \times 5} = 312.5 \text{ kN}$$ (due to left half of span)

$$H_2 = \frac{25}{128}\,\frac{Wl}{h}$$

$$= \frac{25}{128} \times \frac{100 \times 25}{5} = 97.66 \text{ kN}$$ (due to point load at the crown)

$\therefore$ Total "H" $= H_1 + H_2 = 410.16$ kN

"Semi circular-Arch":

$$H = \frac{\int My\, ds}{\int y^2\, ds}$$

$$H = \frac{2\int_0^{\pi/2} \frac{W}{2} R(1-\cos\theta)\, R\sin\theta \cdot Rd\theta}{\int_0^{\pi/2} R^2 \sin^2\theta \cdot Rd\theta}$$

$$= \frac{\frac{W}{2}\int_0^{\pi/2} (1-\cos\theta)\sin\theta\, d\theta}{\int_0^{\pi/2} \sin^2\theta \cdot d\theta}$$

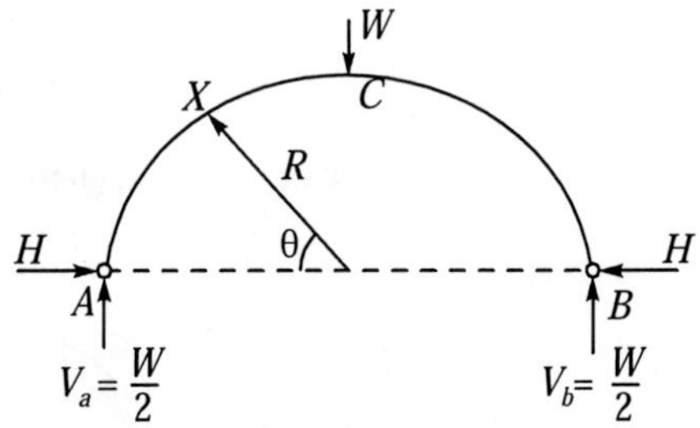

Fig. 18.12

If $(1 - \cos\theta) = u$

$\sin\theta \cdot d\theta = du$

$$\therefore \quad \int (1-\cos\theta)\sin\theta \cdot d\theta = \int u \cdot du = \frac{u^2}{2}$$

$$= \frac{(1-\cos\theta)^2}{2}$$

and $\therefore \int_0^{\pi/2} (1-\cos\theta)\sin\theta \cdot d\theta = \frac{1}{2}$ and, $\int_0^{\pi/2} \sin^2\theta \cdot d\theta = \frac{\pi}{4}$

$$\therefore \quad \text{"}H\text{"} = \frac{W}{2} \times \frac{\frac{1}{2}}{\pi/4} = \frac{W}{\pi}$$

It is important to note that the "*H*" is independent of the magnitude of the radius of the arch.

To find the vertical deflection of the crown:

The actual *B.M.* at any section "*X*" is given by,

$$M_x = \frac{W}{\pi} R(1 - \cos\theta)\, HR\sin\theta$$

$$\therefore \quad M_X = \frac{W}{\pi} R(1-\cos\theta) - \frac{W}{\pi} R\sin\theta = \frac{WR}{2\pi} \quad (\pi - \pi\cos\theta - 2\sin\theta)$$

$\therefore$ Strain energy stored by the arch

$$wi = \int \frac{MX^2 ds}{2\,EI} = \frac{W^2R^2}{4\pi^2}\, 2\int_0^{\pi/2} (\pi - \pi\cos\theta - 2\sin\theta^2) \cdot \frac{Rd\theta}{2\,EI}$$

$$= \frac{W^2R^3}{4\pi^2\,EI}\int_0^{\pi/2} (\pi^2 + \pi^2\cos^2\theta + 4\sin^2\theta - 2\pi^2\cos\theta + 4\pi\sin\theta\cdot\cos\theta - 4\pi\sin\theta\, d\theta)$$

$$= \frac{W^2R^3}{4\pi^2 EI}\left[\pi^2\cdot\frac{\pi}{2}+\pi^2\cdot\frac{\pi}{4}+4\frac{\pi}{4}-2\pi^2+4\pi\frac{1}{2}-4\pi\right]$$

$$= \frac{W^2R^3}{4\pi^2 EI}\left[\frac{3}{4}\pi^3-2\pi^2-\pi\right] = \frac{w^2R^3}{4\pi^2 EI}\times\frac{\pi}{4}\,[3\pi^2-8\pi-4]$$

or $$w_i = \frac{W^2R^3}{16\pi EI}\,(3\pi^2-8\pi-4)$$

By the first theorem of castigliaono, vertical deflection of the crown of the arch = $\delta = \dfrac{\partial\, wi}{\partial\, w}$

$$\text{“}\delta\text{”} = \frac{WR^3}{8\pi EI}\,(3\pi^2-8\pi-4)$$

18.7. CASE OF SEMICIRCULAR ARCH CARRING UDL OVER WHOLE SPAN

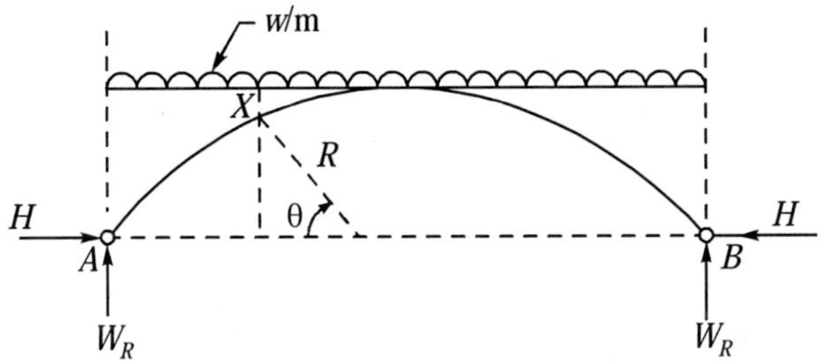

Fig. 18.13

The beam moment at any section (*i.e.*, the *B.M.* ignoring the horizontal thrust)

$$M = wR\cdot R\,(1-\cos\theta) - wR\,(1-\cos\theta)\cdot\frac{R\,(1-\cos\theta)}{2}$$

$$= \frac{wR^2}{2}\,(1-\cos\theta)\,(2-1+\cos\theta)$$

$$= \frac{wR^2}{2}\,(1-\cos^2\theta) = \frac{wR^2}{2}\sin^2\theta$$

The horizontal thrust at the support

$$\text{“}H\text{”} = \frac{\int My\,ds}{\int y^2\,ds} = \frac{2\int_0^{\pi/2}\frac{wR^2}{2}\sin^2\theta\cdot R\sin\theta\cdot Rd\theta}{2\int_0^{\pi/2}R^2\sin^2\theta\cdot Rd\theta}$$

$$= \frac{\frac{wR}{2}\int_0^{\pi/2}\sin^3\theta\cdot d\theta}{\int_0^{\pi/2}\sin^2\theta\cdot d\theta}$$

$$\therefore \quad \boxed{H = \frac{4}{3} \cdot \frac{wR}{\pi}}$$

Example 18.4: Find the Horizontal thrust '*H*' at each support.

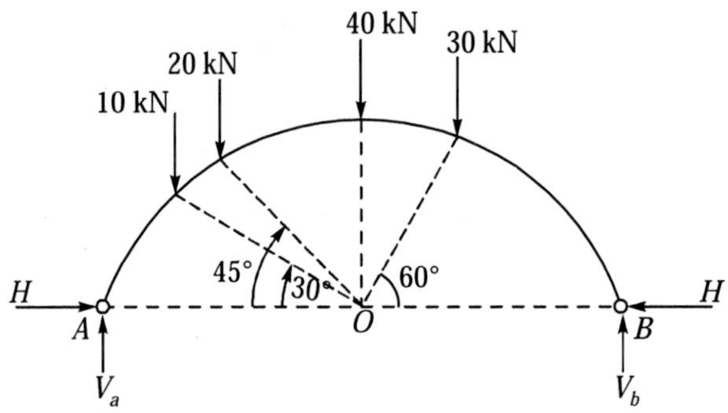

Fig. 18.14

Solution: The horizontal thrust at each support due to a load *W* at a section *X* can angle α is given by,

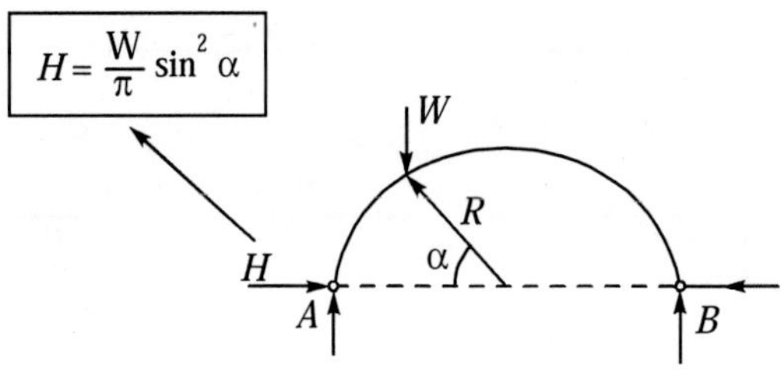

Fig. 18.15

Hence, horizontal thrust at each support

$$H = \sum \frac{W}{\pi}\sin^2\alpha$$

$$= \frac{10}{\pi}\sin^2 30 + \frac{20}{\pi}\sin^2 45 + \frac{40}{\pi}\sin^2 90 + \frac{30}{\pi}\sin^2 60$$

$$= \frac{10}{\pi}\times\frac{1}{4} + \frac{20}{\pi}\times\frac{1}{2} + \frac{40}{\pi}\times 1 + \frac{30}{\pi}\frac{3}{4}$$

$$= \text{“}H\text{”} = \frac{75}{\pi} = 23.873 \text{ kN}$$

18.8. TEMPERATURE EFFECT ON TWO HINGED ARCHES

Figure 18.16(*i*) shows a two hinged arch subjected to a rise in temp. "*T*". Obviously the length of the member tends to increase. If one of the ends say the end *B* had been placed on rollers, there would be a horizontal displacement BB_1 equal to α *T l* Fig. 18.16 (*ii*).

Since such a horizontal displacement cannot take place when the ends are hinged, inwards horizontal thrust *H* will be developed at each support. The hinged end *B* with the horizontal thrust *H* may be looked upon as a roller end *B* with an externally applied horizontal force *H* Fig. 18.16 (*iii*).

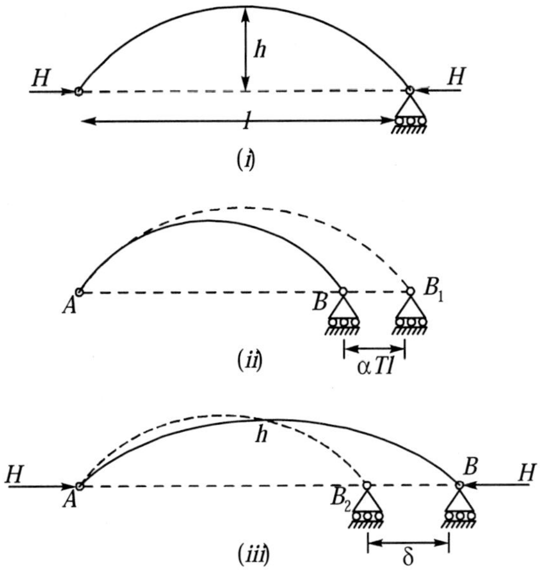

Fig. 18.16

Let 'δ' be the inward horizontal movement due to external force "H".

If "δ" = $\alpha\, T\, l$, then the net displacement of B = 0. Hence the value of H for which $\delta = \alpha\, T\, l$, represents the horizontal thrust developed in the two hinged arch due to rise in temp.

Now, considering the end B as a roller end with an external horizontal force H applied at B the *B.M.* at any section is given by

$$M_X = -Hy$$

$$\text{Strain energy stored by the arch} = wi = \int \frac{MX^2\, ds}{2\, EI} = \int \frac{H^2\, y^2\, ds}{2\, EI}$$

$\therefore$ By the first theorem of castiglianon, Inward horizontal movement of $B = \delta$

$$\therefore \qquad \delta = \frac{\partial\, wi}{\partial\, H} = \int 2H \cdot \frac{y^2\, ds}{2\, EI}$$

$$\delta = H \int \frac{y^2\, ds}{EI}$$

The condition that H may represent the horizontal thrust for the two hinged arch subjected to the rise of temp. is

$$\text{"}\delta\text{"} = \alpha\, T\, l$$

$$\therefore \qquad H \int \frac{y^2\, ds}{EI} = \alpha\, T\, l, \quad \therefore \quad H = \frac{\alpha\, T\, l}{\int \frac{y^2\, ds}{EI}}$$

If the arch section is of uniform flexural rigidity,

$$\boxed{H = \frac{EI\, \alpha\, T\, l}{\int y^2 \cdot ds}}$$

For a semicircular two hinged arch

We know, $$\int y^2 \cdot ds = 2\int_0^{\pi/2} (R\sin\theta^2)\, Rd\theta = 2R^3 \cdot \frac{\pi}{4} = \frac{\pi R^3}{2}$$

Also $$l = 2R$$

$$\therefore \quad H = \frac{EI\,\alpha\,T\,(2R)}{\left(\dfrac{\pi R^3}{2}\right)}$$

$$\therefore \quad \boxed{\text{“}H\text{”} = \frac{4\,EI\,\alpha\,T}{\pi R^2}}$$

For a parabolic two hinged arch,

$$H = \frac{EI_0\,\alpha\,T\,l}{\int y^2\,dx}$$

We knows $$\int y^2\,dx = 2\int_0^{l/2} \left\{\frac{4h}{l^2}\,x\,(l-x)\right\}^2 dx = \frac{8}{15}\,h^2 l$$

$$\therefore \quad \text{“}H\text{”} = \frac{EI_0\,\alpha\,T\,l}{\dfrac{8}{15}\,h^2 l} = \frac{15}{8}\,\frac{EI_0\,\alpha\,T}{h^2}$$

∴ Max *B.M.* due to rise in temperature

$$= M = Hh = \left[\frac{EI\,\alpha\,T\,l}{\int y^2\,ds}\right] h$$

Max. stress due to rise in temperature

Let D be the depth of arch at crown,

∴ Section modulus $$\text{“}Z\text{”} = \frac{I}{\frac{D}{2}} = \frac{2I}{D}$$

∴ Max. stress due to rise in temperature

$$\text{“}f\text{”} = \frac{M}{Z} = \left[\frac{EI\,\alpha\,T\,l}{\int y^2\,ds}\right] h \times \frac{D}{2I}$$

$$\text{“}f\text{”} = \frac{EI\,\alpha\,T\,l\,h\,D}{2\int y^2\,ds}$$

Note: In case of parabolic arch, replace “*ds*” by “*dx*”.

Example 18.5: *A two hinged semi-circular arch of radius 10 mtrs. is subjected to a rise of temp. of 40°C. Find the max. stress due to the rise of temp. Take E = 2 × 10⁵ N/mm² and α = 12 × 10⁻⁶/0°C. The depth of arch = 1000 mm.*

Solution: Horizontal thrust due to rise of temperature

$$= H = \frac{4\,EI\,\alpha\,T}{\pi R^2} = \frac{4 \times 2 \times 10^5 \times 12 \times 10^{-6} \times 40\,I}{\pi \times (10000)^2}$$

$$= \frac{384\,I}{\pi \times 10^8}\ \text{N}$$

Max. *B.M.* due to the rise of temperature

$$M = \frac{384\,I}{\pi \times 10^8} \times 10000 = \frac{384\,I}{\pi \times 10^4}\ \text{Nmm}$$

Section modulus $Z = \frac{I}{500}$

∴ Max. bending stress $= \frac{M}{Z} = \frac{384\,I}{\pi \times 10^4} \times \frac{500}{I} = 6.11\ \text{N/mm}^2$

Example 18.6: *A two hinged parabolic arch of span 40 mtrs. and rise 8 mtr. is subjected to a rise of temperature of 30°C. Find the max. bending stress at the crown due to temp. rise. The rib section is 1000 mm deep. Take E = 2 × 10³ N/mm² of α = 12 × 10⁶ per °C.*

Solution: Horizontal thrust due to rise in temperature

$$= H = \frac{15}{8} \times \frac{EI_0\,\alpha\,T}{h^2} = \frac{15}{8} \times \frac{2 \times 10^5 \times 12 \times 10^{-6} \times 30\,I_0}{(8000)^2}$$

$$= \frac{27\,I_0}{128 \times 10^5}\ \text{N}$$

Max. *B.M.* due to rise in temperature

$$\text{“}M\text{”} = \frac{27\,I_0}{128 \times 10^5} \times 8000 = \frac{27\,I_0}{1600}\ \text{Nmm}$$

Section modulus “Z” $= \frac{I_0}{500}\ \text{mm}^3$

∴ Max. bending stress “f” $= \frac{M}{Z} = \frac{27\,I_0}{1600} \times \frac{500}{I_0} = 8.44\ \text{N/mm}^2$

18.9. EFFECT OF RIB SHORTENING

We know any section of the arch is subjected to a bending moment, a shear force and a normal thrust. The normal thrust causes a shortening of the actual length of the arch.

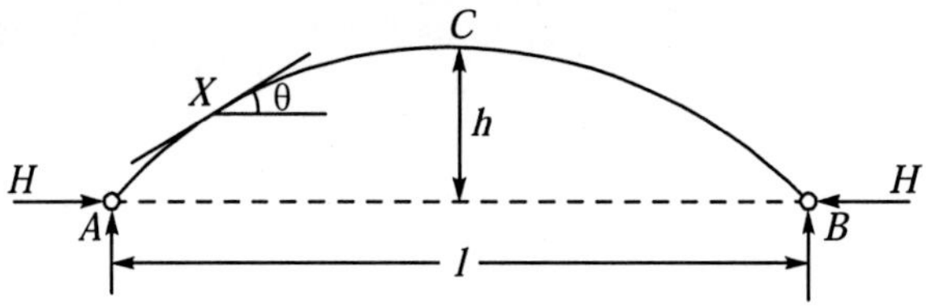

Fig. 18.17

Consider any section "X" of the two hinged arch shown in the Fig. 18.17.

Let the tangent to the arch at X be at α with the horizontal.

Let M_X = B.M. at the section X

V = S.F. at the section as in a corresponding simply supported beam.

N = Normal thrust on the section X of the arch.

But $M_X = M - Hy$, where M = B.M. at the section as in a simply supported beam.

$N = V\sin\alpha + H\cos\alpha$

The strain energy stored by the arch is given by,

$$wi = \int \frac{MX^2\, ds}{2\,EI} + \int \frac{N^2\, ds}{2\,AE}$$

$$wi = \int \frac{(M - Hy)^2\, ds}{2\,EI} + \int \frac{(V\sin\alpha + H\cos\alpha)^2\, ds}{2\,AE}$$

For the condition that strain energy stored is minimum,

$$\frac{\partial wi}{\partial H} = 0$$

$$\frac{\partial wi}{\partial H} = \int \frac{(M - Hy)(-y)\, ds}{EI} + \int \frac{(V\sin\alpha + H\cos\alpha)\cos\alpha \cdot ds}{AE} = 0$$

$$-\int \frac{My\, ds}{EI} + H\int \frac{y^2\, ds}{EI} + \int \frac{V\sin\alpha + H\cos\alpha \cdot ds}{AE}$$

$$+ H\int \frac{\cos^2\alpha \cdot ds}{AE} = 0$$

$$H = \frac{\int \frac{My \cdot ds}{EI} - \int \frac{V\sin\alpha \cdot \cos\alpha \cdot ds}{AE}}{\int \frac{y^2\, ds}{EI} + \int \frac{\cos^2\alpha \cdot ds}{AE}}$$

We know in the practical cases, the sectional area is minimum at the crown and increases towards the supports. The slope α is zero at the crown and increases towards the supports.

Hence the quantity $\frac{A}{\cos\alpha}$ has minimum value at the crown and increases towards supports.

Let A_m = Mean value of $\frac{A}{\cos\alpha}$ and putting $ds\cos\alpha = dx$,

The quantity $\int \frac{\cos^2 \alpha \cdot ds}{AE} = \int \frac{dx}{A_m E} = \frac{l}{A_m E}$

where "l" = Span of arch

We also find that the effect of V in shortening the rib is ignorable,

with above, simplification, we get

$$H = \frac{\int \frac{My\,ds}{EI}}{\int \frac{y^2\,ds}{EI} + \frac{l}{A_m E}} = \frac{\int \frac{My\,ds}{EI_0}}{\int \frac{y^2\,dx}{EI_0} + \frac{l}{A_m E}}$$

Horizontal Thrust due to Temperature Change and Rib Shortening

We know, the horizontal thrust due to temperature rise is given by,

$$H = \frac{\alpha T l}{\int \frac{y^2\,ds}{EI}}$$

To consider the effect of shortening, the "H" is given by,

$$H = \frac{\alpha T l}{\int \frac{y^2\,ds}{EI} + \frac{l}{A_m E}} = \frac{\alpha T l}{\int \frac{y^2\,dx}{EI_0} + \frac{l}{A_m E}}$$

Example 18.7: *A two-hinged parabolic arch of span 36 m and central rise 6 m, has the element of inertia varying as the secant of slope of rib axis. Find the horizontal thrust on the arch, a point load of 100 kN acts at a distance of 9 m from the left end support.* (A.M.I.E.)

Solution: Given: Span (l) = 36 m; Central rise (y_c) = 6 m; Point load (w) = 100 kN and horizontal distance between the load and the left end support (x) = 9 m.

Horizontal thrust on the arch,

$$H = \frac{5}{8} \times \frac{Wx}{y_C l^3}(l - x)(l^2 + lx - x^2)$$

$$= \frac{5}{8} \times \frac{100 \times 9}{6 \times 36^3} (36 - 9)\ (36^2 + (36 \times 9) - 9^2)\ \text{kN}$$

$$= \frac{5 \times 100 \times 9 \times 27 \times 1539}{8 \times 6 \times 36 \times 36 \times 36} = 83.5\ \text{kN} \quad \textbf{Ans.}$$

Example 18.8: *A two hinged semicircular arch of span 20 m is loaded as shown in the Fig. 18.18. below. Find the horizontal thrust.*

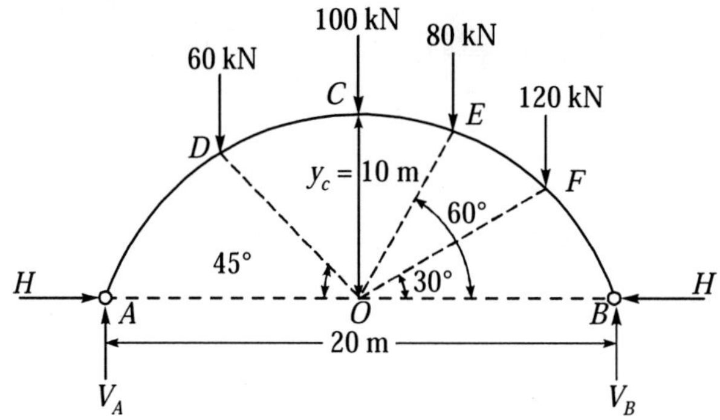

Fig. 18.18.

Solution: Given: Span (l) = 20 m; and central rise y_c = 10 m.

Let H_1 = Horizontal thrust due to load at D

H_2 = Horizontal thrust due to load at C

H_3 = Horizontal thrust due to load at E

H_4 = Horizontal thrust due to load at F

Using the relation,
$$H = \frac{W\cos^2\beta}{\pi}$$

$\therefore$
$$H_1 = \frac{60\cos^2 45°}{\pi} = \frac{60\left(\frac{1}{\sqrt{2}}\right)^2}{\pi} = \frac{30}{\pi} \quad ...(i)$$

Similarly
$$H_2 = \frac{100\cos^2 90°}{\pi} = \frac{100}{\pi} \quad ...(ii)$$

$$H_3 = \frac{80\cos^2 30°}{\pi} = \frac{80\left(\frac{\sqrt{3}}{2}\right)^2}{\pi} = \frac{60}{\pi} \quad ...(iii)$$

and
$$H_4 = \frac{120\cos^2 60}{\pi} = \frac{120\left(\frac{1}{2}\right)^2}{\pi} = \frac{30}{\pi} \quad ...(iv)$$

$\therefore$ Total horizontal thrust

$$H = H_1 + H_2 + H_3 + H_4$$
$$= \frac{30}{\pi} + \frac{100}{\pi} + \frac{60}{\pi} + \frac{30}{\pi} = 70 \text{ kN} \quad \textbf{Ans.}$$

[It may be remembered that β is the angle which the line BD makes with line OC]

Example 18.9: *A two hinged parabolic arch has a span 30 m and a rise of 7.5 m. The M.I. of arch section is proportional to sec θ, where θ is the slope of the arch axis at any pont with the horizontal. Calculate the horizontal thrust caused in the arch due to rise of temperature 25°F. The value of E = 2 × 10^8 kN/m^2 and hte co-efficient of expansion = 6 × 10^{-6} per degree F. The M.I. at crown is 1.25 × 10^{-2} m^4.*

Solution: Given: Span "l" = 30 m

Central rise y_c = 7.5 m

Rise of temp. = 25°F

Modulous of Elasticity, E = 2× 10^8 kN/m^2

co-efficient of Linear expansion α = 6 × 10^{-6}/F°

and M.I. at crown = 1.25 × 10^{-2} m^4.

$$\text{Horizontal thrust due to rise of temp.} = \frac{15 EI_0 \alpha.t}{8\, yc^2}$$

$$= \frac{15 \times 2 \times 10^6 \times 1.25 \times 10^4 \times 25 \times 6 \times 10^{-6}}{8 \times 750 \times 750} = 1250 \text{ kN.}$$

Example 18.10: *A two-hinged parabolic arch of span 20 m has central rise 4 m. The arch is loaded with a triangular load of intensity zero at left abutment and 50 kN/m at right abutment, as shown in Fig. 18.19. Determine the horizontal thrust on the arch.* (A.M.I.E.)

Solution: Given: Span (l) = 20 m and central rise (y_c) = 4 m.

Let H = Horizontal thrust on the arch

For simplification, let us assume another triangle load of intensity 50 kN/m at left abutment and zero at right abutment, to be acting over the arch. We know that as a result of this assumed load*, the horizontal thrust at A and B will be twice the thrust due to given load only.

Now using the relation, $$2H = \frac{wl^2}{80\, y_C} = \frac{50 \times 20^2}{8 \times 4} = 625 \text{ kN.}$$

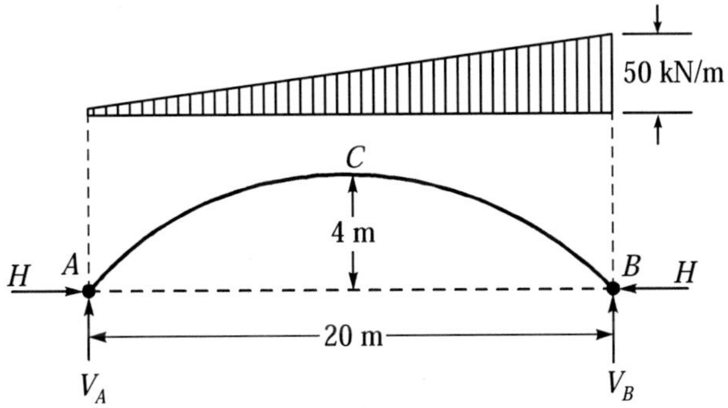

Fig. 18.19

$$\therefore \quad H = \frac{625}{2} = 312.5 \text{ kN} \quad \textbf{Ans.}$$

* Now the arch will be subjected to a uniformly distributed load of 50 kN/m over the entire span.

SUMMARY

- A two hinged arch, as the name indicates, is an arch hinged on its two supports only. Like a three hinged arch, the reactions at the supports of a two hinged arch will consist of a vertical component V_A and V_B as well as horizontal component H.
- Two hinged arch is a statically indeterminate structure to the Ist degree. The expression for horizontal thrust is derived using Castigliano's first theorem as well as by unit load method.
- Two hinged circular arch and parabolic arches are analysed for various loadings. Their theory is extended to linear arches also.
- Depending upon the number of hinges, arches are divided into four classes viz., three hinged, two hinged, single hinged and fixed arches. In bridge construction, especially in rail road bridges, the more frequently used arches are the two hinged and the fixed end ones.
- Horizontal thrust in a two-hinged arch is found out on the following assumptions :
  (*a*) Before loading, the arch is free from stresses.
  (*b*) The supports *A* and *B* are rigidly hinged.
  (*c*) After loading, the horizontal span *AB* of the arch remains unchanged.
- The horizontal thrust in a two hinged arch may be found out either by strain energy method or by the flexural deformation.

SHORT ANSWER QUESTIONS

1. Derive an equation for the horizontal thrust in a two hinged-arch.
2. Show that the horizontal thrust on a semi-circular two hinged arch, when subjected to a load *w* at its crown is $\dfrac{W}{\pi}$.
3. What are the straining actions in a two-hinged arch ? Obtain expressions for them.
4. Obtain an expression for the horizontal thrust caused by the rise of temperature in a two hinged arch.
5. State briefly what do you understand by an Arch ?

CHAPTER

19 Useful-Data

The following data summarises the pervious memory and formulae, the knowledge of which is very essential at this stage.

(a) Algebra

1. $a^0 = 1$; $x^0 = 1$

   (*i.e.,* Anything raised to the power zero is one.)

2. $x^m \times x^n = x^{m+n}$

   (*i.e.,* If the bases are same, in multiplication, the powers are added)

3. $\frac{x^m}{x_n} = x^{m-n}$

   (*i.e.,* If the bases are same, in division, the powers are subtracted)

4. If $ax^2 + bx + c = 0$ (Quadratic equation)

   then $$x = \frac{-b \pm \sqrt{b^2 - 4ac}}{2a}$$

   where a is the coefficient of x^2,

   b is the coefficient of x and c is the constant term.

(b) Trigonometry

In a right-angled triangle ABC as shown in Fig. 17.1

1. $\frac{b}{c} = \sin \theta$

2. $\frac{c}{a} = \cos \theta$

3. $\frac{b}{a} = \frac{\sin \theta}{\cos \theta} = \tan \theta$

4. $\frac{c}{b} = \frac{1}{\sin \theta} = \text{cosec } \theta$

5. $\frac{c}{a} = \frac{1}{\cos \theta} = \sec \theta$

6. $\frac{a}{b} = \frac{\cos \theta}{\sin \theta} = \frac{1}{\tan \theta} = \cot \theta$

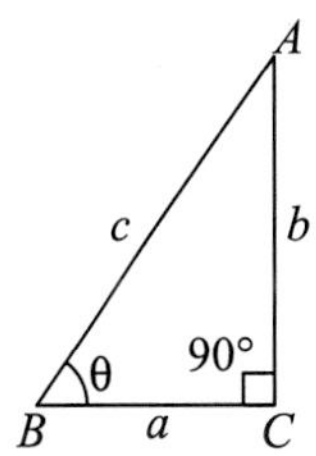

Fig. 19.1

7. The following table shows the values of trigonometrical functions for some typical angles:

Angle	0°	30°	45°	60°	90°
sin	0	$\frac{1}{2}$	$\frac{1}{\sqrt{2}}$	$\frac{\sqrt{3}}{2}$	1
cos	1	$\frac{\sqrt{3}}{2}$	$\frac{1}{\sqrt{2}}$	$\frac{1}{2}$	0
tan	0	$\frac{1}{\sqrt{3}}$	1	$\sqrt{3}$	∞

or in other words, for sin write.

0°	30°	45°	60°	90°
$\frac{\sqrt{0}}{2}$	$\frac{\sqrt{1}}{2}$	$\frac{\sqrt{2}}{2}$	$\frac{\sqrt{3}}{2}$	$\frac{\sqrt{4}}{2}$
0	$\frac{1}{2}$	$\frac{1}{\sqrt{2}}$	$\frac{\sqrt{3}}{2}$	1

For cos write the values in reverse order, for tan divide the value of sin by cos for the respective angle.

8. In the first quadrant (*i.e.*, 0° to 90°) all the trigonometrical ratios are positive.
9. In the second quadrant (*i.e.*, 90° to 180°) only sin θ and cosec θ are positive.
10. In the third quadrant (*i.e.*, 180° to 270°) only tan θ and cot θ are positive.
11. In the fourth quadrant (*i.e.*, 270° to 360°) only cos θ and sec θ are positive.
12. In any triangle *ABC*,

$$\frac{a}{\sin A} = \frac{b}{\sin B} = \frac{c}{\sin C}$$

where *a, b* and *c* are the length of the three sides of a triangle. *A, B* and *C* are opposites angles of the sides *a, b* and *c* respectively.

13. $\sin (A + B) = \sin A \cos B + \cos A \sin B.$
14. $\sin (A - B) = \sin A \cos B - \cos A \sin B.$
15. $\cos (A + B) = \cos A \cos B - \sin A \cos B.$
16. $\cos (A - B) = \cos A \sin B + \sin A \sin B.$
17. $\tan (A + B) = \dfrac{\tan A + \tan B}{1 - \tan A \cdot \tan B}$
18. $\tan (A - B) = \dfrac{\tan A + \tan B}{1 + \tan A \cdot \tan B}$

19. $\sin 2A = 2 \sin A \cos A$

20. $\sin^2 \theta \cos^2 \theta = 1$

21. $1 + \tan^2 \theta = \sec^2 \theta$

22. $1 + \cot^2 \theta = \text{cosec}^2 \theta$

23. $\sin^2 A = \dfrac{1 - \cos 2A}{2}$

24. $\cos^2 A = \dfrac{1 - \cos 2A}{2}$

25. $2 \cos A \sin B = \sin (A + B) - \sin (A - B)$.

26. Rules for the change of trigonometrical ratios:

$$(A)\left\{\begin{aligned} \sin(-\theta) &= -\sin\theta \\ \cos(-\theta) &= \cos\theta \\ \tan(-\theta) &= -\tan\theta \\ \cot(-\theta) &= -\cot\theta \\ \sec(-\theta) &= \sec\theta \\ \text{cosec}(-\theta) &= -\text{cosec}\,\theta \end{aligned}\right.$$

$$(B)\left\{\begin{aligned} \sin(90^\circ-\theta) &= \cos\theta \\ \cos(90^\circ-\theta) &= \sin\theta \\ \tan(90^\circ-\theta) &= \cot\theta \\ \cot(90^\circ-\theta) &= \tan\theta \\ \sec(90^\circ-\theta) &= \text{cosec}\,\theta \\ \text{cosec}(90^\circ-\theta) &= \sec\theta \end{aligned}\right.$$

$$(C)\left\{\begin{aligned} \sin(90^\circ+\theta) &= \cos\theta \\ \cos(90^\circ+\theta) &= -\sin\theta \\ \tan(90^\circ+\theta) &= -\cot\theta \\ \cot(90^\circ+\theta) &= -\tan\theta \\ \sec(90^\circ+\theta) &= -\text{cosec}\,\theta \\ \text{cosec}(90^\circ+\theta) &= -\sec\theta \end{aligned}\right.$$

$$(D)\left\{\begin{aligned} \sin(180^\circ-\theta) &= \sin\theta \\ \cos(180^\circ-\theta) &= -\cos\theta \\ \tan(180^\circ-\theta) &= -\tan\theta \\ \cot(180^\circ-\theta) &= -\cot\theta \\ \sec(180^\circ-\theta) &= -\sec\theta \\ \text{cosec}(180^\circ-\theta) &= \text{cosec}\,\theta \end{aligned}\right.$$

$$(E)\begin{cases} \sin(180° + \theta) = -\sin\theta \\ \cos(180° + \theta) = -\cos\theta \\ \tan(180° + \theta) = \tan\theta \\ \cot(180° + \theta) = \cot\theta \\ \sec(180° + \theta) = -\sec\theta \\ \text{cosec}(180° + \theta) = -\text{cosec}\,\theta \end{cases}$$

Following are the rules to remember the above 30 formulae:

Rule 1: Trigonometrical ratio changes only when the angle is $(90° - \theta)$ or $(90° + \theta)$. In all other cases, trigonometrical ratio remains the same. Following is the law of change:

sin changes into cos and cos changes into sin.
tan changes into cot and cot changes into tan.
sec changes into cosec and cosec changes into sec.

Rule 2: Consider the angle θ to be a small angle and write the proper sign as per formula 8 to 11 above.

(c) Differential Calculus

1. $\frac{d}{dx}$ is the sign of differentiation.

2. $\frac{d}{dx}(x)^n = nx^{n-1}$; $\frac{d}{dx}(x)^8 = 8x^7$, $\frac{d}{dx}(x) = 1$

   (*i.e.*, to differentiate any power of x, write the power before x and subtract one from the power.)

3. $\frac{d}{dx}(C) = 0$; $\frac{d}{dx}(7) = 0$

   (*i.e.*, differential coefficient of a constant is zero)

4. $\frac{d}{dx}(u, v) = u \cdot \frac{dv}{dx} + v \cdot \frac{du}{dx}$

   [*i.e.*, Differential coefficient of product of any two functions] = [(Ist function × Differential coefficient of second function) + (2nd function × Differential coefficient of first function)]

5. $\frac{d}{dx}\left(\frac{u}{v}\right) = \frac{v \cdot \frac{du}{dx} - u \cdot \frac{dv}{dx}}{v^2}$

   [*i.e.*, Differential coefficient of two functions when one is divided by the other] = [((Denominator × Differential coefficient of numerator) − (Numerator × Differential coefficient of denominator)) / Square of denominator]

6. Differential coefficient of trigonometrical functions

$$\frac{d}{dx}(\sin x) = \cos x;$$

$$\frac{d}{dx}(\cos x) = -\sin x;$$

$$\frac{d}{dx}(\tan x) = -\sec^2 x;$$

$$\frac{d}{dx}(\cot x) = \text{cosec}^2 x;$$

$$\frac{d}{dx}(\sec x) = \sec x; \tan x;$$

$$\frac{d}{dx}(\text{cosec } x) = -\text{cosec } x \cdot \cot x$$

Note: The differential coefficient, whose trigonometrical function begins with co, is negative.

7. If the differential coefficient of a function is zero, the function is either maximum or minimum. Conversely, if the maximum or minimum value of a function is required, then differentiate the function and equate it to zero.

(*d*) Integral Calculus

1. $\int dx$ is the sign of integration.

2. $\int x^n \, dx = \frac{x^{n+1}}{n+1}$; $\int x^6 \, dx = \frac{x^7}{7}$

   (*i.e.*, to integrate any power of *x*, add one to the power and divide by the new power).

3. $\int 7 dx = 7 dx$; $\int C dx = Cx$

   (*i.e.*, to integrate any constant, multiply the constant by *x*.)

4. $\int (ax + b)^n \, dx = \frac{(ax+b)^{n+1}}{(n+1) \times a}$

❑❑❑

Index

Win Prizes

ATTENTION READERS

Win Prizes

We request for your frank assessment regarding various aspects of the book as given below :

FUNDAMENTALS OF STRUCTURAL ANALYSIS

By: Col. (Retd.) M.K. Pant

Please Fill up the Sheet in Neat and Clean Writing

(i) What topics of your syllabus, important for your examination point of view, are not covered in this book ?

....

....

....

....

(ii) What are the chapters wherein the treatment of the subject matter is not systematic or organised or updated ?

....

....

....

....

(iii) Have you come across misprints/mistakes/factual inaccuracies in the book? Please specify the chapters and the page numbers.

....

....

....

....

(iv) What other books on the same subject have you found/heard better than the present book? Please specify in terms of price and quality

....

....

....

....

(v) Further suggestions and comments for the improvement of this book.

....

....

....

....

(PS: If need be please attach a separate sheet to write your views)

OTHER DETAILS

(i) Who recommended you this book ? (Please tick (✓) in the box near the option relevant to you.

☐ Teacher ☐ Friends ☐ Bookseller

....

....

....

....

(ii) Name of the recommending teacher, his designation and address :

....

....

....

....

(iii) Name and address of the bookseller you purchased the book from :

....

....

....

....

(iv) Name and address of your institution (Please mention the University or Board as the case may be)

....

....

....

....

(v) Your name and complete postal address :

....

....

....

....

(vi) Course you are enrolled in :

....

....

....

....

The best assessment will be awarded each month: The award will be in the form of our publications as decided by the Editorial Board.

Please mail the filled up coupon at your earliest to :

S.K. KATARIA & SONS®
4885/109, Prakash Mahal, Dr. Subhash Bhargav Lane,
Opposite Delhi Medical Association, Daryaganj, New Delhi–110002
☎: +91-11-43551243, Mobile: +91-9871775858, 7678425209
Email : katariabook@gmail.com, katariabooks@yahoo.com